Information Technology Control and Audit

Information Technology Control and Audit

Fifth Edition

Angel R. Otero

CRC Press
Taylor & Francis Group
Boca Raton London New York

CRC Press is an imprint of the
Taylor & Francis Group, an **informa** business

AN AUERBACH BOOK

CRC Press
Taylor & Francis Group
6000 Broken Sound Parkway NW, Suite 300
Boca Raton, FL 33487-2742

First issued in paperback 2020

ISBN-13: 978-1-4987-5228-2 (hbk)
ISBN-13: 978-0-367-65715-4 (pbk)

Visit the Taylor & Francis Web site at
http://www.taylorandfrancis.com

and the CRC Press Web site at
https://www.crcpress.com

I dedicate this book to my wife, Ghia, my daughter Elizabeth, and my sons, Leonardo and Giancarlo. I also dedicate this book to my parents, Angel and Lydia, and my brothers, Luis Daniel and Carlos.

Contents

SECTION II PLANNING AND ORGANIZATION

SECTION III   AUDITING IT ENVIRONMENT

Preface

Motivation

Throughout the years, organizations have experienced numerous losses, which have had a direct impact on their most valuable asset, information. Attackers are well able and continue to look for opportunities to break traditional/non-traditional computer systems and exploit their vulnerabilities. Attacks like the above, along with enactments of recent laws and regulations—all mainly related to the protection of information—have turned information technology (IT) audit into a critical, dynamic, and challenging profession with a future that brings growth into new and exciting IT-related areas.

I recognized the significance of writing a book that would aid the understanding of the IT audit profession, emphasizing on how IT audit provides organizations and auditors the ability to effectively assess financial information's validity, reliability, and security. Of particular importance was to write about techniques, procedures, and controls that could assist organizations and auditors in determining whether activities related to gathering, processing, storing, and distributing financial information are effective and consistent with the organization's goals and objectives. A further motivating factor to write this book was to cover relevant material typically tested on the Certified Information Systems Auditor (CISA) and Certified Public Accountant (CPA) exams (e.g., internal controls, computer-assisted auditing techniques (CAATs), technological skill sets, etc.). Lastly, I wanted to write a book that would capture the knowledge and experience I have gained throughout my IT audit career.

Purpose

This book recognizes the continuous need for organizations and auditors to effectively manage and examine IT systems in order to meet business goals and objectives. The book provides essential principles, knowledge, and skills on how to control and assess IT systems that will prepare the reader for a successful career in the public practice, private industry, or government. Targeted to both students and industry practitioners in the IT and accounting fields, the book:

- includes IT audit problems, simulations, practical cases, and research assignment opportunities to develop IT audit expertise.
- is designed to meet the increasing needs of IT audit, compliance, security, and risk management professionals.
- represents an ideal resource for those preparing for the CISA and CPA exams.

The book is intended for use in a semester course at undergraduate (senior level) and graduate levels. It is also designed to fit well in an online course. Financial and managerial accounting, accounting information systems, management information systems, and/or management principles are all suggested prerequisites.

New to Fifth Edition

A significant change of this fifth edition is the reduction and/or consolidation of chapters, from 26 (previous edition) to 13 chapters, allowing the faculty to easily fit the chapters with the course term (e.g., 16-week course, 8-week course, etc.). Every chapter has been significantly revised to include up-to-date audit concepts, examples, tools, and/or techniques, including end-of-chapter cases featuring practical IT audit scenarios. These scenarios require students to work individually or in teams, engage in classroom discussions, perform analysis, and come up with decision-making strategies. The fifth edition is tailored to auditing topics, computer techniques, and other technological skill sets and concepts normally tested on both, CISA and CPA exams.

Another significant contribution is the addition or revision of teaching resources and materials that allow faculty to focus their attention on classroom presentation and discussion. Available resources and material include:

- Significantly enhanced PowerPoint presentation slides, with additional relevant information in the notes section.
- Author-certified test banks, ensuring the testing of higher level cognitive skills and not just on text memorization. Test banks include multiple choice, true or false, and essays questions that assess on content, as well as require analysis, synthesis, and evaluation skills.
- Revised instructors manual.
- Syllabus template with suggested class schedules for 16-week and 8-week courses.

Chapter Organization and Content

The book is divided into three major sections:

Section I—Foundation for IT Audit (Four Chapters)

Chapter 1 discusses how technology is constantly evolving and shaping today's business environments. The chapter also talks about the audit profession, with particular interest in IT auditing. Following the discussion of auditing, current trends and needs of IT auditing are described, as well as the various roles of an auditor in the IT field. The chapter ends with explanations of why IT auditing is considered a profession, and the various career opportunities available to IT auditors.

Chapter 2 focuses on federal, state, and international legislation that governs the use, misuse, and privacy of information and its related technology. Legislation has a lasting impact on the online community (government, business, and the public), which is something that those entering the IT audit profession should be knowledgeable of.

Chapter 3 discusses the audit process for IT and the demands that will be placed on the profession in the future. The chapter covers risk assessments, audit planning, required phases of an

IT audit engagement, and the significance of IT audit documentation when supporting financial statement audits.

Chapter 4 defines audit productivity tools and describes how they help the audit process. The chapter then touches upon the various techniques used to document financial application systems. Explanations of CAATs and the role they play in the audit will follow along with descriptions of the various techniques used when defining audit sample size, selecting samples, and reviewing applications. CAATs used in auditing application controls and in operational reviews will then be described followed by explanations of computer forensic tools and techniques.

Section II—Planning and Organization (Four Chapters)

Chapter 5 acknowledges the existing globalization of many industries and financial markets, and stresses the importance of effective governance and controls to the success of organizations. The Sarbanes–Oxley Act of 2002, the Committee of Sponsoring Organizations of the Treadway Commission, the Combined Code on Governance in the United Kingdom, and the Organization for Economic Co-Operation and Development Principles of Corporate Governance in Europe have all set the bar for global corporate governance. For IT, Control Objectives for Information and Related Technology (COBIT) has become the global standard for governance. IT governance provides the structure and direction to achieve the alignment of the IT strategy with the business strategy.

Chapter 6 discusses the process of managing and evaluating risks in an IT environment. Risk management should be integrated into strategic planning, operational planning, project management, resource allocation, and daily operations, as it enables organizations to focus on areas that have the highest impact. Risk assessments, on the other hand, occur at multiple levels of the organization with a focus on different areas. What are the characteristics and components of a risk management process? What are the professional standards of practice for risk assessments? What are examples of risk assessment practices used in varied environments? These are some of the questions that will be answered within the chapter.

Chapter 7 focuses on project management best practices, standards, and methodologies used by managers to effectively and efficiently bring projects to an end. The chapter also discusses factors when conducting effective project management, as well as the auditor's role in project management. Topics such as IT project management and management of big data projects are also discussed.

Chapter 8 discusses the system development life cycle and its common phases, as well as risks and associated controls related to such phases. The chapter also explains common approaches used for software development, and ends with a discussion of the IT auditor's involvement in the system development and implementation process.

Section III—Auditing IT Environment (Five Chapters)

Chapter 9 discusses common risks to various types of application systems, and provides examples of such potential risks. The chapter also touches upon relevant application controls (i.e., input, processing, and output) that can be implemented by organizations in order to mitigate the risks discussed. Lastly, the involvement of IT auditors when examining applications is explained.

Chapter 10 discusses management processes and procedures for change control and organizational change. This chapter also describes configuration management and sample activities conducted as part of a configuration management plan. This chapter ends with a discussion of IT audit involvement in a change control management examination.

Chapter 11 presents an overview of information systems operations as a relevant component of the IT infrastructure. This chapter provides sample objectives and control activities the IT auditor should focus on when examining information systems operations. This chapter also covers end-user computing groups, and provides guidelines when auditing such groups. Lastly, IT audit involvement is discussed when examining information systems operations, including the procedures to follow when auditing data centers and disaster recovery plans, among others.

Chapter 12 describes the importance of information security to organizations, and how information represents a critical asset in today's business environment. This chapter also discusses recent technologies that are revolutionizing organizations and, specifically, the need to implement adequate security to protect the information. Information security threats and risks, and how they continue to affect information systems are also described. Relevant information security standards and guidelines available for organizations and auditors will then be discussed, along with the significance of establishing an information security policy. This chapter continues with a discussion of roles and responsibilities of information security personnel. To end, explanations of information security controls, the significance of selecting, implementing, and testing such controls, and the IT audit involvement in an information security assessment are provided.

Chapter 13 discusses critical success factors when acquiring systems or services from third parties. This chapter also covers service management and expectations for an effective partnership between organizations and suppliers. Outsourcing IT, discussed next, refers to hiring an outside company to handle all or part of an organization's IT processing activities, allowing organizations to focus on what they do best. Lastly, IT audit involvement when examining system acquisitions and outsourced IT services is described.

Key Learning Objectives

- To discuss how technology is constantly evolving and shaping today's IT environments. To explain the IT audit profession, roles of the IT auditor, and career opportunities.
- To describe legislation relevant to IT auditors and its impact on the IT field. To illustrate frequently reported Internet crimes and cyberattacks. To develop audit plans and procedures that assist organizations to comply with relevant laws and regulations.
- To explain the IT audit process, the significance of Control Objectives for Information and Related Technology (COBIT), and the various phases of an IT audit engagement. To develop relevant and practical documentation to perform IT audit work.
- To support the role and significance of tools and computer-assisted audit techniques (CAATs) when performing audit work. To design audit plans that ensure adequate use of tools and technologies when delivering audit work.
- To demonstrate the significance of aligning IT plans, objectives, and strategies with the business (i.e., IT governance).
- To explain risk management, particularly the Enterprise Risk Management—Integrated Framework. To describe risk assessments, and how they form the first step in the risk management methodology.
- To describe project management, as well as project management standards and best practices. To discuss the role of the IT auditor in project management.

- To outline the system development life cycle (SDLC), common approaches, risks, associated controls, and IT auditor's involvement. To develop relevant audit programs listing risks related to SDLC phases, and IT controls and procedures needed to mitigate those risks.
- To discuss risks associated with common types of application systems, as well as application controls and how they are used to safeguard the input, processing, and output of information. To discuss the IT auditor's involvement in an examination of application systems. To develop relevant and practical documentation to perform IT audit work.
- To establish the significance of a change control management process. To illustrate the audit involvement in a change control management examination. To perform actual audit work related to change control management, from completing an understanding of the IT environment through formally communicating audit results to management.
- To demonstrate the importance for organizations of having implemented policies, procedures, and adequate controls related to information systems operations to ensure completeness, accuracy, and validity of information. To describe the audit involvement in an examination of an organization's information systems operations. To design and prepare relevant and practical documentation when performing IT audit work.
- To support the importance of protecting information against security threats and risks, and implementing effective information security policies, procedures, and controls to ensure the integrity of such information. To describe audit involvement in an information security examination.
- To explain the importance of a sourcing strategy as a critical success factor to acquiring IT services or products. To discuss how IT services should be defined to meet organizational objectives and how to measure performance of those IT services.

Acknowledgments

I start by thanking God for all the blessings, life itself, family, friends, and now the opportunity to complete this book. Next, I extend my deepest gratitude to Mr. Richard O'Hanley from CRC Press and Auerbach Publications, who believed in me and trusted that we can perform great work together. I am especially grateful to the editorial development team and faculty reviewers, who enhanced the quality of this book with their priceless suggestions, feedback, and constructive criticism. The depth and sincerity of their reviews clearly evidenced their devotion and passion to the IT auditing field. I was privileged to have their candid and valuable advice, which undoubtedly added significant value to this fifth edition.

To colleagues and friends, thank you for your time, guidance, advice, knowledge, and support provided, all critical in encouraging me to commit to the challenging journey of writing a book.

To my parents, Angel L. Otero and Lydia E. Rivera, my brothers, Dr. Luis Daniel Otero and Dr. Carlos Otero, thank you all for serving as great role models, personally and professionally. Thank you for your constant support, encouragement, and for always challenging me to work at the next level.

To my children, Elizabeth, Leonardo, and Giancarlo, I hope that the achievement I attained today serves as encouragement and motivation for you. Remember to always work hard without harming your neighbor, never settle for less, and condition your mindset for greatness. Be respectful, be humble, and most importantly, always put God first in your lives.

Last, but certainly not least, my most sincere thanks to my lovely wife and best friend, Ghia. Thank you for believing in me since the day I embarked on this endeavor. Thank you for your great help, continuing encouragement, and support with raising our children. Thank you for your unconditional love and the sleepless nights during the past couple of years. I owe the completion of this book to you and all the sacrifices you made. I love you with all the strengths of my heart.

Author

Angel R. Otero, Ph.D., CPA, CISA, CITP, CICA, CRISC is an assistant professor of accounting and program chair for accounting and finance online programs at the College of Business in Florida Institute of Technology (Florida Tech or FIT), Melbourne, FL. Dr. Otero has a B.S. in accounting from Pennsylvania State University, a M.S. in software engineering from Florida Tech, and a Ph.D. in information systems from Nova Southeastern University. He also holds active memberships at the American Institute of Certified Public Accountants (AICPA), ISACA (formerly the Information Systems Audit and Control Association), and the Institute for Internal Controls (IIC) professional organizations.

Dr. Otero has over 20 years of industry experience in the areas of public accounting/auditing, information systems auditing, internal control audits, and information technology consulting. Clients served involve the industries of banking/finance, insurance, government, manufacturing, retail, and wholesale, among others. Before joining FIT, Dr. Otero worked at Deloitte & Touche, LLP for 10 plus years and attained the position of senior manager overseeing offices in the State of Florida and Puerto Rico.

Dr. Otero's research interests involve (1) information systems/technology auditing; (2) accounting information systems; (3) financial audits and internal controls; and (4) information security audits and risk assessments. He has published in multiple peer-reviewed journals and conference proceedings.

FOUNDATION
FOR IT AUDIT

Chapter 1

Information Technology Environment and IT Audit

LEARNING OBJECTIVES

1. Discuss how technology is constantly evolving and shaping today's business (IT) environments.
2. Discuss the auditing profession and define financial auditing.
3. Differentiate between the two types of audit functions that exist today (internal and external).
4. Explain what IT auditing is and summarize its two broad groupings.
5. Describe current IT auditing trends, and identify the needs to have an IT audit.
6. Explain the various roles of the IT auditor.
7. Support why IT audit is considered a profession.
8. Describe the profile of an IT auditor in terms of experience and skills required.
9. Discuss career opportunities available to IT auditors.

Organizations today are more information dependent and conscious of the pervasive nature of technology across the business enterprise. The increased connectivity and availability of systems and open environments have proven to be the lifelines of most business entities. Information technology (IT) is now used more extensively in all areas of commerce around the world.

IT Environment

The need for improved control over IT, especially in commerce, has been advanced over the years in earlier and continuing studies by many national and international organizations. Essentially, technology has impacted various significant areas of the business environment, including the use and processing of information, the control process, and the auditing profession.

- Technology has improved the ability to capture, store, analyze, and process tremendous amounts of data and information, expanding the empowerment of the business decision maker. It has also become a primary enabler to production and service processes. There is a residual effect in that the increased use of technology has resulted in increased budgets, increased successes and failures, and better awareness of the need for **control**.
- Technology has significantly impacted the **control process** around systems. Although control objectives have generally remained constant, except for some that are technology specific, technology has altered the way in which systems should be controlled. Safeguarding **assets**, as a control objective, remains the same whether it is done manually or is automated. However, the manner by which the control objective is met is certainly impacted.
- Technology has impacted the auditing profession in terms of how audits are performed (information capture and analysis, control concerns) and the knowledge required to draw conclusions regarding operational or system effectiveness, efficiency, and reporting **integrity**. Initially, the impact was focused on dealing with a changed processing environment. As the need for **auditors** with specialized technology skills grew, so did the IT auditing profession.

Technology is constantly evolving and finding ways to shape today's IT environment in the organization. The following sections briefly describe various recent technologies that have and will certainly continue to revolutionize organizations, how business is done, and the dynamics of the workplace.

Enterprise Resource Planning (ERP)

According to the June 2016 edition of Apps Run the World, a technology market-research company devoted to the applications space, the worldwide market of ERP systems will reach $84.1 billion by 2020 versus $82.1 billion in 2015. ERP is software that provides standard business functionality in an integrated IT environment system (e.g., procurement, inventory, **accounting**, and human resources [HR]). Refer to Exhibit 1.1 for an illustration of the ERP modular system.

ERPs allow multiple functions to access a common database—reducing storage costs and increasing **consistency** and **accuracy** of data from a single source. Additionally, ERPs:

- Have standard methods in place for automating processes (i.e., information in the HR system can be used by payroll, help desk, and so on).
- Share real-time information from modules (finance, HR, etc.) residing in one common database, hence, **financial statements**, analyses, and reports are generated faster and more frequently.

Some of the primary ERP suppliers today include SAP, FIS Global, Oracle, Fiserv, Intuit, Inc., Cerner Corporation, Microsoft, Ericsson, Infor, and McKesson.

Despite the many advantages of ERPs, they are not much different than purchased or packaged systems, and may therefore require extensive modifications to new or existing business processes. ERP modifications (i.e., software releases) require considerable programming to retrofit all of the organization-specific code. Because packaged systems are generic by nature, organizations may need to modify their business operations to match the vendor's method of processing, for instance. Changes in business operations may not fit well into the organization's culture or other processes, and may also be costly due to training. Additionally, as ERPs are offered by a single

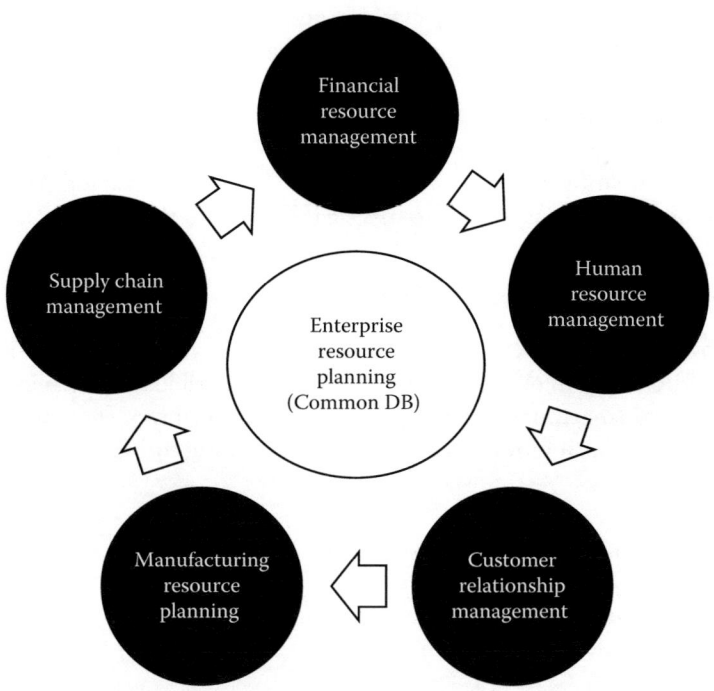

Exhibit 1.1 Enterprise resource planning modular system.

vendor, **risks** associated with having a single supplier apply (e.g., depending on a single supplier for maintenance and support, specific hardware or software requirements, etc.).

Cloud Computing

Cloud computing continues to have an increasing impact on the IT environment. According to ISACA (formerly known as the Information Systems Audit and Control Association), the cloud computing's exponential growth should no longer be considered an emerging technology. Cloud computing has shaped business across the globe, with some organizations utilizing it to perform business critical processes. Based on the July 2015's ISACA Innovation Insights report, cloud computing is considered one of the key trends driving business **strategy**. The International Data Corporation, in its 2015 publication, also predicts that cloud computing will grow at 19.4% annually over the next 5 years. Moreover, Deloitte's 2016 Perspective's Cloud Computing report (report) indicates that for private companies, cloud computing will continue to be a dominant factor.

Cloud computing, as defined by PC Magazine, refers to the use of the Internet (versus one's computer's hard drive) to store and access data and programs. In a more formal way, the National Institute of Standards and Technology (NIST) defines cloud computing as a "model for enabling ubiquitous, convenient, on-demand network access to a shared pool of configurable computing resources (e.g., networks, servers, storage, applications, and services) that can be rapidly provisioned and released with minimal management effort or service provider interaction." NIST also stress that availability is significantly promoted by this particular (cloud) model.

The highly flexible services that can be managed in the virtual environment makes cloud computing very attractive for business organizations. Nonetheless, organizations do not yet feel fully comfortable when storing their information and applications on systems residing outside of their on-site premises. Migrating information into a shared infrastructure (such as a cloud environment) exposes organizations' sensitive/critical information to risks of potential unauthorized access and exposure, among others. Deloitte, one of the major global accounting and auditing firms, also supports the significance of security and privacy above, and added, based in its report, that cloud-stored information related to patient data, banking details, and personnel records, to name a few, is vulnerable and susceptible to misuse if fallen into the wrong hands.

Mobile Device Management (MDM)

MDM, also known as Enterprise Mobility Management, is a relatively new term, but already shaping the IT environment in organizations. MDM is responsible for managing and administering mobile devices (e.g., smartphones, laptops, tablets, mobile printers, etc.) provided to employees as part of their work responsibilities. Specifically, and according to PC Magazine, MDM ensures these mobile devices:

- integrate well within the organization and are implemented to comply with organization policies and procedures
- protect corporate information (e.g., emails, corporate documents, etc.) and configuration settings for all mobile devices within the organization

Mobile devices are also used by employees for personal reasons. That is, employees bring their own mobile (personal) device to the organization (also referred to as bring-your-own-device or BYOD) to perform their work. Allowing employees to use organization-provided mobile devices for work and personal reasons has proved to appeal to the average employee. Nevertheless, organizations should monitor and control the tasks performed by employees when using mobile devices, and ensure employees remain focused and productive. It does represent a risk to the organization's security and a distraction to employees when mobile devices are used for personal and work purposes. Additionally, allowing direct access to corporate information always represents an ongoing risk, as well as raises security and **compliance** concerns to the organization.

Other Technology Systems Impacting the IT Environment

The Internet of Things (IoT) has a potential transformational effect on IT environments, data centers, technology providers, etc. Gartner, Inc. estimates that by the year 2020, IoT will include 26 billion units installed and revenues will exceed $300 billion generated mostly by IoT product and service suppliers.

IoT, as defined by Gartner, Inc., is a system that allows remote assets from "things" (e.g., devices, sensors, objects, etc.) to interact and communicate among them and with other network systems. Assets, for example, communicate information on their actual status, location, and functionality, among others. This information not only provides a more accurate understanding of the assets, but also maximizes their utilization and productivity, resulting in an enhanced decision-making process. The huge volumes of raw data or data sets (also referred to as Big Data) generated as a result of these massive interactions between devices and systems need to be processed and analyzed effectively in order to generate information that is meaningful and useful in the decision-making process.

Big Data, as defined by the TechAmerica Foundation's Federal Big Data Commission (2012), "describes large volumes of high velocity, complex and variable data that require advanced techniques and technologies to enable the capture, storage, distribution, management, and analysis of the information." Gartner, Inc. further defines it as "… high-volume, high-velocity and/or high-variety information assets that demand cost-effective, innovative forms of information processing that enable enhanced insight, decision making, and process automation."

Even though accurate Big Data may lead to more confident decision-making process, and better decisions often result in greater operational efficiency, cost reduction, and reduced risk, many challenges currently exist and must be addressed.

Challenges of Big Data include, for instance, analysis, capture, data curation, search, sharing, storage, transfer, visualization, querying, as well as updating. Ernst & Young, on its EY Center for Board Matters' September 2015 publication, states that challenges for auditors include the limited access to **audit** relevant data, the scarcity of available and qualified personnel to process and analyze such particular data, and the timely integration of analytics into the audit. The IoT also delivers fast-moving data from sensors and devices around the world, and therefore results in similar challenges for many organizations when making sense of all that data.

Other recent technologies listed on the Gartner's 2015 Hype Cycle for Emerging Technologies Report that are currently impacting IT environments include wearables (e.g., smartwatches, etc.), autonomous vehicles, cryptocurrencies, consumer 3D printing, and speech-to-speech translation, among others.

IT Environment as Part of the Organization Strategy

In today's environment, organizations must integrate their IT with business strategies to attain their overall objectives, get the most value out of their information, and capitalize on the technologies available to them. Where IT was formerly viewed as an enabler of an organization's strategy, it is now regarded as an integral part of that strategy to attain profitability and service. At the same time, issues such as IT governance, international information infrastructure, security, and privacy and control of public and organization information have driven the need for self-review and self-assurance.

For the IT manager, the words "audit" and "auditor" send chills up and down the spine. Yes, the auditor or the audit has been considered an evil that has to be dealt with by all managers. In the IT field, auditors in the past had to be trained or provided orientation in system concepts and operations to evaluate IT practices and applications. IT managers cringe at the auditor's ability to effectively and efficiently evaluate the complexities and grasp the issues. Nowadays, IT auditors are expected to be well aware of the organization's IT infrastructure, policies, and operations before embarking in their reviews and examinations. More importantly, IT auditors must be capable of determining whether the **IT controls** in place by the organization ensure data protection and adequately align with the overall organization goals.

Professional associations and organizations such as ISACA, the **American Institute of Certified Public Accountants (AICPA)**, the Canadian Institute of Chartered Accountants (CICA), Institute of Internal Auditors (IIA), Association of Certified Fraud Examiners (ACFE), and others have issued guidance, instructions, and supported studies and research in audit areas.

The Auditing Profession

Computers have been in use commercially since 1952. Computer-related crimes were reported as early as 1966. However, it was not until 1973, when the significant problems at Equity Funding

Corporation of America (EFCA) surfaced, that the auditing profession looked seriously at the lack of controls in computer information systems (IS). In 2002, almost 30 years later, another major **fraud** resulted from corporate and accounting scandals (Enron and WorldCom), which brought **skepticism** and downfall to the financial markets. This time, neither the major accounting firms nor the security- and exchange-regulated businesses in major exchanges were able to avoid the public outrage, lack of investor confidence, and increased government regulation that befell the U.S. economy. Again, in 2008, the U.S. economy suffered as mortgage banking and mortgage investment companies (such as Countrywide, IndyMac, etc.) defaulted from unsound lending strategies and poor risk management.

When EFCA declared bankruptcy in 1973, the minimum direct impact and losses from illegal activity were reported to be as much as $200 million. Further estimates from this major financial fraud escalated to as much as $2 billion, with indirect costs such as legal fees and depreciation included. These losses were the result of a "computer-assisted fraud" in which a corporation falsified the records of its life insurance **subsidiary** to indicate the issuance of new policies. In addition to the insurance policies, other assets, such as **receivables** and **marketable securities**, were recorded falsely. These fictitious assets should have been revealed as non-existent during the corporation's regular year-end audits but were never discovered. As the computer was used to manipulate files as a means of covering the fraud, the accounting profession realized that conventional, manual techniques might not be adequate for audit engagements involving computer application.

In 1973, the AICPA (major national professional organization of certified public accountants), in response to the events at EFCA, appointed a special committee to study whether the auditing standards of the day were adequate in such situations. The committee was requested to evaluate specific procedures to be used and the general standards to be approved. In 1975, the committee issued its **findings**. Even though the special committee found that auditing standards were adequate, and that no major changes were called for in the procedures used by auditors, there were several observations and recommendations issued related to the use of computer programs designed to assist the **examination** of financial statements. Another critical review of the existing auditing standards was started in 1974, when the AICPA created its first standards covering this area. Then, 29 years later, the Enron–Arthur Andersen fiasco of 2002 took us back to 1973.

The issue of "**due professional care**" has come to the forefront of the audit community as a result of major U.S. financial scandals and poor management, including but not limited to, Waste Management (1998), Enron (2001), Worldcom (2002), American Insurance Group (2005), Lehman Brothers (2008), Bernard L. Madoff Securities LLC (2008), MF Global (2011), Anthem Inc. (2015), Wells Fargo (2016), and others. The EFCA scandal of 1973 led to the development of strong state and federal regulation of the insurance industries and corporate creative accounting in the aerospace industry, which provided support for the Foreign Corrupt Practices Act (FCPA) of 1977. Perhaps today, the **Sarbanes–Oxley Act of 2002 (SOX)** will be a vivid reminder of the importance of due professional care. SOX is a major reform package, mandating the most far-reaching changes Congress has imposed on the business world since the FCPA of 1977 and the **Securities and Exchange Commission (SEC)** Act of 1934. Examples of some of these significant changes include the creation of a **Public Company Accounting Oversight Board**,* as well as the increase of criminal penalties for violations of securities laws. SOX will be discussed in more detail in the next chapter.

---

* The PCAOB is a non-for-profit corporation instituted by Congress to oversee the audits of public companies in order to protect the interests of investors and further the public interest in the preparation of informative, accurate, and independent audit reports. http://pcaobus.org/Pages/default.aspx.

Financial Auditing

Financial auditing encompasses all activities and responsibilities concerned with the rendering of an opinion on the fairness of financial statements. The basic rules governing audit opinions indicate clearly that the **scope** of an audit covers all equipment and procedures used in processing significant data.

Financial auditing, as carried out today by the independent auditor, was spurred by legislation in 1933 and 1934 that created the SEC. This legislation mandated that companies whose securities were sold publicly be audited annually by a Certified Public Accountant (CPA). CPAs, then, were charged with attesting to the fairness of financial statements issued by companies that reported to the SEC. The AICPA issued in 1993 a document called *"Reporting on an Entity's Internal Control Structure over Financial Reporting (Statement on Standards for Attestation Engagements 2)"* to further define the importance of **internal control** in the **attestation engagement**.

Within the CPA profession in the United States, two groups of principles and standards have been developed that affect the preparation of financial statements by publicly held companies and the procedures for their audit examination by CPA firms: **Generally Accepted Accounting Principles (GAAP)** and **Generally Accepted Auditing Standards (GAAS)**.

GAAP establishes consistent guidelines for financial reporting by corporate managers. As part of the reporting requirement, standards are also established for the maintenance of financial records on which periodic statements are based. An auditor, rendering an opinion indicating that financial statements are stated fairly, stipulates that the financial statements conform to GAAP. These accounting principles have been formulated and revised periodically by private-sector organizations established for this purpose. The present governing body is the **Financial Accounting Standards Board (FASB)**. Implementation of GAAP is the responsibility of the management of the reporting entity.

GAAS, the second group of standards, was adopted in 1949 by the AICPA for audits. These audit standards cover three categories:

- *General Standards* relate to professional and technical competence, **independence**, and due professional care.
- *Standards of Fieldwork* encompass planning, evaluation of internal control, sufficiency of evidential matter, or documentary evidence upon which findings are based.
- *Standards of Reporting* stipulate compliance with all accepted auditing standards, consistency with the preceding account period, adequacy of disclosure, and, in the event that an opinion cannot be reached, the requirement to state the assertion explicitly.

GAAS provide broad guidelines, but not specific guidance. The profession has supplemented the standards by issuing statements of authoritative pronouncements on auditing. The most comprehensive of these is the SAS series. SAS publications provide procedural guidance relating to many aspects of auditing. In 1985, the AICPA released a codification of the SAS No. 1–49. Today, the number of statements exceeds 120.

A third group of standards, called the **International Financial Reporting Standards (IFRS)**, has been recently created by the **International Accounting Standards Board (IASB)**[*] to respond to the increasing global business environment and address the need to compare financial statements

---

[*] The purpose of the IASB is to develop a single set of high-quality, understandable, enforceable, and globally accepted financial reporting standards based upon clearly articulated principles.

prepared in different countries. The AICPA defines IFRS as the "set of accounting standards developed by the IASB that is becoming the global standard for the preparation of public company financial statements." While many of the global organizations have already migrated to IFRS, the United States has yet to do so. Due to the size of the United States and its significant presence globally, however, U.S. GAAP still has significant global impact. This results in the two major accounting standard-setting efforts in the world: U.S. GAAP and IFRS. Nevertheless, all major nations have now established time lines to converge with or to adopt IFRS standards in the near future.

Internal versus External Audit Functions

There are two types of audit functions that exist today. They have very important roles in assuring the **validity** and integrity of financial accounting and reporting systems. They are the internal and external audit functions.

Internal Audit Function

The IIA defines internal auditing (IA) as "an independent, objective **assurance** and consulting activity designed to add value and improve an organization's operations." IA brings organizations a systematic and disciplined approach to assess and enhance their **risk management**, control, and governance processes, as well as to accomplish their goals and objectives.

IA departments are typically led by a **Chief Audit Executive (CAE)**, who directly reports to the **Audit Committee** of the **Board of Directors**. The CAE also reports to the organization's **Chief Executive Officer (CEO)**. The primary purpose of the IA function is to assure that management-authorized controls are being applied effectively. The IA function, although not mandatory, exists in most private enterprise or corporate entities, and in government (such as federal, state, county, and city governments). The mission, character, and strength of an IA function vary widely within the style of top executives and traditions of companies and organizations. IT audits is one of the areas of support for IA.

The IA group, if appropriately staffed with the resources, performs all year long monitoring and testing of IT activities within the control of the organization. Of particular concern to private corporations is the processing of data and the generation of information of financial relevance or **materiality**.

Given management's large part to play in the effectiveness of an IA function, their concern with the **reliability** and integrity of computer-generated information from which decisions are made is critical. In organizations where management shows and demonstrates concern about internal controls, the role of the IA grows in stature. As the IA function matures through experience, training, and career development, the external audit function and the public can rely on the quality of the internal auditor's work. With a good, continuously improving IA management and staff, the Audit Committee of the Board of Directors is not hesitant to assign additional reviews, consultation, and testing responsibilities to the internal auditor. These responsibilities are often broader in scope than those of the external auditor.

Within the United States, internal auditors from government agencies often come together to meet and exchange experiences through conferences or forums. For example, the Intergovernmental Audit Forum is an example of an event where auditors come together from city, county, state, and federal environments to exchange experiences and provide new information regarding audit techniques and methods. The IIA also holds a national conference that draws an auditor population from around the world, both private and government, to share experiences and discuss new audit methods and techniques.

External Audit Function

The external audit function evaluates the reliability and the validity of systems controls in all forms. The principal objective in such evaluation is to minimize the amount of substantial auditing or testing of transactions required to render an opinion on the financial statements.

External auditors are provided by public accounting firms and also exist in government as well. For example, the Government Accountability Office (GAO) is considered an external reviewer because it can examine the work of both federal and private organizations where federal funds are provided. The Watchdogs of Congressional Spending provide a service to the taxpayer in reporting directly to Congress on issues of mismanagement and poor controls. Interestingly, in foreign countries, an Office of the Inspector General or Auditor General's Office within that country prepares similar functions. Also, the GAO has been a strong supporter of the International Audit Organization, which provides government audit training and guidance to its international audit members representing governments worldwide.

From a public accounting firm standpoint, firms such as Deloitte, Ernst & Young, PricewaterhouseCoopers, and KPMG (altogether referred to as the "Big Four") provide these types of external audit services worldwide. The external auditor is responsible for testing the reliability of client IT systems and should have a special combination of skills and experience. Such an auditor must be thoroughly familiar with the audit **attest** function. The attest function encompasses all activities and responsibilities associated with the rending of an audit opinion on the fairness of the financial statements. Besides the accounting and auditing skills involved in performing the attest function, these external auditors also must have substantial IT audit experience. SOX now governs their role and limits of services that can be offered beyond audit.

What Is IT Auditing?

Before defining what IT auditing is, let us explain the difference between IS and IT. An IS, represented by three components (i.e., people, process, and IT), is the combination of strategic, managerial, and operational activities involved in managing information. The IT component of an IS involves the hardware, software, communication, and other facilities necessary to manage (i.e., input, store, process, transmit, and output) such information. Refer to Exhibit 1.2.

The term audit, according to ISACA, refers to the formal inspection and verification to check whether a standard or set of guidelines is being followed, records are accurate, or efficiency and effectiveness targets are being met. In combining both definitions above, IT auditing can be defined as the *formal, independent, and objective examination of an organization's IT infrastructure to determine whether the activities (e.g., procedures, controls, etc.) involved in gathering, processing, storing, distributing, and using information comply with guidelines, safeguard assets, maintain data integrity, and operate effectively and efficiently to achieve the organization's objectives.* IT auditing provides **reasonable assurance** (never absolute) that the information generated by applications within the organization is accurate, complete, and supports effective decision making consistent with the nature and scope of the engagement previously agreed.

IT auditing is needed to evaluate the adequacy of application systems to meet processing needs, evaluate the adequacy of internal controls, and ensure that assets controlled by those systems are

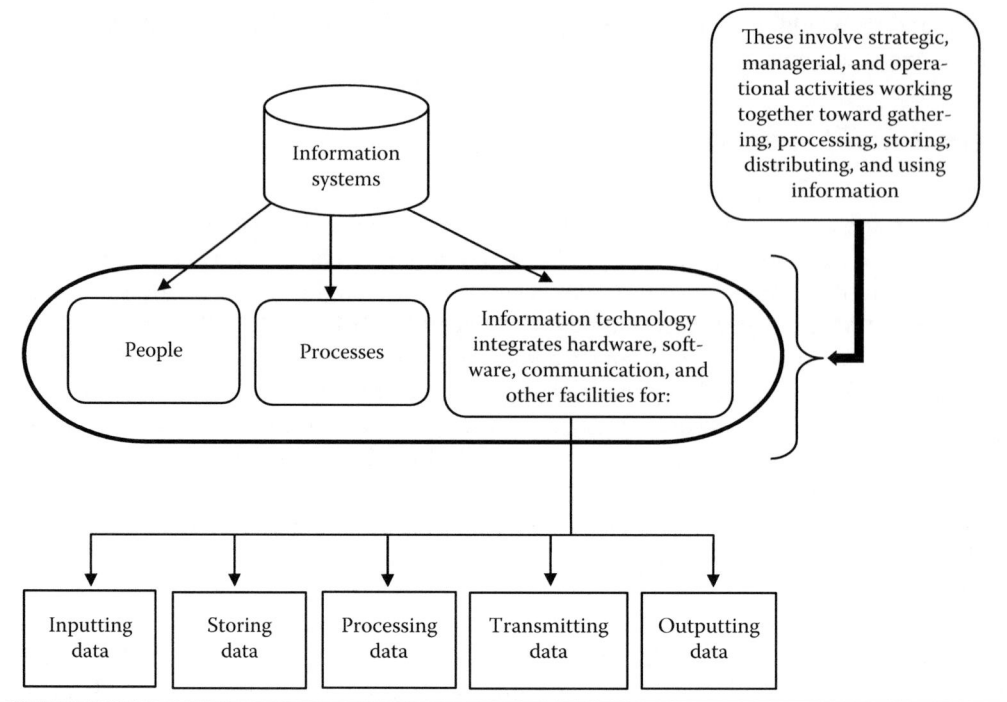

Exhibit 1.2 Information systems versus information technology.

adequately safeguarded. As for the IT auditors of today, their advanced knowledge and skills will progress in two ways. One direction is continued growth and skill in this profession, leading the way in computer audit research and development and progressing up the external and internal audit career paths. The other direction involves capitalizing on a thorough knowledge of organizational systems and moving into more responsible career areas in general management. Today, even in these economic times, the demand for qualified IT auditors exceeds the supply. IT governance has created vast opportunities for the IT auditor.

Learning new ways of auditing is always a priority of internal and external IT auditors. Most auditors want tools or audit methodologies that will aid them in accomplishing their task faster and easier. Almost every large organization or company has some sort of IT audit function or shop that involves an internal audit department. Today, the "Big Four" firms have designated special groups that specialize in the IT audit field. They all have staff that perform these external IT audits. Most of these IT auditors assist the financial auditors in establishing the correctness of financial statements for the companies in which they audit. Others focus on special projects such as Internet security dealing with penetration studies, firewall evaluations, bridges, routers, and gateway configurations, among others.

There are two broad groupings of IT audits, both of which are essential to ensure the continued proper operation of IS. These are as follows:

- *General Computer Controls Audit.* It examines IT general controls ("general controls" or "ITGCs"), including policies and procedures, that relate to many applications and supports the effective functioning of application controls. General controls cover the IT infrastructure and support services, including all systems and applications. General controls commonly include controls over (1) IS operations; (2) information security (ISec); and (3) change control management (CCM) (i.e., system software acquisition, change and maintenance, program change, and application system acquisition, development, and maintenance). Examples of general controls within IS operations address activities such as data backups and offsite storage, job monitoring and tracking of exceptions to completion, and access to the job scheduler, among others. Examples of general controls within ISec address activities such as access requests and user account administration, access terminations, and physical security. Examples of general controls within CCM may include change request approvals; application and database upgrades; and network infrastructure monitoring, security, and change management.

■ *Application Controls Audit.* It examines processing controls specific to the application. Application controls may also be referred to as "automated controls." They are concerned with the accuracy, completeness, validity, and authorization of the data captured, entered, processed, stored, transmitted, and reported. Examples of application controls include checking the mathematical accuracy of records, validating data input, and performing numerical sequence checks, among others. Application controls are likely to be effective when general controls are effective.

Refer to Exhibit 1.3 for an illustration of general and application controls, and how they should be in place in order to mitigate risks and safeguard applications. Notice in the exhibit that the application system is constantly surrounded by risks. Risks are represented in the exhibit by explosion symbols. These risks could be in the form of unauthorized access, loss or theft or equipment and information, system shutdown, etc. The general controls, shown in the hexagon symbols, also surround the application and provide a "protective shield" against the risks. Lastly, there are the application or automated controls which reside inside the application and provide first-hand protection over the input, processing, and output of the information.

IT Auditing Trends

Computing has become indispensable to the activities of organizations worldwide. The Control Objectives for Information and Related Technology (COBIT) Framework was created in 1995 by ISACA. COBIT, now on its fifth edition, emphasizes this point and substantiates the need to research, develop, publicize, and promote up-to-date, internationally accepted IT control objectives. In earlier documents such as the 1993 discussion paper "Minimum Skill Levels in Information Technology for Professional Accountants" and their 1992 final report "The Impact of Information Technology on the Accountancy Profession," the International Federation of Accountants (IFAC) acknowledges the need for better university-level education to address growing IT control concerns and issues.

Reports of information theft, computer fraud, information abuse, and other related control concerns are being heard more frequently around the world. Organizations are more information-conscious, people are scattered due to decentralization, and computers are used more extensively in all areas of commerce. Owing to the rapid diffusion of computer technologies and the ease of information accessibility, knowledgeable and well-trained IT auditors are needed to ensure that more effective controls are put in place to maintain data integrity and manage access to information. The need for better controls over IT has been echoed in the past by prior studies such as the AICPA Committee of Sponsoring Organizations of the Treadway Commission (COSO); International

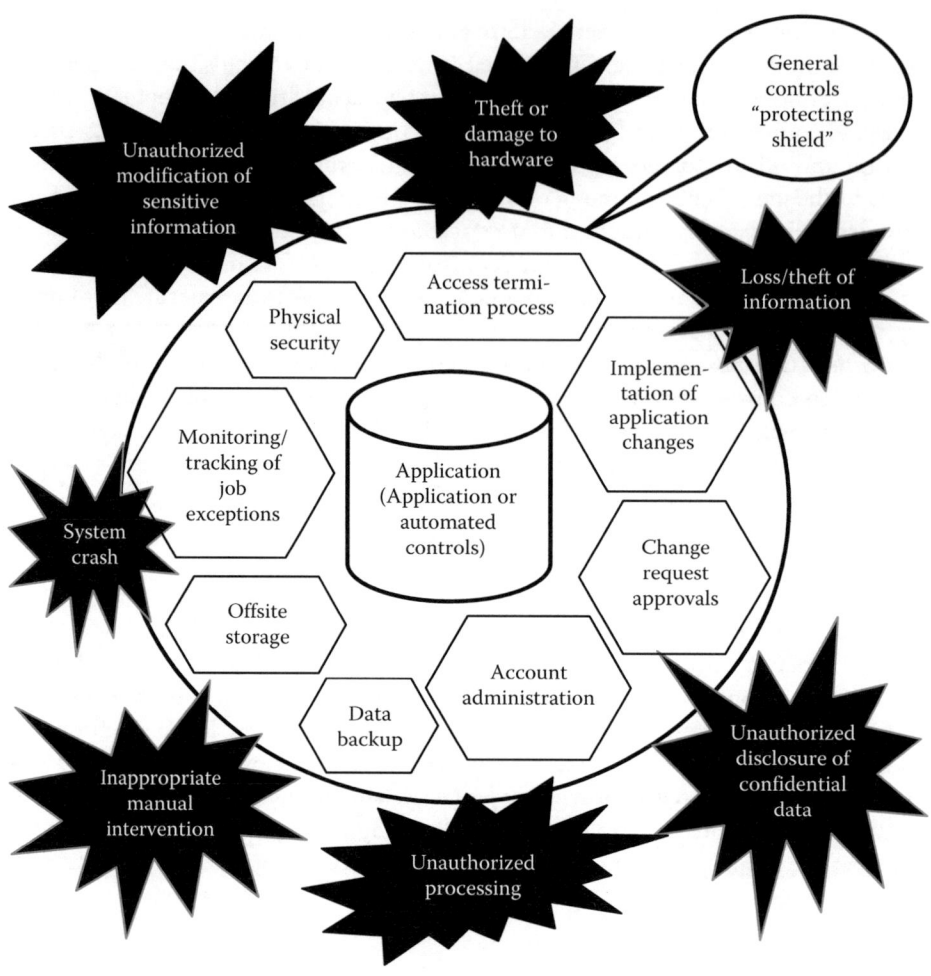

Exhibit 1.3 Relationship between general computer controls and application controls.

Organization for Standardization (ISO) 17799 and 27000; the IIA Systems Auditability and Control Report; Guidelines for the Security of IS by the OECD; the U.S. President's Council on Integrity and Efficiency in Computer Audit Training curriculum; and the United States' National Strategy for Securing Cyberspace released in 2002; among others.

The AICPA's Assurance Services Executive Committee (ASEC) is responsible for updating and maintaining the Trust Services Principles and Criteria (TSPC) and creating a framework of principles and criteria to provide assurance on the integrity of information. TSPC presents criteria for use by practitioners when providing professional attestation or **advisory** services to assess controls relevant to the following principles:

- *Security*: The system is protected against unauthorized access (both physical and logical).
- *Availability*: The system is available for operation and use as committed or agreed.
- *Processing integrity*: System processing is complete, accurate, timely, and authorized.
- *Confidentiality*: Information designated as confidential is protected as committed or agreed.

■ *Privacy*: Personal information is collected, used, retained, disclosed, and destroyed in conformity with the commitments in the entity's privacy notice and with criteria set forth in generally accepted privacy principles issued by the AICPA and CICA.

The theory and methodologies of IT auditing are integrated from five areas: a fundamental understanding of business, traditional auditing, IT management, behavioral science, and IT sciences. Business understanding and knowledge are the cornerstones of the audit process. Traditional auditing contributes knowledge of internal control practices and overall control philosophy within a business enterprise. IT management provides methodologies necessary to achieve successful design and implementation of systems. Behavioral science indicates when and why IT are likely to fail because of people's problems. IT sciences contribute to knowledge about control theory and the formal models that underlie hardware and software designs as a basis for maintaining data integrity.

Ever since the ISACA was formed there has been a growing demand for well-trained and skilled IT audit professionals. The publication *The EDP Auditors Association: The First Twenty-Five Years* documents the early struggles of the association and evolution of IT audit practices in this field.

The area of information assurance has also grown and evolved. The United States in its passage of the Cyber Security Research and Development Act has pledged almost a billion dollars for the development of curriculum, research, and skills for future professionals needed in this field.

Information Assurance

Organizations increasingly rely on critical digital electronic information capabilities to store, process, and move essential data in planning, directing, coordinating, and executing operations. Powerful and sophisticated threats can exploit security weaknesses in many of these systems. **Outsourcing** technological development to countries that could have terrorists on their development staff causes speculation that the potential exists for code to be implanted that would cause disruption, havoc, embezzlement, theft, and so on. These and other weaknesses that can be exploited become **vulnerabilities** that can jeopardize the most sensitive components of information capabilities. However, we can employ deep, layered defenses to reduce vulnerabilities and deter, defeat, and recover from a wide range of threats. From an information assurance perspective, the capabilities that we must defend can be viewed broadly in terms of four major elements: local computing environments, their boundaries, networks that link them together, and their supporting infrastructure. The U.S. National Strategy for Securing Cyberspace is one of those initiatives.

The term "information assurance" is defined as information integrity (the level of confidence and trust that can be placed on the information) and service availability. In all contexts, whether business or government, it means safeguarding the collection, storage, transmission, and use of information. The ultimate goal of information assurance is to protect users, business units, and enterprises from the negative effects of corruption of information or denial of services. The Department of Homeland Security and Supporting Organizations such as the National Security Agency (NSA), Federal Bureau of Investigation (FBI), and Central Intelligence Agency (CIA) have all worked toward supporting this goal.

As the nation's IS and their critical infrastructures are being tied together (government and business), the points of entry and exposure increase, and thus, risks increase. The technological advancement toward higher bandwidth communication and advanced switching systems has reduced the number of communications lines and further centralized the switching functions. Survey data indicates that the increased risk from these changes is not widely recognized. Since 9/11, more coordinated efforts have been made by U.S. defense organizations such as the Defense Information Systems Agency to promulgate standards for the Defense Information Infrastructure and the Global Information Grid, which should have a positive impact on information assurance that will extend beyond the U.S. Department of Defense and impact all segments of the national economy. The NSA has drafted and produced standards for IT security personnel that not only impact federal agencies but also corporate entities who contract IT services in support of the federal government. NIST, for example, has generated security guidance for Health Insurance Portability and Accountability Act compliance that impacts the medical profession and all corporations/business servicing the health field who handle medical information. A similar example includes the Payment Card Industry Data Security Standards (PCI DSS), maintained, managed, and promoted by the PCI Security Standards Council (Council) worldwide. The Council was founded in 2006 by major credit card companies, such as, American Express, Discover, JCB International, MasterCard, and Visa, Inc. These companies share equally in governance, execution, and compliance of the Council's work. PCI DSS refer to technical and operational requirements applicable specifically to entities that store, process, or transmit cardholder data, with the intention of protecting such data in order to reduce credit card fraud.

Need for IT Audit

Initially, IT auditing (formerly called electronic data processing [EDP], computer information systems [CIS], and IS auditing) evolved as an extension of traditional auditing. At that time, the need for an IT audit came from several directions:

- Auditors realized that computers had impacted their ability to perform the attestation function.
- Corporate and information processing management recognized that computers were key resources for competing in the business environment and similar to other valuable business resource within the organization, and therefore, the need for control and auditability were critical.
- Professional associations and organizations, and government entities recognized the need for IT control and auditability.

The early components of IT auditing were drawn from several areas. First, traditional auditing contributes knowledge of internal control practices and the overall control philosophy. Another contributor was IS management, which provides methodologies necessary to achieve successful design and implementation of systems. The field of behavioral science provided such questions and analysis to when and why IS are likely to fail because of people problems. Finally, the field of computer science contributes knowledge about control concepts, discipline, theory, and the formal models that underlie hardware and software design as a basis for maintaining data validity, reliability, and integrity.

IT auditing became an integral part of the audit function because it supports the auditor's judgment on the quality of the information processed by computer systems. Auditors with IT audit skills were viewed as the technological resource for the audit staff. The audit staff often looked to them for technical assistance. The IT auditor's role evolved to provide assurance that adequate and appropriate controls are in place. Of course, the responsibility for ensuring that adequate internal controls are in place rests with management. The audit's primary role, except in areas of management advisory services, is to provide a statement of assurance as to whether adequate and reliable internal controls are in place and are operating in an efficient and effective manner. Management's role is to ensure and the auditors' role is to assure.

There are several types of needs within IT auditing, including organizational IT audits (management control over IT), technical IT audits (infrastructure, data centers, data communication), and application IT audits (business/financial/operational). There are also development/implementation IT audits (specification/requirements, design, development, and post-implementation phases), and compliance IT audits involving national or international standards.

When auditing IT, the breadth and depth of knowledge required are extensive. For instance, auditing IT involves:

- Application of risk-oriented audit approaches
- Use of computer-assisted audit tools and techniques
- Application of standards (national or international) such as the ISO* to improve and implement quality systems in software development and meet IT security standards
- Understanding of business roles and expectations in the auditing of systems under development as well as the purchase of software packaging and project management
- **Assessment** of information security, confidentiality, privacy, and availability issues which can put the organization at risk
- Examination and verification of the organization's compliance with any IT-related legal issues that may jeopardize or place the organization at risk
- Evaluation of complex systems development life cycles (SDLC) or new development techniques (i.e., prototyping, end-user computing, rapid systems, or application development)
- Reporting to management and performing a follow-up review to ensure actions taken at work

The auditing of IT and communications protocols typically involves the Internet, intranet, extranet, electronic data interchange, client servers, local and wide area networks, data communications, telecommunications, wireless technology, integrated voice/data/video systems, and the software and hardware that support these processes and functions. Some of the top reasons to initiate an IT audit include the increased dependence on information by organizations, the rapidly changing technology with new risks associated with such technology, and the support needed for financial statement audits.

SOX also requires the assessment of internal controls and makes it mandatory for SEC registrants. As part of the process for assessing the effectiveness of internal controls over financial reporting, management needs to consider controls related to the IS (including technologies) that support relevant business and financial processes. These controls are referred to as ITGCs (or IT general controls). As mentioned earlier, ITGCs are IT processes, activities, and/or procedures that are performed within the IT environment and relate to how the applications and systems are developed, maintained, managed, secured, accessed, and operated. Exhibit 1.4 illustrates other top reasons to have IT audits.

---

* Examples of ISO standards include ISO/IEC 27002, ISO/IEC 27000, and ISO 17799.

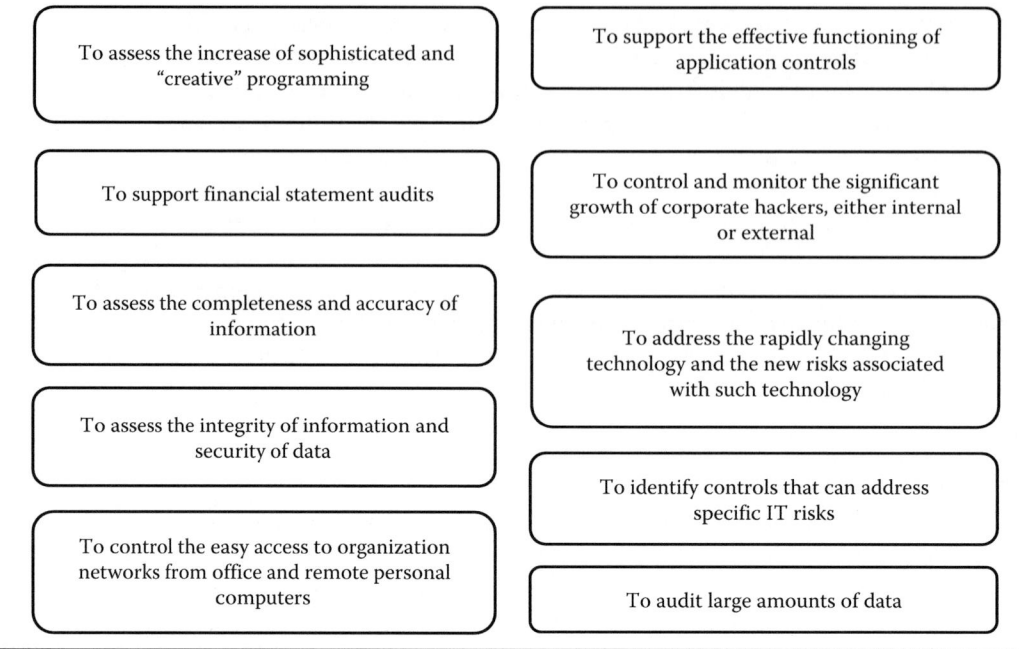

Exhibit 1.4 Top reasons for having an IT audit.

IT Governance

There have been many changes in the way enterprises address IT issues, resulting in a renewed focus on the concepts of IT governance. CEOs, **Chief Financial Officers**, **Chief Operating Officers**, **Chief Technology Officers**, and **Chief Information Officers** agree on the founding principles of IT governance, which focus on strategic alignment between IT and enterprise objectives. This, in turn, creates changes to tactical and day-to-day operational management of IT in the organization.

IT governance is the process by which an enterprise's IT is directed and controlled. As defined earlier, IT refers to the hardware, software, communication, and other facilities used to input, store, process, transmit, and output data in whatever form. Effective IT governance helps ensure that IT supports business goals, maximizes business investment in IT, and appropriately manages IT-related risks. IT governance also helps ensure achievement of critical success factors by efficiently and effectively deploying secure, reliable information, and applied technology.

Because IT impacts the operation of an entire organization, everyone within the organization should have an interest and role in governing its use and application. This growing awareness has led organizations to recognize that, if they are to make the most of their IT investment and protect that investment, they need a formal process to govern it. Reasons for implementing an IT governance program include:

- Increasing dependence on information and the systems that deliver the information
- Increasing vulnerabilities and a wide spectrum of threats
- Scale and cost of current and future investments in information and IS
- Potential for technologies to dramatically change organizations and business practices to create new opportunities and reduce costs

As long as these factors remain a part of business, there will be a need for effective, interdependent systems of enterprise and IT governance.

An open-standard IT governance tool that helps nontechnical and technical managers and auditors understand and manage risks associated with information and related IT is COBIT, developed by the IT Governance Institute and the Information Systems Audit and Control Foundation. COBIT is a comprehensive framework of control objectives that helps IT auditors, managers, and executives discharge fiduciary responsibilities, understand the IT systems, and decide what level of security and control is adequate. COBIT provides an authoritative, international set of generally accepted IT practices for business managers and auditors. COBIT is discussed in Chapter 3.

Role of the IT Auditor

The auditor evaluating today's complex systems must have highly developed technical skills to understand the evolving methods of information processing. Contemporary systems carry risks such as non-compatible platforms, new methods to penetrate security through communication networks (e.g., the Internet), and the rapid decentralization of information processing with the resulting loss of centralized controls.

As the use of IT in organizations continues to grow, auditing computerized systems must be accomplished without many of the guidelines established for the traditional auditing effort. In addition, new uses of IT introduce new risks, which in turn require new controls. IT auditors are in a unique position to evaluate the relevance of a particular system to the enterprise as a whole. Because of this, the IT auditor often plays a role in senior management decision making.

The role of IT auditor can be examined through the process of IT governance and the existing standards of professional practice for this profession. As mentioned earlier, IT governance is an organizational involvement in the management and review of the use of IT in attaining the goals and objectives set by the organization.

IT Auditor as Counselor

In the past, users have abdicated responsibility for controlling computer systems, mostly because of the psychological barriers that surround the computer. As a result, there are few checks and balances, except for the IT auditor. IT auditors must take an active role in assisting organizations in developing policies, procedures, standards, and/or best practices on safeguarding of the information, auditability, control, testing, etc. A good information security policy, for instance, may include:

- Specifying required security features
- Defining "reasonable expectations" of privacy regarding such issues as monitoring people's activities
- Defining access rights and privileges and protecting assets from losses, disclosures, or damages by specifying acceptable use guidelines for users
- Providing guidelines for external communications (networks)
- Defining responsibilities of all users
- Establishing trust through an effective password policy
- Specifying recovery procedures
- Requiring violations to be recorded

- Acknowledging that owners, custodians, and clients of information need to report irregularities and protect its use and dissemination
- Providing users with support information

The SANS Institute provides general information security policy templates on its Website, which can be downloaded and be a great starting point for any organization. A good computer security policy will differ for each organization, corporation, or individual depending on security needs. An information security policy will not guarantee a system's security or make the network completely safe from possible attacks from cyberspace. Nevertheless, a security policy, helped by effective security products and a plan for recovery, may help targeting potential losses to levels considered "acceptable," and minimize the leaking of private information. The IT auditor is part of an institutional team that helps create shared governance over the use, application, and assurance over IT within the organization.

An IT audit staff in a large corporation can make a major contribution to computer system control by persuading user groups to insist on a policy of comprehensive testing for all new systems and all changes to existing systems. By reviewing base-case results, user groups can control the accuracy of new or changed systems by actually performing a complete control function. Auditors must convince users and IT personnel of the need for a controlled IT environment.

Insisting that all new systems be reviewed at predefined checkpoints throughout the system's development life cycle can also enhance control of IT. The prospect of audit review should prompt both user and systems groups to define their objectives and assumptions more carefully. Here, too, IT auditors can subtly extend their influence.

IT Auditor as Partner of Senior Management

Although the IT auditor's roles of counselor and skilled technician are vital to successful company operation, they may be irrelevant if the auditor fails to view auditing in relation to the organization as a whole. A system that appears well controlled may be inconsistent with the operation of a business.

Decisions concerning the need for a system traditionally belonged to management, but because of a combination of factors (mostly the complex technology of the computer), computer system audits were not successfully performed. When allocating funds for new systems, management has had to rely on the judgment of computer personnel. Although their choices of new and more effective computer systems cannot be faulted, computer personnel have often failed to meet the true business needs of the organization.

Management needs the support of a skilled computer staff that understands the organization's requirements, and IT auditors are in such a position to provide that information. They can provide management with an independent assessment of the effect of IT decisions on the business. In addition, the IT auditor can verify that all alternatives for a given project have been considered, all risks have been accurately assessed, the technical hardware and software solutions are correct, business needs will be satisfied, and costs are reasonable.

IT Auditor as Investigator

As a result of increased legislation and the use of computer evidence within the courts, the ability to capture and document computer-generated information related to criminal activity is critical for purposes of prosecution. The awareness and use of computer-assisted tools and techniques in performing forensic support work have provided new opportunities for the IT auditor, IT security personnel, and those within law enforcement and investigation. For the IT audit professional, **computer forensics** is an exciting, developing field. The IT auditor can work in the field of computer forensics or work side by side with a computer forensics specialist, supplying insight into a particular system or network. The specialists can ask the IT audit professionals questions pertaining to the system and get responses faster than having to do research and figure everything out on their own. Although the specialist is highly trained and can adapt to almost any system or platform, collaboration can make the jobs of the forensic specialist and the IT professional easier and more efficient.

Since its birth in the early 1970s, computer forensics has continuously evolved into what is now a very large field. New technologies and enhancements in protocols are allowing engineers and developers to create more stable and robust hardware, software, and tools for the specialist to use in computer-related criminal investigations. As computers become more advanced and more abundant, so do criminal activities. Therefore, the computer forensics niche is also in constant progression along with the technological advancements of computers.

IT Audit: The Profession

With the passage of the Homeland Security Act, the Patriot Act, and SOX, the role of the auditor (internal and external) is more critical to the verification and validation of the financial infrastructure. The profession of IT auditing can provide a person with exposure to the way information flows within an organization and give its members the ability to assess its validity, reliability, and security. IT auditing involves people, technology, operations, and systems. It is a dynamic and challenging profession with a future that brings growth into new areas such as IT security and computer forensics, to name a few.

Today, IT auditors interact with managers, users, and technicians from all areas of most organizations. They must have interpersonal skills to interact with multiple levels of personnel and technical skills to understand the variety of technology used in information processing activity—especially technology used in generating and/or processing the company's financial information (e.g., financial statements, etc.). The IT auditor must also gain an understanding of and be familiarized with the operational environment to assess the effectiveness of the **internal control structure**. Finally, the IT auditor must understand the technological complexities of existing and future systems and the impact they have on operations and decisions at all levels.

IT auditing is a relatively new profession, and employment opportunities are present in all sectors of private industry, public accounting, and government worldwide. A profession is more than just an occupation. A profession has certain special characteristics, including a common body of knowledge, certification, continuing education, professional associations and ethical standards, and educational curriculum.

A Common Body of Knowledge

Since 1975, there have been various studies identifying a common body of knowledge for the IT audit profession. A common body of knowledge consists of clearly identified areas in which a person must attain a specific level of understanding and competency necessary to successfully practice within the profession. These areas are categorized into core areas. Organizations such as ISACA, AICPA, IIA, CICA, ISSA, InfoSec, and others around the world have issued major studies and papers on the topic of the knowledge, skills, and abilities needed to audit computer systems. Students, especially the ones with business and computer majors, receive a degree of base-level training in (1) auditing concepts and practices; (2) management concepts and practices; (3) computer systems, telecommunications, operations, and software; (4) computer information processing techniques; and (5) understanding of business on local and international scales. These are some of the major core areas of competency identified by the various independent studies for the individual who enters the IT audit, control, and security field.

Certification

Certification is a vital component of a profession. As you prepare for entry into your profession, whether it is accounting, IS, or other business fields, certification will be the measure of your level of knowledge, skills, and abilities in the profession. For example, attainment of the CPA designation is an important career milestone for the practicing accountant. In IT auditing, the Certified Information Systems Auditor (CISA) is one of the main levels of recognition and attainment. There are certain requirements for candidates to become CISA certified, such as:

■ Passing a rigorous written examination
■ Evidencing a minimum of 5 years of professional IS auditing, control or security work experience
■ Adhering to the ISACA's Code of Professional Ethics and the Information Systems Auditing Standards as adopted by ISACA
■ Agreeing to comply with the CISA Continuing Education Policy

The CISA examination covers areas (or domains) within the process of auditing IS; governance and management of IT; IS acquisition, development and implementation; IS operations, maintenance and service management; and the protection of information assets. Thus, university education plays an important part in providing the groundwork toward the certification process. Other licenses and certifications relevant to the IT auditor include the following: CPA, Certified Chartered Accountant (CA), Certified Internal Auditor (CIA), Certified Computer Professional (CCP), Certified Government Financial Manager (CGFM), Certified Information Systems Security Professional (CISSP), Certified Information Security Manager (CISM), Certified in Risk and Information Systems Control (CRISC), AICPA's Certified Information Technology Professional (CITP), and Certified Fraud Examiner (CFE).

Certification is important and a measure of skill attainment within the profession. Attainment of more than one certification will enhance your knowledge, skills, and abilities within the audit domain. Proficiency in skill application comes from experience and continuing education. The dynamic changes in business (commerce), IT, and world events continue to shape the future for this exciting profession.

Continuing Education

Certification requires continuing education so that those who are certified maintain a level of proficiency and continue their certification. Continuing education is an important element for career growth. As graduates enter their profession, they will find that their academic education is the foundation for continued development of career-enhancing knowledge, skills, and abilities. A continuing education requirement exists to support the CISA program. The IT auditor of the future will constantly face change with regard to existing systems and the dynamics of the environment (i.e., reorganization, new technology, operational change, and changing requirements).

The breadth and depth of knowledge required to audit IT is extensive. For example, IT auditing involves the application of risk-oriented audit approaches; the use of computer-assisted audit tools and techniques (e.g., EnCase, CaseWare, Idea, ACL, Guardant, eTrust, CA-Examine, etc.); the application of national or international standards (i.e., ISO 9000/3, ISO 17799, ISO 27000, and related amendments to improve and implement quality systems in software development); the auditing of systems under development involving complex SDLC or new development techniques (e.g., prototyping, end-user computing, rapid systems development, etc.); and the auditing of complex technologies involving electronic data interchange, client servers, local and wide area networks, data communications, telecommunications, and integrated voice/data/video systems.

Because the organizational environment in which the IT auditor operates is a dynamic one, it is important that new developments in the profession be understood so that they may be appropriately applied. Thus, the continuing education requirement helps the CISA attain new knowledge and skills to provide the most informed professional opinion. Training courses and programs are offered by a wide variety of associations and organizations to assist in maintaining the necessary skills that they need to continue to improve and evolve. Methods for receiving such training may even be global with video teleconferencing and telecommuting and with the Internet playing a major role in training delivery.

Professional Associations and Ethical Standards

As a manager at any level, one must remember that auditors, whether internal or external, have standards of practice that they must follow. Like IT professionals, auditors may belong to one or more professional associations and have code of ethics and professional standards of practices and guidance that help them in performing their reviews and audits. If they are seen not performing their work to "standards of practice" for their profession, they know they could be open to a potential lawsuit or even "decertified." Some of the organizations that produced such standards of practice are the AICPA, IIA, IFAC, CICA, GAO, and ISACA.

ISACA, created in 1969, is the leading IT governance, assurance, as well as security and control professional association today. ISACA:

- provides knowledge and education on areas like IS assurance, information security, enterprise governance, IT risk management, and compliance.
- offers globally known certifications/designations, such as, CISA, CISM, Certified in the Governance of Enterprise IT (CGEIT), and Certified in Risk and CRISC.
- develops and frequently updates international IS auditing and control standards, such as, the COBIT standard. COBIT assist both, IT auditors and IT management, in performing their daily duties and responsibilities in the areas of assurance, security, risk and control, and deliver value to the business.

To act as an auditor, one must have a high standard of moral ethics. The term *auditor* is Latin for one that hears complaints and makes decisions or acts like a judge. To act as a judge, one definitely must be morally ethical or it defeats the purpose. Ethics are a very important basis for our culture as a whole. If the auditor loses favor in this area, it is almost impossible to regain the trust the auditor once had with audit management and auditees. Whether an auditor is ethical in the beginning or not, they should all start off with the same amount of trust and good favor from the client or auditee. If the bond is not broken, the auditor establishes a good name as someone who can be trusted with sensitive material.

In today's world economy, trust is an unheard-of word. No one can trust anyone these days and for this reason it is imperative that high ethics are at the top of the manager's list of topics to cover with new audit teams. Times are changing and so are the clients requesting audit services. Most managers will state that they cherish this aspect called ethics because it distinguishes them from others without it.

For example, say a budget calls for numerous hours. It is unethical to put down hours not worked. It is also unethical to overlook something during the audit because the client says it is not important. A fine line exists between what is ethical and what is legal. Something can be ethically wrong but still legal. However, with that being said, some things initially thought to be unethical become illegal over time. If there is a large enough population opposed to something ethically incorrect, you will see legislation introduced to make it illegal.

When IT auditors attain their CISA certification, they also subscribe to a Code of Professional Ethics. This code applies to not only the professional conduct but also the personal conduct of IT auditors. The code is actually not in conflict with codes of ethics from other audit/assurance related domains (e.g., IIA, AICPA, etc.). It requires that the ISACA standards are adhered to, confidentiality is maintained, any illegal or improper activities are reported, the auditor's competency is maintained, due care is used in the course of the audit, the results of audit work are communicated, and high standards of conduct and character are maintained.

Educational Curricula

IT auditing is a profession with conduct, aims, and qualities that are characterized by worldwide technical and ethical standards. It requires specialized knowledge and often long and intensive academic preparation. Most accounting, auditing, and IT professional societies believe that improvements in research and education will definitely provide a "better-developed theoretical and empirical knowledge base for the IT audit function." They feel that emphasis should be placed on education obtained at the college level.

The academic communities both in the United States and abroad have started to incorporate portions of the common body of knowledge and the CISA examination domains into courses taught at the university level. Several recent studies indicate the growth of computer audit courses emerging in university curricula worldwide.

Various universities have developed curricula tailored to support the profession of IT auditing. Although the curricula at these universities constantly evolve, they currently exist at institutions such as Bentley University (Massachusetts), Bowling Green State University (Ohio), California State Polytechnic University, University of Mississippi, University of Texas, Georgia State University, University of Maryland, University of Tennessee, National Technological University (Argentina), University of British Columbia (Canada), York University (Canada), and the Hong Kong University of Science and Technology, among others. Graduates from these programs qualify for 1 year work experience toward their CISA certification.

A Model Curriculum for undergraduate and graduate education in IS and IT audit education was initially issued in March 1998 and updated in 2004, 2009, and 2011 by the IS Audit and Control Association and Foundation. The purpose of the Model is to provide colleges, universities, and/or educational institutions the necessary tools to educate students, and prepare them to enter the IT audit profession. Education through the Model focuses on fundamental course components of IT audit and control, as well as keeps up with the rapid pace of technological change. Such education is also in line with recent events, government regulations, and changes in business processes, all of which have affected the role of IT audit and the methodologies used by IT auditors.

IT Auditor Profile: Experience and Skills

Experience in IT audit is a definite must. Nothing in this world can compare to actual on-the-job, real-world experiences. Theory is also valuable, and for the most part an IT auditor should rely on theory to progress through an audit. For example, if IT auditors wish to demonstrate their commitment and knowledge level of the field, they can select an area to be tested. A number of professional certifications exist that can benefit the auditor. In the IT audit area, for instance, to pass the CISA exam, one must know, understand, and be able to apply the theory of modern IT auditing to all exam questions posed. There are other relevant licenses and certifications, as mentioned earlier, that can be very useful to an IT auditor's career and future plans.

The understanding of theory is definitely essential to the successful IT auditor. However, theory can only take one so far. This textbook and others available should be viewed as a guide. In this field, due to the technology complexity and situation, there comes a time when an IT auditor has to rely on experience to confront a new, never before encountered situation. Experience in the field is a definite plus, but having experience in a variety of other fields can sometimes be more beneficial. For example, an IT audit manager working for a Big Four public accounting firm is going to be exposed to a wide variety of IT audit situations and scenarios. Such experience will help broaden horizons and further knowledge in the IT audit field. Another example would be an Internal Audit Supervisor that has performed risk-focused and compliance audits for all departments within an organization. Such ample experience is nothing but a plus, and likely will allow the auditor to add significant, above-and-beyond value to the organization's operations.

Direct entry into the profession, as is the situation today, may change with entry-level requirements, including experience in business processes, systems, and technology, as well as sound knowledge of general auditing theory supplemented by practical experience. Additionally, IT auditors may require specific industry expertise such as banking, telecommunications, transportation, or finance and insurance to adequately address the industry-specific business/technology issues. This book provides current information and approaches to this complex field, which can help the practitioners and those wanting to learn more.

Experience comes with time and perseverance, as is well known, but auditors should not limit themselves to just one industry, software, or operating system. They should challenge themselves and broaden their horizons with a multitude of exposure in different environments, if possible. The broader and well rounded the IT auditor is, the better the chance for a successful audit career.

In addition to the experience, effective IT auditors must possess a variety of skills that enable them to add value to their organizations or clients. The finest technical experience or training does not necessarily fully prepare auditors for the communication and negotiation skills that are required for success.

Many of the nontechnical or supplemental skills are concerned with gathering information from and, of comparable importance, presenting information to people. As such, these supplemental skills are readily transferable to other disciplines, for example, finance, management, and marketing. The final product auditors create is an audit report. If the information within the audit report is not effectively and efficiently delivered via solid oral and written communication skills, all value accruing from the **audit process** could potentially be lost.

Having a diverse set of supplemental or "soft" skills never hurts when one is working with an **auditee**. For example, a senior IT auditor was recently conducting an audit in which she was faced with a client/auditee that was not very cooperative. During the questioning process, the senior IT auditor established a rapport with the client by using people skills or "soft skills." The role of an auditor is not an easy one when we are asked to review, question, and assess the work of others. Many times, the auditee must have a clear understanding of our role and that the auditor's focus is not to be critical of the individual but of the organizational policies, procedures, and process. The audit objectives focus on both the organization's goals and objectives.

Career Opportunities

There are a number of career opportunities available to the individual seeking an opportunity in IT audit. For the college graduate with the appropriate entry-level knowledge, skills, and abilities, this career provides many paths for growth and development. Further, as a career develops and progresses, IT audit can provide mobility into other areas as well. Today's IT auditors are employed by public accounting firms, private industries, management consulting firms, and the government.

Public Accounting Firms

Public accounting firms offer individuals an opportunity to enter the IT auditing field. Although these firms may require such individuals to begin their careers in financial audits to gain experience in understanding the organization's audit methodologies, after initial audit experience the individual who expresses interest in a particular specialization (e.g., forensics, security, etc.) will be transferred to such specialty for further training and career development. Many who have taken this career path have been successful, and several have become partners, principals, or directors within the firm. The primary sources for most public accounting firms are college recruitment and development within. However, it is not uncommon for a firm to hire from outside for specialized expertise (e.g., computer forensics, telecommunication, database systems, etc.).

Private Industry

Like public accounting firms, private industry offers entry-level IT audit professional positions. In addition, IT auditors gain expertise in more specialized areas (i.e., telecommunications, systems software, and systems design), which can make them candidates for IT operations, IT forensics, and IT security positions. Many CEOs view audit experience as a management training function. The IT auditor has particular strengths of educational background, practical experience with corporate IS, and understanding of executive decision making. Some companies have made a distinction between IT auditors and operational and financial auditors. Others require all internal auditors to be capable of auditing IT systems. Sources for persons to staff the IT audit function within a company generally may come from college recruitment, internal transfers, promotions, and/or outside hiring.

Management Consulting Firms

Another area of opportunity for IT audit personnel is management consulting. This career area is usually available to IT auditors with a number of years' experience. Many management consulting practices, especially those that provide services in the computer IS environment, hire experienced IT auditors. This career path allows these candidates to use their particular knowledge, skills, and abilities in diagnosing an array of computer and management information issues and then assist the organization in implementing the solutions. The usual resources for such positions are experienced personnel from public accounting CPA firms, private industries, and the government. IT forensics is another growing area in management consulting services.

Government

The government offers another avenue for one to gain IT audit experience. In the United States, federal, state, county, and city governments employ personnel to conduct IT audit-related responsibilities. Federal organizations such as the NSA, FBI, Department of Justice, and the CIA employ personnel who have IT audit experience, computer security experience, and IT forensics experience. Governments worldwide also employ personnel to conduct IT audits.

Government positions offer training and experience to personnel responsible for performing IT audit functions. Sources for government IT auditors are college recruits and employees seeking internal promotion or transfer. There are occasions when experienced resources may be hired from the outside as well.

Conclusion

Business operations are changing at a rapid pace because of the fast continuing improvement of technology. Technology has impacted various areas of the business environment, including the use and processing of information, existing control processes, and how audits are performed to draw conclusions regarding operational or system effectiveness, efficiency, and reporting integrity. It is also noted that technology constantly changes and identifies ways to shape today's IT environments in the organization. There were several recent technologies described that have and certainly will continue to revolutionize organizations, in particular how business is done and the dynamics of the workplace.

Because of major corporate and accounting fraud and scandals, the auditing profession, both internal and external functions, now looks seriously at the lack of controls in computer information systems. Within financial auditing, for instance, there are principles and standards that rule the CPA profession in the United States (i.e., GAAP and GAAS). These look for accurate preparation of financial statements as well as effective procedures for their audit examinations. A different type of auditing, IT auditing, has become an integral part of the audit function because it supports the auditor's judgment on the quality of the information processed by computer systems. IT auditing provides reasonable assurance (never absolute) that the information generated by applications within the organization is accurate, complete, and supports effective decision making consistent with the nature and scope agreed. There are two broad groupings of IT audits (i.e., General Computer Controls Audit and Application Controls Audit), both essential to ensure the continued proper operation of IS.

For the IT auditor, the need for audit remains critical and continues to be a demanding one. There are many challenges ahead; everyone must work together to design, implement, and safeguard the integration of new and existing technologies in the workplace. Given the various role hats IT auditors can wear, they must keep updated with reviews and changes in the existing laws governing the use of computers and the Internet. IT auditors can provide leverage in helping organizations understand the risks they face and the potential for consequences.

Review Questions

1. Technology has impacted the business environment in three areas. Summarize those areas.
2. Differentiate between internal and external auditors in terms of their roles and responsibilities.
3. How is IT auditing defined?
4. General Computer Controls Audit and Application Controls Audit are the two broad groupings of IT audits. Summarize both audits and provide specific examples supporting the controls evaluated within each type of audit.
5. The TSPC, maintained by the AICPA's ASEC, presents criteria for use by practitioners when providing professional attestation or advisory services to assess controls relevant to five principles. Describe in your own words these principles.
6. Explain what information assurance is.
7. One of the roles of the IT auditor is to act as a Counselor to organizations. As a Counselor, IT auditors can assist organizations in developing policies, procedures, standards, and/or best practices, such as an information security policy. Using the characteristics of a good information security policy listed in the chapter, develop five information security policies you would share with your client.
8. Explain why IT audit is considered a profession. Describe the requirements for candidates to become CISA certified.
9. What is ISACA and how does it helps the IT audit profession?
10. Where are the current career opportunities for the IT auditor? Search the Internet and identify at least one job profile/description for each career opportunity identified above. For each job profile identified, list the following in a table form:
    a. Job description
    b. Duties, tasks, and responsibilities required
    c. Minimum job requirements (or qualifications)
    d. Minimum education and/or certification requirements
    e. Knowledge, skills, and abilities required, etc.

Exercises

1. After reading this chapter, you should feel comfortable about the general roles and responsibilities of an IT auditor.
    a. Describe in your own words what do IT auditors do.
    b. Why should they be part of the overall audit team when performing the annual financial audit of a client?
2. List five Websites you can go to for information about:
    a. IT auditing
    b. IT security and privacy issues
3. Visit the Websites of four external audit organizations: two private and two government sites. Provide a summary of who they are and their roles, function, and responsibilities.
4. Interview an IT auditor and gather the following information:
    a. Position and company?
    b. Number of years of experience in IT auditing?
    c. Degree(s) and professional certifications?
    d. Career path?

  e. Why did he or she join IT auditing?

  f. Likes and dislikes about IT auditing?

  g. Where do they see themselves 5 years from now?

 5. You are asked by your IT audit manager to:

  a. Prepare a list of at least five professional certifications/designations that would be helpful for the IT audit staff to have. In a three-column table format, document the name of the professional certification or designation, name of the issuance professional organization, reasons why you think it would be relevant for the IT auditor, and the source link of the Website or source examined.

Further Reading

1. AICPA IFRS Resources. *What Is IFRS?* www.ifrs.com/ifrs_faqs.html#q1 (accessed October 2016).
2. American Institute of Certified Public Accountants (AICPA). (2011). *Top Technology Initiatives*, www.aicpa.org/InterestAreas/InformationTechnology/Resources/TopTechnologyInitiatives/Pages/2011TopTechInitiatives.aspx
3. Chen, Y., Paxson, V., and Katz, R. H. (2010). *What's New about Cloud Computing Security?* Technical report UCB/EECS-2010-5, EECS Department, University of California, Berkeley, 2010, www.eecs.berkeley.edu/Pubs/TechRpts/2010/EECS-2010-5.html
4. Deloitte. *Cloud Computing in 2016-Private Company Issues and Opportunities*, www2.deloitte.com/us/en/pages/deloitte-growth-enterprise-services/articles/private-company-cloud-computing.html (accessed October 2016).
5. EY Center for Board Matters. (September 2015). *EY Big Data and Analytics in the Audit Process*, www.ey.com/Publication/vwLUAssets/ey-big-data-and-analytics-in-the-audit-process/$FILE/ey-big-data-and-analytics-in-the-audit-process.pdf (accessed December 2015).
6. NIST. Final version of NIST cloud computing definition published, www.nist.gov/news-events/news/2011/10/final-version-nist-cloud-computing-definition-published (accessed October 2011).
7. Gallegos, F. (2002). Due professional care. *Inf. Syst. Control J.*, 2, 25–28.
8. Gallegos, F. (2003). IT auditor careers: IT governance provides new roles and opportunities. *IS Control J.*, 3, 40–43.
9. Gallegos, F. and Carlin, A. (July 2007). IT audit: A critical business process. *Comput. Mag.*, 40(7), 87–89.
10. Gartner IT Glossary. (n.d.). www.gartner.com/it-glossary/big-data/ (accessed October 2016).
11. Gartner's 2015 hype cycle for emerging technologies identifies the computing innovations that Organizations Should Monitor, www.gartner.com/newsroom/id/3114217 (accessed July 2015).
12. Gartner says the Internet of Things will transform the data center, www.gartner.com/newsroom/id/2684616 (accessed October 2014).
13. High Technology Crime Investigation Association. HTCIA.org
14. Ibrahim, N. IT Audit 101: Internal audit is responsible for evaluating whether IT risks are appropriately understood, managed, and controlled. *Internal Auditor*, http://go.galegroup.com/ps/i.do?id=GALE%7CA372553480&sid=googleScholar&v=2.1&it=r&linkaccess=fulltext&issn=00205745&p=AONE&sw=w&authCount=1&u=melb26933&selfRedirect=true (accessed June 2014).
15. IDC. Worldwide public cloud services spending forecast to reach $266 billion in 2021, according to IDC. USA, www.idc.com/getdoc.jsp?containerId=prUS42889917 (accessed July 2017).
16. Information Systems Audit and Control Foundation. *COBIT*, 5th Edition. Information Systems Audit and Control Foundation, Rolling Meadows, IL, www.isaca.org/Knowledge-Center/COBIT/Pages/Overview.aspx (accessed June 2012).
17. Information Systems Audit and Control Association. (2011). *CISA Examination Domain*, ISACA Certification Board, Rolling Meadows, IL.
18. ISACA. Innovation insights: Top digital trends that affect strategy. www.isaca.org/knowledge-Center/Research/Pages/isaca-innovation-insights.aspx (accessed March 2015).

19. ISACA. ISACA innovation insights, www.isaca.org/knowledge-center/research/pages/cloud.aspx (accessed September 2016).

20. ISACA. ISACA innovation insights, www.isaca.org/knowledge-Center/Research/Pages/isaca-innovation-insights.aspx (accessed September 2016).

21. ISACA. ISACA's glossary, www.isaca.org/Pages/Glossary.aspx?tid=1095&char=A (accessed October 2016).

22. ISACA. ISACA's glossary, www.isaca.org/Pages/Glossary.aspx?tid=1490&char=I (accessed October 2016).

23. ISACA. ISACA's glossary, www.isaca.org/Pages/Glossary.aspx?tid=1489&char=I (accessed October 2016).

24. ISACA. The code of professional ethics, Information Systems Audit Control Association Website, www.isaca.org

25. ISACA. ISACA's programs aligned with the model curriculum for IS audit and control, http://www.isaca.org/Knowledge-Center/Academia/Pages/Programs-Aligned-with-Model-Curriculum-for-IS-Audit-and-Control.aspx (accessed October 2016).

26. Nelson, B., Phillips, A., and Steuart, C. (2010). *Guide to Computer Forensics and Investigations*, Course Technology, Cengage Learning, Boston, MA.

27. Otero, A. R. (2015). Impact of IT auditors' involvement in financial audits. *Int. J. Res. Bus. Technol.*, 6(3), 841–849.

28. PCI Security. PCI Security Standards Council, www.pcisecuritystandards.org/pci_security/ (accessed October 2016).

29. SANS' Information Security Policy Templates. www.sans.org/security-resources/policies/general (accessed October 2016).

30. Senft, S., Gallegos, F., and Davis, A. (2012). *Information Technology Control and Audit*. CRC Press/Taylor & Francis, Boca Raton, FL.

31. Singleton, T. (2003). The ramifications of the Sarbanes–Oxley. *IS Control J.*, 3, 11–16.

32. AICPA. Statements on auditing standards, www.aicpa.org/research/standards/auditattest/pages/sas.aspx#SAS117 (accessed October 2016).

33. Takabi, H., Joshi, J. B. D., and Ahn, G. (2011). Security and privacy challenges in cloud computing environments. *IEEE Secur. Priv.*, 8(6), 24–31.

34. TechAmerica Foundation Federal Big Data Commission. (2012). Demystifying big data: A practical guide to transforming the business of government, https://bigdatawg.nist.gov/_uploadfiles/M0068_v1_3903747095.pdf (accessed December 2012).

35. The best mobile device management (MDM) solutions of 2016. *PC Magazine*, www.pcmag.com/article/342695/the-best-mobile-device-management-mdm-software-of-2016 (accessed November 2016).

36. Comprehensive National Cybersecurity Initiative. www.whitehouse.gov/cybersecurity/comprehensive-national-cybersecurity-initiative (accessed July 2012).

37. Institute of Internal Auditors. Definition of internal auditing, www.iia.org.au/aboutIIA/definition-OfIA.aspx (accessed October 2016).

38. Top 10 ERP software vendors and market forecast 2015–2020. Apps run the world. www.appsruntheworld.com/top-10-erp-software-vendors-and-market-forecast-2015-2020/ (accessed October 2016).

39. U.S. Securities and Exchange Commission. *SEC Announces Financial Fraud Cases*. Press Release, www.sec.gov/news/pressrelease/2016-74.html (accessed October 2016).

40. What is cloud computing? *PC Magazine*, www.pcmag.com/article2/0,2817,2372163,00.asp (accessed November 2016).

41. Worldwide public cloud services spending forecast to double by 2019, according to IDC, https://www.informationweek.com/cloud/infrastructure-as-a-service/idc-public-cloud-spending-to-double-by-2019/d/d-id/1324014 (accessed October 2016).

Chapter 2

Legislation Relevant to Information Technology

LEARNING OBJECTIVES

1. Discuss IT crimes and explain the three main categories of crimes involving computers.
2. Define cyberattack, and illustrate recent cyberattacks conducted on U.S. companies.
3. Summarize the Sarbanes–Oxley Act of 2002 federal financial integrity legislation.
4. Describe and discuss federal financial security legislation relevant to IT auditors.
5. Describe and discuss privacy-related legislation relevant to IT auditors.
6. Discuss state laws relevant to IT auditors.
7. Discuss international privacy laws relevant to IT auditors.

The Internet has grown exponentially from a simple linkage of a relatively few government and educational computers to a complex worldwide network that is used by businesses, governments, and the public—almost everyone from the computer specialist to the novice user and everyone in between. Today, common uses for the Internet include everything from accounting, marketing, sales, and entertainment purposes to e-mail, research, commerce, and virtually any other type of information sharing. As with any breakthrough in technology, advancements have also given rise to new legislation. This chapter focuses on legislation that governs the use and misuse of IT. It is believed that legislation has had a lasting impact on the online community (government, business, and the public), which is something that those entering the IT audit and information assurance profession should be knowledgeable of.

IT Crimes and Cyberattacks

The IT explosion has opened up many new gateways for criminals, requiring organizations to take the necessary precautions to safeguard their intellectual assets against computer crime. According to the 2016 Internet Crime Report, the FBI's Internet Crime Complaint Center (IC3) received a total of 298,728 complaints with reported losses in excess of $1.3 billion. In 2015, the FBI received

127,145 complaints from a total of 288,012 concerning suspected Internet-facilitated criminal activity actually reported having experienced a loss. Total losses reported on 2015 amounted to $1,070,711,522 (or almost a 134% increase from the 2014 total reported loss of $800,492,073). In 2014, there were 123,684 complaints received (from a total of 269,422) by the FBI that actually reported a loss from online criminal activity. In 2015, most of the continuing complaints received by the FBI involved criminals hosting fraudulent government services websites in order to acquire **personally identifiable information (PII)** and to collect fraudulent fees from consumers. Other notable ones from 2014 to 2016 involved "non-payment" (i.e., goods/services shipped or provided, but payment never rendered); "non-delivery" (i.e., payment sent, but goods/services never received); identity theft; personal data breach; extortion; and others. Some of the most frequently reported Internet crimes from 2014 to 2016 are listed in Exhibit 2.1.

There are three main categories of crimes involving computers. These crimes may be committed as individual acts or concurrently. The first of these is where the computer is the target of the crime. Generally, this type of crime involves the theft of information that is stored in the computer. This also covers unauthorized access or modification of records. The most common way to gain unauthorized access is for the criminal to become a "super-user" through a backdoor in the system. The backdoor in the system is there to permit access should a problem arise. Being a super-user is equivalent to being the system's manager and it allows the criminal access to practically all areas and functions within the system. This type of crime is of the greatest concern to industry.

The next general type of computer crime occurs when the computer is used as an instrument of the crime. In this scenario, the computer is used to assist the criminal in committing the crime. This category covers fraudulent use of automatic teller machine (ATM) cards, credit cards, telecommunications, and financial fraud from computer transactions.

In the third category, the computer is not necessary to commit the crime. The computer is incidental and is used to commit the crime faster, process greater amounts of information, and make the crime more difficult to identify and trace. The most popular example of this crime is child pornography. Owing to increased Internet access, child pornography is more widespread, easier to access, and harder to trace. IT helps law enforcement prosecute this crime because the incriminating information is often stored in the computer. This makes criminal prosecution easier. If the criminal is savvy, the computer is programmed to encrypt the data or erase the files if it is not properly accessed. Thus, the fields of computer forensics and computer security are opening new job opportunities for audit and security professionals who use their skills to capture the evidence.

A notoriously computer crime that organizations commonly deal with, and which also may involve all three types of computer crimes just explained, are cyberattacks. The Oxford Dictionary defines cyberattack as "an attempt by hackers to damage or destroy a computer network or system."[*] Another definition for a cyberattack is the deliberate and malicious exploitation of computer networks, systems, and (computer-generated) data by individuals or organizations to obtain valuable information from the users through fraudulent means.[†] Valuable, confidential, and/or sensitive information may take the form of passwords, financial details, classified government information, etc. Cyberattacks can be labeled as either **cyber campaigns**, **cyberwarfare**, or **cyberterrorism** depending on their context. Cyberterrorism, for instance, is discussed in a later section within this chapter.

---

[*] https://en.oxforddictionaries.com/definition/cyberattack.
[†] www.britannica.com/topic/cyberwar#ref1085374.

Exhibit 2.1   Frequently Reported Internet Crimes from the FBI's Internet Crime Complaint Center (IC3) from 2014 to 2016

Internet Crime	Description
Business Email Compromise (BEC)	Sophisticated scam targeting businesses working with foreign suppliers and/or businesses who regularly perform wire transfer payments.
Ransomware	Ransomware is a form of malware targeting both human and technical weaknesses in an effort to deny the availability of critical data and/or systems.
Tech Support Fraud	Tech support fraud occurs when the subject claims to be associated with a computer software or security company, or even a cable or Internet company, offering technical support to the victim.
Auto Fraud	Typical automobile fraud scam involves selling a consumer an automobile (listed on a legitimate Website) with a price significantly below its fair market value. The seller (fraudster) tries to rush the sale by stating that he/she must sell immediately due to relocation, family issues, need of cash, or other personal reasons. The seller does not allow for inspecting the automobile nor meet with the consumer face to face. The seller then asks the consumer to wire payment to a third-party agent, and to fax the payment receipt back to him or her as proof of payment. The seller keeps the money and never gets to deliver the automobile.
Government Impersonation E-mail Scam	This type of Internet crime involves posing as government, law enforcement officials, or simply someone pretending to have certain level of authority in order to persuade unaware victims to provide their personal information.
Intimidation/Extortion Scam	This type of crime utilizes demands for money, property, assets, etc. through undue exercise of authority (i.e., threats of physical harm, criminal prosecution, or public exposure) in order to extort and intimidate.
Real Estate Fraud	Similar to Auto Fraud. The seller (fraudster) tries to rush the sale of a house (with a price significantly below its market rental rates) by stating that he/she must sell immediately due to relocation, new employment, family issues, need of cash, or other personal reasons. Such significant price reduction is used to attract potential victims. The seller will then ask the consumer to provide personal identifying information and to wire payment to a third-party. Upon receiving payment, the seller is never found.
Confidence Fraud/ Romance Scam	This type of crime refers to schemes designed to look for companionship, friendship, or romance via online resources.

Cyberattacks have become increasingly common in recent years. Some of the most recent and infamous cyberattacks conducted on U.S. companies are listed in Exhibit 2.2. Let us now discuss current (federal, state, and international) legislation in place to deal with these computer crimes and attacks, and which are relevant to the IT auditor.

Exhibit 2.2   Recent Cyberattacks Conducted on U.S. Companies

Company / Industry	Cyberattack Description
Verizon (2017) / Telecommunications	Verizon, the major telecommunications provider, suffered a data security breach with over 6 million U.S. customers' personal details exposed on the Internet.[a]
Yahoo! (2016) / Internet Computer Software	Considered by many the largest data breach discovered in the history of the Internet. Breach took place in late 2014, where hackers stole information associated with at least 500 million Yahoo! user accounts, including names, e-mail addresses, telephone numbers, encrypted or unencrypted security questions and answers, dates of births, and encrypted password. Yahoo! publicly disclosed the data breach 2 years later on September 22, 2016.[b]
Experian PLC (2015) / Business Services	Servers storing credit assessment information (e.g., names, addresses, social security numbers, etc.) of more than 15 million customers were attacked by hackers.[c]
WhatsApp Inc. (2015) / Communications	Up to 200,000 users were either at risk of a cyberattack or have already had personal information compromised reported the cross-platform messaging application. Through internet connection, WhatsApp provides texting services, replacing the regular SMS text messages.[b]
Anthem, Inc. (2015) / Managed Health Care	Considered the largest healthcare breach to date, the cyberattack on Anthem affected up to 80 million current and former customers. Anthem, Inc.'s president and CEO Joseph Swedish stated that "Anthem was the target of a very sophisticated external cyberattack."[d] Hackers gained access to Anthem's computer system and got information including names, birthdates, medical IDs, Social Security numbers, street addresses, e-mail addresses, and employment information, including income data.
Chick-Fil-A, Inc. (2014) / Restaurant	Cyberattacks on point-of-sale systems during 10 months at numerous Chick-Fil-A restaurants resulted in around 9,000 credit cards being compromised.[b]
Staples, Inc. (2014) / Retail	Malware (software that damages or disables computer systems) was detected in the point-of-sale systems of 115 stores, affecting around 1.16 million credit cards.[b]

(Continued)

Exhibit 2.2 (*Continued*) Recent Cyberattacks Conducted on U.S. Companies

Company / Industry	Cyberattack Description
Sony Pictures Entertainment Inc. (2014) / Entertainment	A cyberattack on Sony Pictures Entertainment's computer networks stole significant amounts of private and of confidential data and also released them to the public. Hackers were believed to be linked to the North Korean government, which was extremely angry at the major Hollywood movie studio for producing a movie (i.e., The Interview) that portrayed North Korea in and negative way, and depicted the assassination of their leader.[e]
Target Corporation (2014) / Retail	Cyberattack during the 2013 Christmas holiday season compromised Target's computer systems and stole data from up to 40 million customers' credit and debit cards. Considered the second-largest breach reported by a U.S. retailer.[f]

[a] http://money.cnn.com/2017/07/12/technology/verizon-data-leaked-online/.
[b] www.cnbc.com/2016/09/22/yahoo-data-breach-is-among-the-biggest-in-history.html.
[c] www.heritage.org/research/reports/2015/11/cyber-attacks-on-us-companies-since-november-2014.
[d] www.usatoday.com/story/tech/2015/02/04/health-care-anthem-hacked/22900925/.
[e] www.vox.com/2014/12/14/7387945/sony-hack-explained.
[f] www.reuters.com/article/us-target-breach-idUSBRE9BH1GX20131219.

Federal Financial Integrity Legislation—Sarbanes–Oxley Act of 2002

It has been more than a decade since the Enron–Arthur Andersen LLP financial scandal (2001), but it still continues to plague today's financial market as the trust of the consumer, the investor, and the government to allow the industry to self-regulate have all been violated. The reminder of the Enron fiasco is today's scandals in the mortgage and mortgage investment market and the domino effect it has had on government, private industry, and the public.

Therefore, the Sarbanes–Oxley Act (SOX) of 2002, which changed the world of financial audit dramatically, will be a vivid reminder of the importance of due professional care. SOX prohibits all registered public accounting firms from providing audit clients, contemporaneously with the audit, certain non-audit services including internal audit outsourcing, financial-information-system design and implementation services, and expert services, among others. These scope-of-service restrictions go beyond existing Security and Exchange Commission (SEC) independence regulations. All other services, including tax services, are permissible only if preapproved by the issuer's audit committee and all such preapprovals must be disclosed in the issuer's periodic reports to the SEC. Issuers refer to a legal entity (e.g., corporations, etc.) that registers and sells securities in order to finance its operations.

SOX discusses requirements for the Board of Directors (board), including composition and duties. The board must (1) register public accounting firms; (2) establish or adopt, by rule, auditing, quality control, ethics, independence, and other standards relating to the preparation of audit reports for issuers; (3) conduct inspections of accounting firms; (4) conduct investigations and disciplinary proceedings, and impose appropriate sanctions; (5) perform such other duties or functions as necessary or appropriate; (6) enforce compliance with the act, the rules of the board, professional standards, and the securities laws relating to the preparation and issuance of audit reports and the obligations and liabilities of accountants with respect thereto; and (7) set the budget and manage the operations of the board and the staff of the board.

SOX is a major reform package mandating the most far-reaching changes. Congress has imposed on the business world since the Foreign Corrupt Practices Act of 1977 and the SEC Act of the 1930s. It seeks to thwart future scandals and restore investor confidence by, among other things, (1) creating the Public Company Accounting Oversight Board (PCAOB); (2) revising auditor independence rules and corporate governance standards; and (3) significantly increasing the criminal penalties for violations of securities laws. These are described below:

PCAOB

To audit a publicly traded company, a public accounting firm must register with the PCAOB. The PCAOB shall collect a registration fee and an annual fee from each registered public accounting firm in amounts that are sufficient to recover the costs of processing and reviewing applications and annual reports. The PCAOB shall also establish a reasonable annual accounting support fee to maintain its operations.

Annual quality reviews must be conducted for public accounting firms that audit more than 100 issuers; all others must be conducted every 3 years. The SEC and the PCAOB may order a special inspection of any registered audit firm at any time. The PCAOB can impose sanctions if the firm fails to reasonably supervise any associated person with regard to auditing or quality control standards.

It is unlawful for a registered public accounting firm to provide any non-audit service to an issuer during the same time with the audit. These non-audit services are listed below:

- **Bookkeeping** or other services related to the accounting records or financial statements of the audit client
- Financial IS design and implementation
- Appraisal or valuation services, fairness opinions, or contribution-in-kind reports
- **Actuarial services**
- Internal audit outsourcing services
- Management functions or human resources
- Broker or dealer, investment adviser, or investment banking services
- Legal services and expert services unrelated to the audit

The PCAOB may, on a case-by-case basis, exempt from the prohibitions listed above any person, issuer, public accounting firm, or transaction, subject to review by the commission. However, the SEC has oversight and enforcement authority over the PCAOB. The PCAOB, in its rulemaking process, is to be treated as if it were a registered securities association.

It will not be unlawful to provide other non-audit services if the audit committee preapproves them in the following manner. SOX allows an accounting firm to engage in any non-audit service, including tax services that are not listed previously, only if the audit committee of the issuer preapproves the activity. The audit committee will disclose to investors in periodic reports its decision to preapprove non-audit services. Statutory insurance company regulatory audits are treated as an audit service, and thus do not require preapproval.

Auditor Independence Rules and Corporate Governance Standards

For independence acceptance, SOX requires auditor (not audit firm) rotation. The lead audit or coordinating partner and the reviewing partner must rotate off of the audit every 5 years. SOX provides no distinction regarding the capacity in which the lead audit partner or the concurring review partner provided such audit services. Any services provided as a manager or in some other capacity appear to count toward the 5-year period. The provision starts as soon as the firm is registered, therefore, absent guidance to the contrary, the lead audit partner and the concurring review partner must count back 5 years starting with the date in which registration occurs. Also, the accounting firm must report to the audit committee all critical accounting policies and practices to be used, all alternative methods to Generally Accepted Accounting Principles (GAAP) that have been discussed with management, and ramifications of the use of such alternative disclosures and methods.

Another audit independence compliance issue is that the chief executive officer (CEO), controller, chief financial officer (CFO), chief accounting officer, or person in an equivalent position cannot be employed by the company's audit firm during the 1-year period preceding the audit. In addition and in order to comply with Section 302: Corporate Responsibility for Financial Reports, for example, both the CEO and the CFO of the company shall:

- prepare and sign off a statement (accompanying the audit report) to certify to **stakeholders** that the company's financial statements and all supplemental disclosures contained within the report are truthful, reliable, and fairly present, in all material respects, the operations and financial condition of the company.
- state that they are indeed responsible for implementing and maintaining the internal control structure.
- support that they have implemented all necessary steps to ensure that the disclosure processes and controls within the company consistently generate financial information that can be relied on by stakeholders.
- present conclusions about the effectiveness of the internal control structure resulting from their evaluation (such evaluation to occur within 90 days prior to the issuance of the report).
- identify for the company's external auditors:
  - any **deficiencies** (significant or not) in the design or operation of internal controls which could adversely affect the company's ability to record, process, summarize, and report financial information;
  - any **material weaknesses** in internal controls;
  - any fraud (material or not) that involves any company personnel who have a significant role in the company's internal controls; and
  - any significant changes implemented that could materially affect internal controls subsequent to the date of their evaluation.

A violation of this section must be knowing and intentional to give rise to liability. It shall be unlawful for any officer or director of an issuer to take any action to fraudulently influence, coerce, manipulate, or mislead any auditor engaged in the performance of an audit for the purpose of rendering the financial statements materially misleading. Another critical and related section of SOX is Section 404: Management Assessment of Internal Controls, which requires that the company's external auditors report on how reliable is the assessment of internal controls performed by management. For this, the annual financial report package that is prepared by the external auditors must include a report (i.e., internal control report) stating that management is responsible for implementing and maintaining an adequate internal control structure. Such report must also include the evaluation performed by management to support the effectiveness of the control structure. Any faults, deficiencies, or weaknesses identified as a result of the assessment must also be reported. The external auditors must further attest to the accuracy of the company management assertion that internal accounting controls are in place and operating effectively.

Increasing Criminal Penalties for Violations of Securities Laws

SOX penalizes executives for nonperformance. If an issuer is required to prepare a restatement due to material noncompliance with financial reporting requirements, the CEO and the CFO must reimburse the issuer for any bonus or other incentive- or equity-based compensation received during the 12 months following the issuance. SOX also prohibits the purchase or sale of stock by officers and directors and other insiders during **blackout periods**. Any profits resulting from sales in violation of this will be recoverable by the issuer.

Each financial report that is required to be prepared in accordance with GAAP shall reflect all material-correcting adjustments that have been identified by a registered accounting firm. Each annual and quarterly financial report shall disclose all material off-balance sheet transactions and other relationships with unconsolidated entities that may have a material current or future effect on the financial condition of the issuer. Also, directors, officers, and 10% or more owners must report designated transactions by the end of the second business day following the day on which the transaction was executed. SOX requires each annual report of an issuer to contain an internal control report. The SEC shall issue rules to require issuers to disclose whether at least one member of its audit committee is a financial expert. Also, the issuers must disclose information on material changes in the financial condition or operations of the issuer on a rapid and current basis.

SOX identifies as a crime for any person to corruptly alter, destroy, mutilate, or conceal any document with the intent to impair the object's integrity or availability for use in an official proceeding or to otherwise obstruct, influence, or impede any official proceeding, such a person being liable for up to 20 years in prison and a fine.

The SEC is also authorized to freeze the payment of an extraordinary payment to any director, officer, partner, controlling person, agent, or employee of a company during an investigation of possible violations of securities laws. Finally, the SEC may prohibit a person from serving as an officer or director of a public company if the person has committed securities fraud.

Federal Security Legislation

It appears that traditional security methods and techniques are simply not working. In fact, the literature argues that the current use of information security tools and technologies (e.g., encryption, firewalls, access management, etc.) alone is not sufficient to protect the information and address information security challenges. Similarly, current security legislation, although addressing issues of unwanted entry into a network, may allow for ways by which criminals can escape the most severe penalties for violating authorized access to a computer system. The computer networking industry is continually changing. Because of this, laws, policies, procedures, and guidelines must constantly change with it; otherwise, they will have a tendency to become outdated, ineffective, and obsolete.

The private industry has in the past been reluctant to implement these U.S. federal government laws because of the fear of the negative impact it could bring to a company's current and future earnings and image to the public. Following are descriptions of some of the U.S. Federal Government laws that regulate IT security.

Computer Fraud and Abuse Act of 1984

The Computer Fraud and Abuse Act (CFAA) was first drafted in 1984 as a response to computer crime. The government's response to network security and network-related crimes was to revise the act in 1994 under the Computer Abuse Amendments Act to cover crimes such as trespassing (unauthorized entry) into an online system, exceeding authorized access, and exchanging information on how to gain unauthorized access. Although the Act was intended to protect against attacks in a network environment, it does also have its fair share of faults.

The Act requires that certain conditions needed to be present for the crime to be a violation of the CFAA. Only if these conditions are present will the crime fall under violation of the CFAA. The three types of attacks that are covered under the Act and the conditions that have to be met include:

- *Fraudulent trespass.* This is when a trespass is made with the intent to defraud that results in both furthering the fraud and the attacker obtaining something of value.
- *Intentional destructive trespass.* This is a trespass along with actions that intentionally cause damage to a computer, computer system, network, information, data, or program, or results in **denial of service** and causes at least $1,000 in total loss in the course of a year.
- *Reckless destructive trespass.* This is when there is the presence of trespass along with reckless actions (although not deliberately harmful) that cause damage to a computer, computer system, network, information, data, or program, or results in denial of service and causes at least $1,000 in total loss in the course of a year.

Each of the definitions above is geared toward a particular type of attack. Fraudulent trespass was a response against crimes involving telephone fraud that is committed through a computer system, such as using a telephone company computer to obtain free telephone service. This condition helps prosecute individuals responsible for the large financial losses suffered by companies such as AT&T. Telephone toll fraud has snowballed into over a billion dollars a year problem for the phone companies. The other two usually apply to online systems and have been implemented to address problems of hackers or crackers, worms, viruses, and virtually any other type of intruder that can damage, alter, or destroy information. These two attacks are similar in many ways, but the key in differentiating the two are the words "intentional," which would, of course, mean a deliberate attack with the intent to cause damage, whereas "reckless" can cover an attack in which damage was caused due to negligence. Penalties under Section 1030(c) of the CFAA vary from 1-year imprisonment for reckless destructive trespass on a nonfederal computer to up to 20 years for an intentional attack on a federal computer where the information obtained is used for "the injury of the United States or to the advantage of any foreign nation" (i.e., cases of espionage).

Computer Security Act of 1987

Another act of importance is the Computer Security Act of 1987, which was drafted due to congressional concerns and public awareness on computer security-related issues and because of disputes on the control of unclassified information. The general purpose of the Act was a declaration from the government that improving the security of sensitive information in federal computer systems is in the public interest. The Act established a federal government computer security program that would protect sensitive information in federal government computer systems. It would also develop standards and guidelines for unclassified federal computer systems and facilitate such protection.*

The Computer Security Act of 1987 also assigned responsibility for developing government-wide computer system security standards, guidelines, and security training programs to the National Bureau of Standards (now the NIST). It further established a Computer System Security and Privacy Advisory Board within the Department of Commerce, and required federal agencies to identify computer systems containing sensitive information and develop security plans for those systems. Finally, it provided periodic training in computer security for all federal employees and contractors who managed, used, or operated federal computer systems.

The Computer Security Act of 1987 is particularly important because it is fundamental to the development of federal standards of safeguarding unclassified information and establishing a balance between national security and other non-classified issues in implementing security policies within the federal government. It is also important in addressing issues concerning government control of **cryptography**.

Homeland Security Act of 2002

The terrorist attack events of September 11, 2001 prompted the passage of the Homeland Security Act of 2002, whose purpose was to prevent terrorist attacks within the United States and to reduce the vulnerability of the United States to terrorism. It plays a major role in the security of cyberspace because it enforces many limitations and restrictions to users of the Internet. For example, one goal of the Act is to establish an Internet-based system that will only allow authorized persons the access to certain information or services. Owing to this restriction, the chances for vulnerability and attacks may decrease. The impact of this Act will definitely contribute to the security of cyberspace because its primary function is to protect the people of the United States from any form of attack, including Internet attacks. The passage of the Homeland Security Act of 2002 and the inclusion of the Cyber Security Enhancement Act (CSEA) within that Act makes the need to be aware and practice **cybersecurity** everyone's business.

The CSEA (H.R. 3482) was incorporated into the Homeland Security Act of 2002. The CSEA demands life sentences for those hackers who recklessly endanger lives. The Act also included provisions that seek to allow Net surveillance to gather telephone numbers, Internet Protocol (IP) addresses, and universal resource locaters (URLs) or e-mail information without recourse to a court where an "immediate threat to a national security interest" is suspected. Finally, Internet Service Providers (ISPs) are required to hand over users' records to law enforcement authorities, overturning current legislation that outlaws such behavior.

The Homeland Security Act of 2002 added phrasing that seeks to outlaw the publication anywhere of details of tools such as Pretty Good Privacy, which encode e-mails so that they cannot be read by snoops. This provision allows police to conduct Internet or telephone eavesdropping randomly with no requirement to ask a court's permission first. This law has a provision that calls for punishment of up to life in prison for electronic hackers who are found guilty of causing death to others through their actions. Any hacker convicted of causing injuries to others could face prison terms up to 20 years under cybercrime provisions, which are in Section 225 of the CSEA provision of the Homeland Security Act.

---

* Office of Technology Assessment, Issue Update on Information Security and Privacy in Networked Environments, p. 105.

Payment Card Industry Data Security Standards of 2004

Payment Card Industry Data Security Standards (PCI DSS) refer to technical and operational requirements applicable to entities that store, process, or transmit cardholder data, with the intention of protecting such data in order to reduce credit card fraud. PCI DSS are maintained, managed, and promoted by the PCI Security Standards Council (Council) worldwide to protect cardholder data. The Council was founded in 2006 by major credit card companies, such as American Express, Discover, JCB International, MasterCard, and Visa, Inc. These companies share equally in governance, execution, and compliance of the Council's work.

All merchants that either accept or process payment through cards must comply with the PCI DSS. Some specifics goals and requirements of PCI DSS include the following:

- Building and maintaining a secure network—implement a strong firewall configuration; avoid using vendor-supplied defaults for system passwords which are easy to decipher
- Protecting stored cardholder data—employ encryption techniques on all transmissions of cardholder data
- Maintaining a vulnerability management program—develop stronger, secure systems; implement (and update as necessary) anti-virus software or programs
- Implementing strong access control measures—assign unique IDs; configure access to cardholder data to the minimal level possible consistent with business needs, related tasks, and responsibilities (i.e., principle of least privilege); restrict physical access to cardholder data
- Monitoring and testing networks—monitor all access to network resources where cardholder data are being transmitted; regularly test the security systems transmitting and processing cardholder data
- Maintaining an information security policy—specify required security features and acceptable use guidelines for users; define user expectations, responsibilities, and access rights and privileges

Federal Information Security Management Act of 2002

The Federal Information Security Management Act (FISMA) was enacted as part of the E-Government Act of 2002 to "provide a comprehensive framework for ensuring the effectiveness of information security controls over information resources that support Federal operations and assets and to provide for development and maintenance of minimum controls required to protect Federal information and information systems." In other words, FISMA requires federal agencies to develop, document, and put in place information security programs with the purpose of protecting both, the information and the systems implemented to support the operations and assets of the agencies, including those provided or managed by another agency, contractor, or other source. Specifically, FISMA requires Federal agencies to:

- ensure that appropriate officials (e.g., Chief Information Officer, etc.) are assigned security responsibility and authority to ensure compliance with the requirements imposed by FISMA
- plan and implement information security programs
- develop and maintain inventories of the agency's major information systems
- have annual independent evaluations (i.e., free of any bias or influence) of their information security program and practices
- report on the adequacy and effectiveness of their information security controls, policies, procedures, and practices

It is critical that agencies understand and, most importantly, implement the tasks listed above in order to mitigate risks to acceptable levels and other factors that could adversely affect their missions. Agencies must constantly monitor and assess their information security programs to safeguard the information (and systems generating it) from events that may result from unauthorized access, as well as the use, disclosure, disruption, modification, or destruction of information.

Electronic Signature Laws—Uniform Electronic Transactions Act of 1999 and Electronic Signatures in Global and National Commerce Act of 2000

An area of concern for many companies involves electronic signatures. Similar to online storage, electronic signatures can significantly improve business operations even though "care must be taken to avoid compromising sensitive customer data and/or violating government regulations on the subject."

There are at least two main pieces of legislation with respect to electronic signature laws that companies should know about: Uniform Electronic Transactions Act (UETA) and Electronic Signatures in Global and National Commerce Act (ESIGN). With these two laws, companies can significantly speed business transaction turnaround times by stating their agreement to contractual terms with only a click of a mouse (i.e., replacing traditional paper signature documents with electronic forms).

UETA is one of the several U.S. Uniform Acts proposed by the National Conference of Commissioners on Uniform State Laws. It exists to harmonize state laws concerning retention of paper records (especially checks) and the validity of electronic signatures. UETA was introduced in 1999 and has been adopted by 47 U.S. states, as well as the District of Columbia, Puerto Rico, and the U.S. Virgin Islands. Simply put, UETA makes electronic signatures valid and in compliance with law requirements when parties ready to enter into a transaction have agreed to proceed electronically.

ESIGN, on the other hand, is a federal law passed by the U.S. Congress in 2000. Like UETA, ESIGN recognizes electronic signatures and records granted all contract parties opt to use electronic documents and sign them electronically. In other words, with ESIGN, documents with electronic signatures and records are equally as good as their standard paper equivalents, and therefore subject to the same legal examination of authenticity that applies to traditional paper documents and wet ink signatures.

For an electronic signature to be recognized as valid under U.S. law (ESIGN and UETA), the following must take place:

- There must be a clear intent to sign by all involved parties.
- Parties to the transaction must consent to do business electronically.
- The application system used to capture the electronic signature must be configured and ready to retain (for validation purposes) all processing steps performed in generating the electronic signature, as well as the necessary electronic signature records for accurate and timely reproduction or restoration, if needed.

Privacy Legislation

On the subject of privacy, in 2009, the California Department of Public Health (CDPH) found that a Children's Hospital of Orange County sent patient records by mistake to an auto shop.

The auto shop business received six faxes containing healthcare information, including information that identified the patient's name, date of birth, and details about the visits. Hospital staff told the CDPH that a test fax should have been sent first, per hospital policy. This is an example of a privacy breach. Privacy, as defined by ISACA, involves the "freedom from unauthorized intrusion or disclosure of information about an individual." Privacy focuses on protecting personal information about customers, employees, suppliers, or business partners. Organizations have an ethical and moral obligation to implement controls to protect the personal information that they collect.

Privacy of information has also been accessed by criminals within the online world. Some of the legislation passed does protect the user against invasion of privacy. However, some of the laws observed contain far too many exceptions and exclusions to the point that their efficacy suffers. In addition, the government continues to utilize state-of-the-art techniques for the purpose of accessing information for the sake of "national security" justified currently under the Homeland Security Act. New bills and legislation continue to attempt to find a resolution to these problems, but new guidelines, policies, and procedures need to be established, and laws need to be enforced to their full extent if citizens are to enjoy their right to privacy as guaranteed under the constitution.

Privacy Act of 1974

In addition to the basic right to privacy that an individual is entitled to under the U.S. Constitution, the government also enacted the Privacy Act of 1974. The purpose of this is to provide certain safeguards to an individual against an invasion of personal privacy. This act places certain requirements on federal agencies, such as permitting individuals to[*]:

- determine what personal records are collected and maintained by federal agencies
- prevent personal records that were obtained for a particular purpose from being used or made available for another purpose without consent
- gain access to their personal information in federal agency records and to correct or amend them

The Act also requires federal agencies to collect, maintain, and use any private information in a manner that assures that such action is for a necessary and lawful purpose, that the information is current and accurate, and that safeguards are provided to prevent misuse of the information.

Although the Privacy Act of 1974 is an important part of safeguarding individual privacy rights, it is important for the IT auditor to recognize that there are many exemptions under which it may be lawful for certain information to be disclosed. This could, in some cases, allow federal and nonfederal agencies the means by which they can obtain and disclose information on any individuals simply because they may fall under one of the many exemptions that the Privacy Act allows. For example, the subsequent Freedom of Information Act provides the federal government a way to release historical information to the public in a controlled fashion. The Privacy Act of 1974 has also been updated over time through the amendment process.

Electronic Communications Privacy Act of 1986

In the area of computer networking, the Electronic Communications Privacy Act of 1986 is one of the leading early pieces of legislation against violation of private information as applicable to

---

[*] CSR Privacy/Information Archive, Privacy Act of 1974 and Amendments.

online systems. The Act specifically prohibits interception and disclosure of wire, oral, or electronic communications, as well as the manufacture or possession of intercepting devices.

Communications Decency Act of 1996

The Communication Decency Act (CDA) of 1996 bans the making of "indecent" or "patently offensive" material available to minors through computer networks. The Act imposes a fine of up to $250,000 and imprisonment for up to 2 years. The CDA does specifically exempt from liability any person who provides access or connection to or form a facility, system, or network that is not under the control of the person violating the Act. The CDA also states that an employer shall not be held liable for the actions of an employee unless the employee's conduct is within the scope of his or her employment. More recent application of this law has been used to protect minor's use of social networks and falling prey to predators/stalkers.

Children's Online Privacy Protection Act of 1998

This is another act passed by Congress following the CDA, effective April 2000. The Children's Online Privacy Protection Act (COPPA) of 1998 applies to the online collection of personal information from children under 13. The new rules spell out what a Website operator must include in a privacy policy when and how to seek verifiable consent from a parent, and what responsibilities an operator has to protect children's privacy and safety online. Operators or owners of a commercial Website or an online service directed to children under 13 years must comply with the COPPA when collecting personal information from such children.

To determine whether a Website is directed to children, the Federal Trade Commission (FTC) considers several factors, including the subject matter; visual or audio content; the age of models on the site; language; whether advertising on the Website is directed to children; information regarding the age of the actual or intended audience; and whether a site uses animated characters or other child-oriented features.

To determine whether an entity is an "operator" with respect to information collected at a site, the FTC will consider who owns and controls the information, who pays for the collection and maintenance of the information, what the preexisting contractual relationships are in connection with the information, and what role the Website plays in collecting or maintaining the information.

In 2008, Congress amended this Act and included it as Title II "Protecting the Children" of the Broadband Data Improvement Act of 2008, Public Law 110-385—October 10, 2008. The amendment specifically defines personal information for a child. Personal information is defined as individually identifiable information about a child that is collected online, such as full name, home address, e-mail address, telephone number, or any other information that would allow someone to identify or contact the child. The Act also cover other types of information—for example, hobbies, interests, and information collected through cookies or other types of tracking mechanisms—when they are tied to individually identifiable information.

Health Insurance Portability and Accountability Act of 1996

National standards for electronic transactions encourage electronic commerce in the healthcare industry and ultimately simplify the processes involved. This results in savings from the reduction in administrative burdens on healthcare providers and health plans. Today, healthcare providers and health plans that conduct business electronically must use many different formats for electronic transactions. For example, about 400 different formats exist today for healthcare claims. With a national standard for electronic claims and other transactions, healthcare providers are able to submit the same transaction to any health plan in the United States and the health plan must accept it. Health plans are also able to send standard electronic transactions such as remittance advices and referral authorizations to healthcare providers. These national standards make electronic data interchange a viable and preferable alternative to paper processing for providers and health plans alike.

Health Insurance Portability and Accountability Act (HIPAA) of 1996, the first ever federal privacy standards to protect patients' medical records and other health information provided to health plans, doctors, hospitals, and other healthcare providers took effect on April 14, 2003. Developed by the Department of Health and Human Services, these new standards provide patients with access to their medical records and more control over how their personal health information is used and disclosed. They represent a uniform, federal floor of privacy protections for consumers across the country. State laws providing additional protections to consumers are not affected by this new rule.

HIPAA calls for stringent security protection on electronic health information while being maintained and transmitted. For IT directors, complying with HIPAA's privacy requirements is primarily a matter of computer security protecting the confidentiality of medical patient information and standardizing the reporting and billing processes for all health- and medical-related information. Confidentiality refers to the protection of any type of sensitive information from unauthorized access. It is critical for an organization's reputation and also to comply with privacy regulations. Risks associated with confidentiality include allowing unauthorized access or disclosure of sensitive and valuable organization data (e.g., corporate strategic plans, policyholder information, etc.). From an organization's stand point, sensitive and/or critical information may include:

- Strategic plans
- Trade secrets
- Cost information
- Legal documents
- Process improvements

To comply with HIPAA, the following must occur:

- Any connection to the Internet or other external networks or systems occurs through a **gateway** or **firewall**.
- Strong authentication is used to restrict the access to critical systems or business processes and highly sensitive data.
- Assessments of vulnerability, reliability, and the threat environment are made at least annually.

The Health Information Technology for Economic and Clinical Health of 2009

The Health Information Technology for Economic and Clinical Health (HITECH) Act of 2009 was enacted as part of the American Recovery and Reinvestment Act of 2009. HITECH promotes the adoption and meaningful use of health IT in the United States. HITECH provides the U.S. Department of Health and Human Services with the authority to establish programs to improve healthcare quality, safety, and efficiency through the "meaningful use" and promotion of health IT, including electronic health records and private and secure electronic health information exchange. Meaningful use refers to minimum U.S. government standards for using electronic health records, and for exchanging patient clinical data between healthcare providers, healthcare providers and insurers, and healthcare providers and patients. Sections within HITECH include the following:

- Subtitle A—Promotion of Health IT
- Subtitle B—Testing of Health IT
- Subtitle C—Grants and Loans Funding
- Subtitle D—Privacy

Subtitle A's goals include the protection and safeguarding of each patient's health information consistent with the law; improvement of healthcare quality; and reduction of medical errors and healthcare costs resulting from inefficiency; among others. Subtitle B lists descriptions and requirements for: (1) testing and implementing Health Information Technology (HIT) standards; (2) testing of HIT infrastructure (e.g., technical test beds, etc.); and (3) assisting higher-education institutions to establish multidisciplinary Centers for Health Care Information Enterprise Integration. Subtitle C implements grants, loans, and demonstration programs as incentives for utilizing health IT. Lastly, Subtitle D deals with privacy and security concerns tied to electronic transmissions of health information.

Both, HITECH and HIPAA, although separate and unrelated laws, supplement each other in some ways. For instance, HITECH demands that its technologies and IT-related standards do not compromise HIPAA privacy and security laws. HITECH also stipulates that physicians and hospitals attesting to meaningful use, must have previously performed a security risk assessment, as HIPAA requires. HITECH further establishes notification rules for data breach instances, which are also mirrored by HIPAA.

Gramm–Leach–Bliley Act of 1999

The Gramm–Leach–Bliley Act of 1999 requires financial institutions to protect consumer financial privacy. Financial institutions are companies that offer consumers financial products or services like loans, financial or investment advice, or insurance. Under the Gramm–Leach–Bliley Act of 1999, financial institutions are required to explain their information-sharing practices to their customers and protect their sensitive data.

In order to comply with the Act, financial institutions must assess, manage, and control risk; oversee service providers; and adjust security programs as needed based on changing risk. One specific provision requires financial institutions to identify internal and/or external threats which can potentially result in unauthorized disclosures, as well as misuse, destruction, or manipulation of customer's sensitive information.

Uniting and Strengthening America by Providing Appropriate Tools Required to Intercept and Obstruct Terrorism Act (USA PATRIOT Act) of 2001

The purpose of the USA PATRIOT Act of 2001 is to deter and punish terrorist acts in the United States and around the world, to enhance law enforcement investigatory tools, and other purposes, some of which include:

- To strengthen U.S. measures to prevent, detect, and prosecute international **money laundering** and financing of **terrorism**
- To subject to special scrutiny foreign jurisdictions, foreign financial institutions, and classes of international transactions or types of accounts that are susceptible to criminal abuse
- To require all appropriate elements of the financial services industry to report potential money laundering
- To strengthen measures to prevent use of the U.S. financial system for personal gain by corrupt foreign officials and facilitate repatriation of stolen assets to the citizens of countries to whom such assets belong

Sadly terrorism still occurs and there are not many signs that it is going away anytime soon. For example, Congress must continuously monitor the U.S. counterterrorism enterprise and determine whether other measures are needed in order to improve it. As mentioned in a 2015 article from The Heritage Foundation, and relevant to the topics discussed within this textbook, cyber-investigation capabilities should be prioritized by Congress. With so much terrorism-related activity occurring on the Internet, law enforcement must be capable of identifying, monitoring, and tracking such violent activity effectively and on a timely basis. Severe cyberattacks, such as cyberterrorism, are capable of shutting down nuclear centrifuges, air defense systems, and electrical grids. To some, these types of attacks should be treated as acts of war as they pose a serious threat to national security. Some of the recent major notable cyberterrorism on the United States include:

- Cyberattacks and cyber espionage conducted by China targeting and exploiting the U.S. government and private networks (2016)
- Cyberattacks and cyber breaches conducted by Russia targeting reporters at the New York Times and other U.S. news organizations (2016)
- Cyberattacks orchestrated by Russia against the Democratic National Committee and the Democratic Congressional Campaign Committee, to disrupt or discredit the presidential election (2016)
- Cyberattacks against U.S. financial institutions (e.g., American Express, JP Morgan Chase, etc.) instigated by Iran and North Korean governments (2013)
- Cyberattacks and cyber breaches claimed by a Syrian hacker group on media organizations (New York Times, Twitter, and the Huffington Post) (2013)

The above represents only a few of cyberterrorism attacks perpetrated on U.S. government and private entities. A summary of all the federal U.S. laws described within this section is included in Exhibit 2.3.

State Laws

IT legislation at the State level is equally relevant to IT auditors tasked with examining applications, data, networks, and controls, and the risk associated with a failure to comply. Described below are examples of these State laws, which include Security Breach Notification Laws, Cybersecurity Legislation, State Social Media Privacy Laws, and others.

At present, 47 states, the District of Columbia, Guam, Puerto Rico, and the Virgin Islands have all enacted security breach notification legislation requiring entities in the private, governmental,

Exhibit 2.3 Summary of U.S. Federal Laws Relevant to IT Auditors

Legislation Type	U.S. Law to Combat IT Crimes	Description
Federal Financial Integrity	Sarbanes–Oxley Act (SOX) of 2002	• Major reform package mandating the most far-reaching changes Congress has imposed on the business world since the Foreign Corrupt Practices Act of 1977 and the SEC Act of the 1930s. • Establishes requirements for organizations' Board of Directors, including its composition and duties. • Requires auditor (not audit firm) rotation. • Created the Public Company Accounting Oversight Board (PCAOB). • Prohibits registered public accounting firms from providing audit clients, contemporaneously with the audit, certain non-audit services. • Requires an internal control report to be included as part of the annual report. • Require that CEO and CFO confirm and assert to stakeholders that SEC disclosures are (1) truthful and reliable; and that (2) management has taken the necessary steps to ensure that the company's disclosure processes and controls are capable of generating accurate and consistent financial information, reliable by stakeholders (Section 302). SEC disclosures include the company's financial statements as well as all supplemental disclosures. The company's external auditor must also report on the reliability of management's assessment of internal control (Section 404). • Penalizes CEOs, CFOs, etc. for nonperformance.
Federal Security	Computer Fraud and Abuse Act of 1984	• Protects against trespassing (unauthorized entry); exceeding authorized access; intentional and reckless destructive trespass; and exchanging information on how to gain unauthorized access.
Federal Security	Computer Security Act of 1987	• Protects sensitive information in federal government systems. • Established a federal government computer security program, NIST, to assist in developing government wide, computer-system-security standards, guidelines, and security training.

(Continued)

Exhibit 2.3 (*Continued*)   Summary of U.S. Federal Laws Relevant to IT Auditors

Legislation Type	*U.S. Law to Combat IT Crimes*	*Description*

Federal Security	Homeland Security Act of 2002	• Prevents terrorist attacks in the United States; reduces the vulnerability of the United States to terrorism; and includes the Cyber Security Enhancement Act, which:
		  – demands life sentences for hackers that recklessly endanger lives.
		  – allows for net surveillance to gather personal and private data (phone #s, IP addresses, URLs, e-mails, etc.) without a court order.
		  – requires Internet Service Providers (ISPs) to hand users' records over to law enforcement authorities, overturning current legislation.

Federal Security	Payment Card Industry Data Security Standards (PCI DSS) of 2004	• PCI DSS are technical and operational requirements that apply to entities that store, process, or transmit cardholder data. • The main purpose of the PCI DSS is to protect cardholder data in order to reduce credit card fraud. • All merchants that either accept or process payment through cards must comply with the PCI DSS. • PCI DSS are maintained, managed, and promoted worldwide by a PCI Security Standards Council. • The Council is formed by major credit card companies (i.e., American Express, Discover, JCB International, MasterCard, and Visa, Inc.). These companies share equally in governance, execution, and compliance of the Council's work.
Federal Security	Federal Information Security Management Act (FISMA) of 2002	• FISMA requires federal agencies to develop, document, and put in place information security programs with the purpose of protecting both, the information and the systems/applications implemented to support the operations and assets of the agencies, including those provided or managed by another agency, contractor, or other source.

(*Continued*)

Exhibit 2.3 (*Continued*) Summary of U.S. Federal Laws Relevant to IT Auditors

Legislation Type	U.S. Law to Combat IT Crimes	Description
Federal Security	Electronic Signature Laws of 1999 and 2000	• Two main pieces of legislation with respect to electronic signature laws are: Uniform Electronic Transactions Act (UETA) and Electronic Signatures in Global and National Commerce Act (ESIGN). • UETA, a state law introduced in 1999 and already adopted by 47 U.S. states, the District of Columbia, Puerto Rico, and the U.S. Virgin Islands, makes electronic signatures valid and in compliance with law requirements when parties ready to enter into a transaction have agreed to proceed electronically. • ESIGN, a federal law passed in 2000 by the U.S. Congress, recognizes electronic signatures and records granted all contract parties opt to use electronic documents and sign them electronically.
Privacy	Privacy Act of 1974	• Safeguards individuals against invasion of their personal privacy by: 　– allowing individuals to determine what information is collected. 　– guaranteeing that information collected is only used for one purpose. 　– assuring information is current and accurate.
Privacy	E-Communications Privacy Act of 1986	• Prohibits (1) interception/disclosure of wire, oral, or electronic communications; and (2) manufacture or possession of intercepting devices.
Privacy	Communications Decency Act of 1996	• Prohibits the making of indecent or patently offensive material available to minors via computer networks. • Employers are not liable for actions of an employee unless it is within the scope of their employment.
Privacy	Children's Online Privacy Protection Act of 1998	• Applies to the online collection of personal information from children under 13. • The act specifically defines what personal information is for a child.

(*Continued*)

Exhibit 2.3 (*Continued*)   Summary of U.S. Federal Laws Relevant to IT Auditors

Legislation Type	U.S. Law to Combat IT Crimes	Description
Privacy	Health Insurance Portability & Accountability Act (HIPAA) of 1996	• Protects confidentiality and security of healthcare information (e.g., patients' medical records, health information provided to doctors, hospitals). • Restricts insurers to reject workers based on preexisting health conditions; requires security and privacy to protect personal information.
Privacy	The Health Information Technology for Economic and Clinical Health of 2009	• Promotes the adoption and meaningful use of health IT in the United States. • Provides the U.S. Department of Health and Human Services with the authority to establish programs to improve healthcare quality, safety, and efficiency through the "meaningful use" and promotion of health IT. • Meaningful use refers to minimum government standards for using electronic health records and exchanging patient data between healthcare providers, healthcare providers and insurers, and healthcare providers and patients.
Privacy	Gram-Leach-Bliley Act of 1999	• Requires financial institutions to assess, manage, and control risk; oversee service providers; and adjust security programs based on risk.
Privacy	USA Patriot Act of 2001	• Deters and punishes terrorist acts in the United States and around the world. • Enhances law enforcement investigatory tools, among others.

and/or educational industries to provide notification to individuals of security breaches involving their PII. Normally, security breach laws involve the following provisions:

- Who needs to comply with the law (e.g., businesses, data/information brokers, government entities, etc.)
- How is "personal information" defined (e.g., name combined with social security numbers, driver's license or state ID, account numbers, etc.)
- What constitutes a security breach (e.g., unauthorized acquisition of data)

- Requirements for notification (e.g., timing or method of notice, who must be notified)
- Exemptions (e.g., for encrypted information)

Cybersecurity legislation is another example of State laws in place to protect against cyber threats and their vast implications for government and private industry security, economic prosperity, and public safety. States have employed legislation with a significant number of approaches to deal specifically with cybersecurity. Examples of these methods include requiring government agencies to establish security practices as well as offering incentives to the cybersecurity industry. Additional procedures to combat cybersecurity include providing exemptions from public records laws for security information and creating cybersecurity commissions, studies, or task forces to promote the education of cybersecurity.

Other common and relevant State laws include State Social Media Privacy Laws, Data Disposal Laws, and Computer Crime Statutes. Regarding social media, State legislation was introduced in 2012 to prevent employers from requesting passwords to personal Internet accounts (including social media accounts) to obtain or maintain a job. Similar legislation prohibits colleges and universities from requiring access to students' social media accounts. The rationale here was that both, employees and students, considered such requests to be an invasion of their privacy. Data disposal laws have been enacted in at least 31 states and Puerto Rico requiring businesses and government entities to dispose all PII collected and stored, both electronically and on paper. Specifically and in order to comply with established laws, businesses and the government must "destroy, dispose, or otherwise make personal information unreadable or undecipherable." Lastly, computer crime statutes are in place to prohibit "actions that destroy or interfere with normal operation of a computer system," such as hacking, gaining unauthorized access without consent, and setting or transmitting malicious computer instructions (e.g., computer viruses, malware, etc.) to modify, damage, or destroy records of information within a computer system or network without the permission of the owner.

International Privacy Laws

Data protection is all about safeguarding such private information, which gets either collected, processed, and stored by electronic means, or intended to form part of filing systems. Data protection laws must be in place to control and shape such activities especially those performed by companies and governments. These institutions have shown time after time that unless rules and laws restrict their actions, they will attempt to collect, examine, and keep all such data without adequately notifying the individuals about such use and purpose of accessing their data.

As of 2014, over 100 countries around the world have enacted data protection or privacy laws, and several other countries are in the process of passing such legislation. For example, the European Union has implemented some of the strongest and most comprehensive data privacy laws (i.e., The 1995 Data Protection Directive). Canada is another leading example with legislation at both, national and provincial levels (e.g., Personal Information Protection and Electronic Documents Act of 2000, etc.). Exhibit 2.4 summarizes these and other relevant international data protection and privacy laws like Mexico's Law on the Protection of Personal Data Held by Private Parties of 2010, and the Safe Harbor Act of 1998.

Exhibit 2.4 Summary of International Privacy Laws Relevant to IT Auditors

International Privacy Legislation	*Brief Description*
Personal Information Protection and Electronic Documents Act of 2000 (PIPED Act, or PIPEDA)—Canada	One of the main purposes of PIPEDA is to support and promote electronic commerce by "protecting personal information that is collected, used or disclosed in certain circumstances."[a] The following 10 principles, established by PIPEDA, govern the collection, use, and disclosure of personal information[b]: 1. Accountability 2. Identifying Purposes 3. Consent 4. Limiting Collection 5. Limiting Use, Disclosure, and Retention 6. Accuracy 7. Safeguards 8. Openness 9. Individual Access 10. Challenging Compliance
Law on the Protection of Personal Data Held by Private Parties of 2010—Mexico	The law requires Mexican business organizations (as well as any company that operates or advertises in Mexico or uses Spanish-language call centers and other support services located in Mexico) to have either consent or legal obligation for/when collecting, processing, using, and disclosing personally identifiable information (PII). Organizations dealing with PII must inform individuals about such use and, most importantly, provide notification to all affected persons in the event of a security breach.[b] The law also include eight general principles that Mexican business organizations must follow when handling personal data[c]: • Legality • Consent • Notice • Quality • Purpose Limitation • Fidelity • Proportionality • Accountability
European Union Data Protection Directive of 1995	The Directive establishes rigorous limits on the collection and use of personal data, and demands that each member state institute an independent national body responsible for the protection of such data.[b] The Directive impacts European businesses (as well as non-European companies to which data is exported), and includes the seven governing principles described below[d]:

(*Continued*)

Exhibit 2.4 (*Continued*)   Summary of International Privacy Laws Relevant to IT Auditors

International Privacy Legislation	*Brief Description*
European Union Data Protection Directive of 1995	1. *Notice* should be given to all affected data subjects when their data is being collected. 2. Data should only be used for the *purpose* stated. 3. Data should not be disclosed without the subject's *consent*. 4. Collected data should be kept *secure* from any potential abuses. 5. *Disclosure* of who is collecting the data should be provided to all affected data subjects. 6. Data subjects should be allowed to *access* their data and make corrections to any inaccurate data. 7. Data subjects should have an available method to hold data collectors *accountable* for following these six principles above.
Safe Harbor Act of 1998	Under the Act, transferring personal data to non-European Union nations (e.g., U.S. companies) not complying with the European "adequacy" standard for privacy protection (established by the European Union Data Protection Directive) is prohibited. The Act (specifically related to U.S. companies doing business in Europe) was intended to bridge the different privacy approaches of the United States and Europe, thus enabling U.S. companies to safely engage in trans-Atlantic transactions without facing interruptions or even prosecution by European authorities.[b] Some key requirements or provisions of the Act include[h]: • Companies participating in the safe harbor will be deemed adequate, and data flows to those companies will continue. • Member state requirements for prior approval of data transfers either will be waived or approval will be automatically granted. • Claims brought by European citizens against U.S. companies will be heard in the United States, subject to limited exceptions.
	Note: A list of other relevant international privacy laws, segregated by country and region, can be found at https://informationshield.com/free-security-policy-tools/international-data-privacy-laws/

[a] www.parl.gc.ca/HousePublications/Publication.aspx?Pub=Bill&Doc=C-6&Language=E&Mode=1&Parl=36&Ses=2&File=35.

[b] www.csoonline.com/article/2126072/compliance/the-security-laws--regulations-and-guidelines-directory.html?page=4.

[c] www.dof.gob.mx/nota_detalle.php?codigo=5150631&fecha=05/07/2010.

[d] http://eur-lex.europa.eu/LexUriServ/LexUriServ.do?uri=OJ:L:2001:008:0001:0022:EN:PDF.

[e] http://europa.eu/rapid/pressReleasesAction.do?reference=IP/00/865&format=HTML&aged=1&language=EN&guiLanguage=en.

Conclusion

Business operations are changing in a rapid pace because of the fast continuing improvement of technology. The Internet in particular now includes everything from marketing, sales, and entertainment purposes to e-mail, research, and commerce, and virtually any other type of information sharing. As with any breakthrough in technology, advancements have also given rise to various new problems and IT crimes issues that must be addressed. These problems are often being brought to the attention of the IT audit and control specialist. Legislation must be in place to govern the use and misuse of IT, including the security and privacy of the information.

It is believed that federal legislation has had a lasting impact on the online community (government, business, and the public), which is something that those entering the IT audit and information assurance profession should be knowledgeable of. Federal financial integrity legislation, such as the SOX of 2002, changed the world of financial audit dramatically, stressing the significance of due professional care. Similarly, federal security legislation was implemented to prevent criminals from escaping penalties for violating authorized access to a computer system. Regarding privacy legislation, the U.S. government enacted the Privacy Act of 1974 to provide certain safeguards to an individual against an invasion of personal privacy. The act also places certain requirements on federal agencies.

Laws at both, state and international levels, cannot be ignored by the IT auditor. IT auditors must be familiarized with these specific legislation and ensure procedures are in place when examining applications, data, networks, and controls, for instance, in order to address risks and comply with these relevant IT laws.

Review Questions

1. Summarize the three main categories of crimes involving computers.
2. What does the Sarbanes–Oxley Act (SOX) of 2002 prohibit? What does SOX require from the Board of Directors?
3. What is the Computer Fraud and Abuse Act (CFAA) of 1984?
4. What is the purpose of the Computer Security Act of 1987, and what does it protect?
5. Why was the Homeland Security Act created? Can hackers who cause injury or death to others be prosecuted under this act? Please elaborate.
6. Differentiate between the Uniform Electronic Transactions Act (UETA) and the Electronic Signatures in Global and National Commerce Act (ESIGN). Provide examples of specific transactions where an electronic signature can be valid under the U.S. law. Hint: For this, you should review the requirements for a valid electronic signature listed in the chapter.
7. What is the Privacy Act of 1974? What requirements does it place on federal agencies?
8. What is the Electronic Communications Privacy Act of 1986, and what does it prohibit?
9. What does the Children's Online Privacy Protection Act of 1998 apply to? Which factors does the Federal Trade Commission (FTC) consider to determine whether a Website is directed to children?
10. What does HIPAA stand for and what does it protect? List the three factors that must be in place to comply with HIPAA.
11. Why was the USA PATRIOT Act of 2001 implemented?

Exercises

1. Using an Internet web browser, search and examine five websites on each of the topics below. In a three-column table format, document the name of the Website examined in the first column, the source link in the second column, and a brief summary of the information provided by the Website in the third column.
   a. Computer crime
   b. Computer privacy
   c. Computer law

2. Explain why you think is important for IT auditors to know about each type of legislation below. Your explanation for each type of legislation should take no less than three paragraphs and incorporate supporting examples. You are also highly encouraged to look for outside sources.
   a. Federal legislation
   b. State legislation
   c. International legislation

3. You were engaged as a consultant by a client that just started doing business. Some of the services your client provides include storing, processing, and/or transmitting credit card data. You client is unaware of any laws or regulations related to the aforementioned services. You know right from the start that your client must comply with PCI DSS standards. Using a memo format, prepare communication to your client including the following:
   a. Summarize what PCI DSS are and why are they relevant to your client. You are highly encouraged to look for outside sources.
   b. Using the six goals and requirements (bullet points) of PCI DSS listed in the chapter as objectives, develop a plan to meet each objective. Your plan must include the specific objective along with a brief explanation of the activity or procedure that you will advise your client to implement in order to comply with the specific objective. For example, for one of the goals or objectives, "Protecting stored cardholder data," you should explain how specifically will the cardholder data be protected and what encryption techniques should be put in place (you may want to elaborate here since your client had expressed to you that she is not well-familiarized with technology). Ultimately, your communication should bring comfort to your client and ensure that all transmissions of cardholder data are indeed safeguarded.

4. Identify two recent cyberattacks (not mentioned in the book) conducted either in the United States or internationally. Summarize both cyberattacks consistent with Exhibit 2.2 (i.e., Company, Industry, and Cyberattack Description), and be ready to present them to the class in a 5-minute presentation.

Further Reading

1. Author unknown. (March 2016). *2009 Internet Crime Report*, Internet Crime Complaint Center IC3, p. 14, www.ic3.gov/media/annualreport/2009_IC3Report.pdf (accessed June 2, 2010).
2. Author unknown. (March 2016). *The Department of Justice's Efforts to Combat Identity Theft*, US Department of Justice Office of the Inspector General, www.justice.gov/oig/reports/plus/a1021.pdf (accessed June 2, 2010).
3. CIPHER—Electronic Newsletter of the Technical Committee on Security and Privacy, A Technical Committee of the Computer Society of the IEEE, www.ieee-security.org/cipher.html

4. Computer Security Division, Computer Security Resource Center. NIST, http://csrc.nist.gov/groups/SMA/fisma/overview.html (accessed October 2016).
5. Computer Crime Statutes. National Conference of State Legislatures, www.ncsl.org/research/telecommunications-and-information-technology/computer-hacking-and-unauthorized-access-laws.aspx (accessed January 2017).
6. Confidentiality breach: Hospital sent patient records to auto shop. Questex LLC, www.fiercehealthcare.com/story/confidentiality-breach-hospital-sent-patient-records-auto-shop/2010-06-28 (accessed January 2017).
7. Cybersecurity Legislation. (2016). National Conference of State Legislatures, www.ncsl.org/research/telecommunications-and-information-technology/cybersecurity-legislation-2016.aspx (accessed October 2016).
8. China continuing cyber attacks on U.S. networks. *The Washington Free Beacon*, http://freebeacon.com/national-security/china-continuing-cyber-attacks-on-u-s-networks/ (accessed October 2016).
9. Data disposal laws. National Conference of State Legislatures, www.ncsl.org/research/telecommunications-and-information-technology/data-disposal-laws.aspx (accessed October 2016).
10. Data protection. Privacy International, www.privacyinternational.org/topics/data-protection (accessed November 2016).
11. Federal Bureau of Investigation. Criminals host fake government services web sites to acquire personally identifiable information and to collect fraudulent fees, Public Service Announcement. www.ic3.gov/media/2015/150407-2.aspx (accessed December 2015).
12. Gallegos, F. (2001). *Federal Laws Affecting IS Audit and Control Professionals*, EDP Auditing Series #72-10-20, Auerbach Publishers, Boca Raton, FL, pp. 1–20.
13. Health IT legislation and regulations. HealthIT.gov, https://www.healthit.gov/topic/laws-regulation-and-policy/health-it-legislation (accessed October 2016).
14. Hathaway, O. A., Crootof, R., Levitz, P., Nix, H., Nowlan, A., Perdue, W., and Spiegel, J. The law of cyber-attack. *Cal. L. Rev.*, 100(4), 817–885. www.jstor.org/stable/23249823
15. Herath, T., and Rao, H. R. (2009). Encouraging information security behaviors in organizations: Role of penalties, pressures, and perceived effectiveness. *Decis. Support Syst.*, 47(2), 154–165.
16. HITECH Act Enforcement Interim Final Rule. U.S. Department of Health & Human Services, www.hhs.gov/hipaa/for-professionals/special-topics/HITECH-act-enforcement-interim-final-rule/ (accessed November 2016).
17. HG.org Legal Resources. Information technology law—Guide to IT law, www.hg.org/information-technology-law.html#1 (accessed October 2016).
18. Federal Trade Commission, (2002). How to comply with the privacy of consumer financial information rule of the Gramm–Leach–Bliley Act. www.ftc.gov/tips-advice/business-center/guidance/how-comply-privacy-consumer-financial-information-rule-gramm#whois
19. Inserra, D. *69th Islamist terrorist plot: Ongoing spike in terrorism should force congress to finally confront the terrorist threat*. The Heritage Foundation, www.heritage.org/research/reports/2015/06/69th-islamist-terrorist-plot-ongoing-spike-in-terrorism-should-force-congress-tofinally-confront-the-terrorist-threat
20. Meaningful use. Tech Target, http://searchhealthit.techtarget.com/definition/meaningful-use (accessed October 2016).
21. Medical privacy—National standards to protect the privacy of personal health information, www.hhs.gov/ocr/hipaa/
22. New York Times, Twitter hacked by Syrian group. *The Daily Star*, www.thedailystar.net/news/new-york-times-twitter-hacked-by-syrian-group (accessed February 2016).
23. PCI security. PCI Security Standards Council, www.pcisecuritystandards.org/pci_security/ (accessed December 2016).
24. Privacy. ISACA® Glossary of Terms, www.isaca.org/Knowledge-Center/Documents/Glossary/glossary.pdf
25. Privacy Act of 1974 and Amendments, Document from the CPSR Privacy/Information Archive, https://epic.org/privacy/1974act/
26. Privacy Guidelines for NII Review of Proposed Principles of the Privacy Working Group, compiled by Electronic Privacy Information Center, www.epic.org

27. Russia suspected in cyberattacks on US news outlets. *New York Post*, http://nypost.com/2016/08/23/russia-suspected-in-cyber-attacks-on-us-news-outlets/ (accessed August 2016).

28. Sarbanes–Oxley-101. Section 302: Corporate Responsibility for Financial Reports, www.sarbanes-oxley-101.com/SOX-302.htm (accessed August 2016).

29. Sarbanes–Oxley-101. Section 404: Management Assessment of Internal Controls, www.sarbanes-oxley-101.com/SOX-404.htm (accessed August 2016).

30. Security Breach Notification Laws. National Conference of State Legislatures, www.ncsl.org/research/telecommunications-and-information-technology/security-breach-notification-laws.aspx#1 (accessed October 2016).

31. Senft, S., Gallegos, F., and Davis, A. (2012). *Information Technology Control and Audit*, CRC Press/Taylor & Francis, Boca Raton, FL.

32. National Conference of State Legislatures. State social media privacy laws, www.ncsl.org/research/telecommunications-and-information-technology/state-laws-prohibiting-access-to-social-media-usernames-and-passwords.aspx (accessed October 2016).

33. UETA and ESIGN Act. DocuSign Inc, www.docusign.com/learn/esign-act-ueta (accessed December 2016).

34. U.S. Congress. Computer Security Act of 1987, compiled by Electronic Privacy Information Center, www.epic.org/crypto/csa

35. U.S. Department of Health and Services. Office for Civil Rights—HIPAA, http://aspe.hhs.gov/admnsimp/pL104191.htm

36. U.S. Department of Justice, Federal Bureau of Investigation. 2016. Internet crime report, https://pdf.ic3.gov/2016_IC3Report.pdf (accessed November 2016).

37. U.S. Department of Justice, Federal Bureau of Investigation. 2015. Internet crime report, https://pdf.ic3.gov/2015_IC3Report.pdf (accessed December 2015).

38. U.S. Department of Justice, Federal Bureau of Investigation. 2014. Internet crime report, https://pdf.ic3.gov/2014_IC3Report.pdf (accessed December 2015).

39. U.S. Department of Justice, Office of Justice Programs, Bureau of Justice Assistance. E-Government Act of 2002, www.it.ojp.gov/PrivacyLiberty/authorities/statutes/1287 (accessed June 2013).

40. United States formally accuses Russian hackers of political cyberattacks. Reuters, www.reuters.com/article/us-usa-cyber-russia-idUSKCN12729B (accessed December 2016).

41. U.S. Government Accountability Office. (February 14, 2008). Testimony before Congressional Subcommittees on Information Security.

Chapter 3

The IT Audit Process

LEARNING OBJECTIVES

1. Describe what audit universe is, and illustrate example.
2. Define control objectives for information and related technology and explain why they are useful for organizations and auditors.
3. Explain what a risk assessment is and its significance to the audit function. Illustrate an example of a risk assessment following the National Institute of Standards and Technology methodology.
4. Describe an audit plan and its components. Illustrate examples of IT audit documentation supporting a financial statement audit.
5. Define the audit process and describe the phases of an IT audit engagement.
6. Discuss other types of audits conducted in IT.

The role of IT audit continues to be a critical mechanism for ensuring the integrity of information systems and the reporting of organization finances to prevent future financial fiascos such as Enron (2001) and WorldCom (2002). Unfortunately, these fiascos continue to occur. Global economies are more interdependent than ever and geopolitical risks impact everyone. Electronic infrastructure and commerce are integrated in business processes around the globe. The need to control and audit IT has never been greater.

Today's IT auditor is faced with many concerns about the exposure of information systems to a multitude of risks. From these concerns arise the objectives for the audit process and function. This chapter looks at the audit process for IT and the demands that will be placed on the profession in the future.

Audit Universe

One of the best practices for an audit function is to have an audit universe. The audit universe is an inventory of all the potential audit areas within an organization. Basic functional audit areas within an organization include sales, marketing, customer service, operations, research and development, finance, human resource, information technology, and legal. An audit universe documents the key business processes and risks of an organization. Documenting processes and, particularly, risks have proved to be a best practice for organizations. The IIA's Performance Standard 2010 encourages the establishment of risk-based plans to determine the priorities for internal audit activity.

An audit universe includes the basic functional audit area, organization objectives, key business processes that support those organization objectives, specific audit objectives, risks of not achieving those objectives, and controls that mitigate the risks. Tying the audit universe to organizational objectives links the entire audit process to business objectives and risks, making it easier to communicate the impact of control deficiencies. Exhibit 3.1 shows an example of an audit universe related to the IT area of an organization.

The audit universe is also an essential building block to a properly risk-based internal audit process. Typically, internal audit groups prepare annual audit schedules to determine the number of hours available and the number of audits that can be performed. The audit universe is an ongoing process; as an organization changes, new risks arise or existing risks change, and new regulations are introduced. Organizations can either remove lower-priority audits from the schedule or hire external auditors to supplement internal staff.

IT audits, for example, have specific IT processes to include in the audit universe. Control Objectives for Information and Related Technology (COBIT) provides a comprehensive list of critical IT processes, which can be used as a starting point.

COBIT

COBIT is an authoritative, international set of generally accepted IT practices or control objectives that help employees, managers, executives, and auditors in: understanding IT systems, discharging fiduciary responsibilities, and deciding adequate levels of security and controls.

COBIT supports the need to research, develop, publicize, and promote up-to-date internationally accepted IT control objectives. The primary emphasis of the COBIT framework issued by Information Systems Audit and Control Foundation in 1996 is to ensure that technology provides businesses with relevant, timely, and quality information for decision-making purposes. The COBIT framework, now on its fifth edition (COBIT 5), has evolved over the years and each time there are major changes to the framework, the framework is numbered to its current version.

The benefit of a standard framework for IT controls, such as COBIT, is that it allows management to benchmark its environment and compare it to other organizations. IT auditors can also use COBIT to substantiate their internal control assessments and opinions. Because the framework is comprehensive, it provides assurances that IT security and controls exist.

COBIT 5, which can be downloaded from www.isaca.org, helps organizations create optimal value from IT by maintaining a balance between realizing benefits and optimizing risk levels and resource use. COBIT 5 is based on five principles (see Exhibit 3.2). COBIT 5 considers the IT needs of internal and external stakeholders (Principle 1), while fully covering the organization's governance and management of information and related technology (Principle 2). COBIT 5 provides an integrated framework that aligns and integrates easily with other frameworks (e.g., Committee of Sponsoring Organizations of the Treadway Commission-Enterprise Risk Management (COSO-ERM), etc.), standards, and best practices used (Principle 3). COBIT 5 enables IT to be governed and managed in a holistic manner for the entire organization (Principle 4) through:

a. Establishing principles, policies, and practical guidance for daily management.
b. Implementing processes to achieve overall IT-related goals and objectives.

Exhibit 3.1   Example of an Audit Universe Related to the IT Area of an Organization

Basic Functional Audit Area: Information Technology Organization's Objective: To provide secure access to financial information, technology, and services for all authorized employees.			
Key Business Process	*IT Audit Objective*	*IT Risk*	*IT Mitigating Control*
Access Control Management	System's security is appropriately implemented, administered, and logged to safeguard against unauthorized access to or modifications of programs and data, that result in incomplete, inaccurate, or invalid processing or recording of financial information.	Users possess privileges that are not consistent with their job functions, thus allowing unauthorized or incorrect modifications to financial or accounting data, which could cause segregation of duties conflicts, unprevented or undetected errors, incorrect financials, or management decisions based upon misleading information.	User access privileges are periodically reviewed by application owners to verify access privileges remain appropriate and consistent with job requirements.
Change Control Management	Programs and systems are appropriately implemented in a manner supporting the accurate, complete, and valid processing and recording of financial information.	Developers or programmers have the ability to promote incorrect or inappropriate modifications or changes to financial data, programs, or settings into the production processing environment, thus resulting in invalid accounting data and/or fraud.	Application systems, databases, networks, and operating systems are developed, modified, and tested in an environment separate from the production environment. Access to the development and test environments is appropriately restricted.
Management of Data Center, Network, and Support	Data are appropriately managed to provide reasonable assurance that financial data remain complete, accurate, and valid throughout the update and storage process.	Financial reporting information and accounting data cannot be recovered in the event of system failure, impacting the entity's ability to report financial information according to established reporting requirements.	Backups are archived off-site to minimize risk that data are lost.

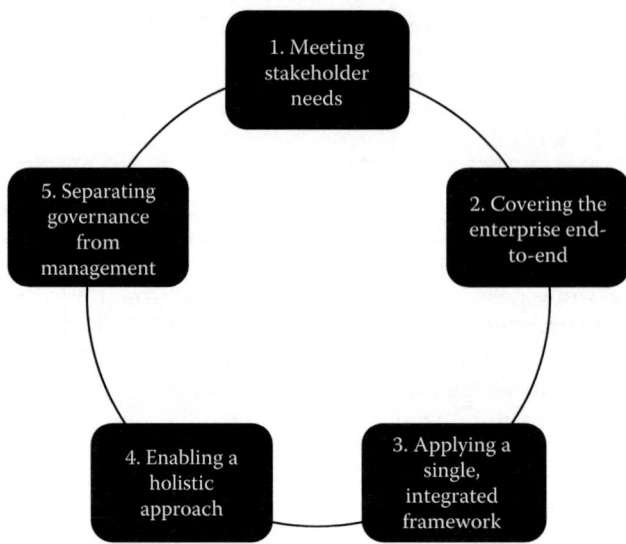

Exhibit 3.2 Principles of the COBIT 5 framework. (Adapted from http://www.isaca.org/cobit/pages/cobit-5-framework-product-page.aspx.)

  c. Putting in place organizational structures with key decision-making capabilities.
  d. Promoting good culture, ethics, and behavior in the organization.
  e. Recognizing that information is pervasive throughout any organization, and often the key product of the organization itself.
  f. Taking into account the infrastructure, technology, and applications that provide the organization with IT processing and services.
  g. Recognizing that people, skills, and competencies are required for successful completion of all activities and correct-decision making.

COBIT 5 assists organizations in adequately separating governance from management objectives (Principle 5). Both governance and management are described below.

  a. *Governance*—optimizes the use of organizational resources to effectively address risks. Governance ensures that the Board of Directors ("board"):
    i. evaluates stakeholder needs to identify objectives,
    ii. guides management by prioritizing objectives, and
    iii. monitors overall management's performance.
  b. *Management*—plan, build, run, and monitor the activities and processes used by the organization to pursue the objectives established by the board.

COBIT 5's framework is valuable for all size types organizations, including commercial, not-for-profit, or in the public sector. The comprehensive framework provides a set of control objectives that not only helps IT management and governance professionals manage their IT operations, but also IT auditors in their quests for examining those objectives.

The COBIT processes can be customized to the organization's environment. IT auditors can help audit management identify the applications associated with the critical business and financial processes, as well as controls that are necessary to make the area being audited free from significant exposures to risk. This objective also encompasses validating adherence of the application systems under examination to appropriate standards (e.g., financial accounting should conform to GAAP, etc.).

The next step in the planning process is to perform a risk assessment for each universe item from Exhibit 3.1. The risk assessment will analyze exposures and help prioritize "high risk" audit projects.

Risk Assessment

Risk assessments are considered the foundation of the audit function as they assist in developing the process for planning individual audits. Specifically, risk assessments:

- improve the quality, quantity, and accessibility of planning data, such as risk areas, past audits and results, and budget information;
- examine potential audit projects in the audit universe and choose those that have the greatest risk exposure to be performed first; and
- provide a framework for allocating audit resources to achieve maximum benefits.

Given the high number of potential audits that can be performed and often the limited amount of audit resources, it is important to focus on the right audits. The risk assessment approach provides explicit criteria for systematically evaluating and selecting these audits.

In today's environment, it is difficult to keep pace with organization and regulatory changes to provide timely information on internal controls. Change increases the audit universe, the number of business partners (i.e., vendors), and the number of projects where an objective and independent perspective is needed. An effective risk assessment planning process allows auditing to be more flexible and efficient to meet the needs of a changing organization, such as:

- identifying new risk areas
- identifying changes in existing risk areas
- accessing current regulatory and legal information
- taking advantage of information gathered during the audit process to improve risk assessment

Audit areas can be evaluated using a weighted scoring mechanism. However, audit management must evaluate the results using their knowledge of the organization objectives and environment to make sure the priorities reflect reality. Audit areas may also be grouped to improve audit efficiency when reviewing similar processes. The auditing function is cyclical in that it uses historical and current information for risk assessment, evaluates controls, communicates results, and incorporates those results back into the risk assessment.

In an IT risk assessment, for instance, financial applications are common audits/projects to be ranked. Their risks can be identified, assessed, and prioritized. Controls (safeguards) are also identified to be put in place to address and mitigate such risks. IT risks surrounding financial applications can be identified through:

- Audits, reviews, inspections
- Reading flowcharts of operations

- Using risk analysis questionnaires
- Analyzing financial statement trends
- Completing insurance policy checklists

Absolute security from threads and risks in today's technology environments is unrealistic. Risk assessments, according to the National Institute of Standards and Technology (NIST) Special Publication 800-30, are used to assist organizations determine the extent of potential threats and the risks associated with IT systems and applications. The results of the above assist management in identifying and implementing appropriate IT controls for reducing or eliminating those threats and risks during the mitigation process. NIST recommends that for a risk assessment, it is important that organizations follow these steps:

1. Have a process in place to identify or characterize assets (e.g., financial applications, etc.).
2. Define vulnerabilities on those assets and the threat-sources that can trigger them.
3. Determine the likelihood or probability levels (e.g., very high, high, medium, etc.) that vulnerabilities may be exercised. For example, probabilities of very high = 1.00, high = 0.75, medium = 0.50, low = 0.25, and very low = 0.10 may be assigned for each vulnerability based on the organization's estimate of their likelihood level.
4. Assign a magnitude of impact to determine how sensitive the asset may be against successfully exercised threats. Magnitudes of impact and impact level values are typically assigned by management for every successful threat that may exercise a vulnerability.
5. Associate assets with correspondent IT and/or business risks.
6. Compute risk rating by multiplying the probability assigned from Step 3 above (e.g., 1.00, 0.75, etc.) times the impact level value assigned in Step 4.
7. Recommend the controls that are needed to mitigate the risks according to their priority or ranking.

It is up to the organization to determine how to deal with the risks they have identified: take a chance and live with them or take action to protect their assets. At the same time, they must consider the costs associated with implementing controls, their impact on users, the manpower required to implement and manage them, and the scope of the action. Exhibit 3.3 shows an example of an IT risk assessment performed to identify and prioritize risks within financial applications. Risk assessment is covered in more detail in a later chapter.

Audit Plan

The audit function should formulate both long-range and annual plans. Planning is a basic function necessary to describe what must be accomplished, include budgets of time and costs, and state priorities according to organizational goals and policies. The objective of audit planning is to optimize the use of audit resources. To effectively allocate audit resources, internal audit departments must obtain a comprehensive understanding of the audit universe and the risks associated with each universe item. Failure to select appropriate items can result in missed opportunities to enhance controls and operational efficiencies. Internal audit departments that develop and maintain audit universe files provide themselves with a solid framework for audit planning.

The intent of the audit plan is to provide an overall approach within which audit engagements can be conducted. It provides the guidance for auditing the organization's integral processes.

Exhibit 3.3 Risk Assessment Example for the IT Functional Audit Area

Financial Application	IT Area / Vulnerability	Threat-Source	Likelihood Determination		Impact		Risk	Risk Rating[a]	Recommended Control	Action Priority
			Likelihood Level	Probability Assigned	Magnitude of Impact	Impact Level Value				
Financial Application #1 (FA1)	IS Operations / There is no offsite storage for data backups to provide reasonable assurance of availability in the event of a disaster.	Hurricanes, system failures, unexpected shutdowns	Medium	0.50	High	75	FA1 information cannot be recovered in the event of system failure, impacting the Company's ability to report financial information according to established reporting requirements.	37.5	Backups of FA1 financial data are archived off-site to minimize risk that data are lost.	Medium
	Information Security / Several of the Company's logical security settings (i.e., passwords) configured for FA1 are not consistent with industry best practices.	Unauthorized users (hackers, terminated employees, and insiders)	High	0.75	High	75	Security parameters are not appropriately configured, allowing for potential unauthorized user access to FA1.	56.25	The identity of users is authenticated to FA1 through passwords consistent with industry best practices minimum security values. Passwords must incorporate configuration for minimum length, periodic change, password history, lockout threshold, and complexity.	High

(Continued)

Exhibit 3.3 (*Continued*) Risk Assessment Example for the IT Functional Audit Area

Financial Application	IT Area / Vulnerability	Threat-Source	Likelihood Determination		Impact		Risk	Risk Rating[a]	Recommended Control	Action Priority
			Likelihood Level	Probability Assigned	Magnitude of Impact	Impact Level Value				
Financial Application #2 (FA2)	Information Security / FA2 owners do not periodically review user access privileges.	Unauthorized users (hackers, terminated employees, and insiders)	Very High	1.00	High	75	Users possess privileges that are not consistent with their job functions, allowing unauthorized or incorrect modifications to FA2's data, which could cause management decisions based upon misleading information.	75	User access privileges within FA2 are periodically reviewed by application owners to verify access privileges remain appropriate and consistent with job requirements.	Very High
	Information Security / Terminated user accounts are not removed from FA2.	Unauthorized users (terminated employees)	Very High	1.00	High	75	Terminated users can gain access to FA2 and view or modify its financial information.	75	The security administrator is notified of employees who have been terminated. Access privileges of such employees are immediately changed to reflect their new status.	Very High

(Continued)

Exhibit 3.3 (*Continued*)   Risk Assessment Example for the IT Functional Audit Area

| Financial Application | IT Area / Vulnerability | Threat-Source | Likelihood Determination | | Impact | | Risk | Risk Rating[a] | Recommended Control | Action Priority |
			Likelihood Level	Probability Assigned	Magnitude of Impact	Impact Level Value				
	Change Control Management / Test results for FA2 upgrades are not approved by management, prior to their implementation into production.	Unauthorized application changes and modifications	Low	0.25	High	75	FA2 changes are not properly authorized. Implementation of such changes could result in invalid or misleading data.	18.75	Changes to FA2 are tested and approved by management prior to their implementation in production in accordance with test plans and results.	Low

[a] Computed by multiplying the "Probability Assigned" and the "Impact Level Value."

The organization and its management must participate in and support this effort fully. Commitment can be gained if participants recognize that a good plan can help pinpoint problems in a highly dynamic, automated IT environment, for instance. Thus, it should be the responsibility of all participants not only to help pinpoint such problems, but also to assist in the measurement and quantification of problems.

Identifying, measuring, and quantifying problems in the IT area are difficult. The IT field is technologically complex and has a language of its own. Participants in the formulation of an IT audit plan, and particularly the IT auditors themselves, must have sufficient experience and training in technical matters to be able to grasp key concepts and abstractions about application systems. For example, abstractions about IT might include significant aspects that are susceptible to naming, counting, or conceptualizing. Understanding the systems at this level can lead to the identification of major problem areas. Audit concentration, then, may be directed to the major problem areas most likely to yield significant results.

Based on this identification of problems, the IT auditor determines what additional data might be required to reach evaluation decisions. The audit process, therefore, must be flexible enough to combine skilled personnel, new technology, and audit techniques in new ways to suit each situation. However, this flexibility of approach requires documentation in planned, directed steps. Systems that are understood poorly (or that have been designed without adequate controls) can result in lost revenues, increased costs, and perhaps disaster or fraud.

During the audit planning phase, the IT audit manager should meet with the chief information officer (CIO) and senior members of IT management to gain their input and concurrence with the risk assessment of the IT processes in the audit universe. If there is an IT steering committee, the audit universe should be reviewed with it as well. This will help ensure alignment between IT, business, and audit on the key risk areas. The meeting with the CIO and IT managers must also introduce the audit staff and communicate the scope, objectives, schedule, budget, and communication process to be used throughout the engagement. This is also an opportunity for an open discussion of IT management's perception of risk areas, significant changes in the area under review, and identification of appropriate contacts in IT.

An IT audit plan partitions the audit into discrete segments that describe application systems as a series of manageable audit engagements and steps. At the detailed planning or engagement level, these segments will have objectives that are custom-tailored to implement organizational goals and objectives within the circumstances of the audit. Thus, IT auditing does not call for "canned" approaches. There is no single series of detailed steps that can be outlined once and then repeated in every audit. The audit plan, therefore, is an attempt to provide an orderly approach within which flexibility can be exercised. At a minimum, an IT audit plan, after gathering a comprehensive understanding of the audit universe and the risks associated with each universe item, should:

1. List the audit objectives and describe the context
2. Develop the audit schedule
3. Create the audit budget and define scope
4. List audit team members, describe audit tasks, determine deadlines

Objectives and Context

The objective and context of the work are key elements in any audit environment and should not be overlooked. They are simply the basis by which all audits should be approached. The *objective* is what is trying to be accomplished. The *context* is the environment in which the work will be performed. Thus, everything ultimately depends on both the objective and the context of the work to be performed. That is, the decisions made about the scope, nature, and timing of the audit work depends on what the auditor's trying to do (e.g., gain assurance of an Accounts Receivable balance, ensure that a newly-implemented financial application will work correctly, assess whether a client Website is secure, etc.) and the environment he/she is working in (e.g., a large versus a small company, a domestic organization with a centralized system versus a multinational with multiple divisions, a New York-based organization versus one based in North Dakota, etc.).

Keep in mind what works well for one organization, may not work as well in another based on many combinations of objective and context. For example, if the IT auditor has a General Controls Assessment, the audit objectives may be to verify that all controls surrounding financial applications and related to the data center, information systems operations, information security, and change control management are adequate. Therefore, the IT auditor needs to verify the controls because the financial auditors were relying on such financial computer system to provide them with the correct financial information. The context is where the auditor's true analytical skills come into play. Here, the environment is for the most part always different from shop to shop. The auditor must assess the context for which he or she has entered and make a decision as to how the environment should be addressed (e.g., big company, small company, large staff, small staff, etc.).

By defining appropriate objectives and context of the work, management can ensure that the audit will verify the correct functioning and control of all key audit areas. A common objective/context set for IT audits is to support financial statement audits.

IT Audits Conducted to Support Financial Statement Audits

Once the auditor has gained a general familiarity with the client's accounting and financial procedures, specific areas of audit interest must be identified. The auditor must decide what applications will have to be examined at a more detailed level. For applications used to support significant business processes, the auditor must determine their sophistication and extent of use. This preliminary study goes just deep enough for the auditor to evaluate the complexity and sophistication of the applications and determine the procedures to be followed in evaluating their internal controls.

Understanding financial applications and determining whether IT controls are in place to effectively secure them and the information generated represent a significant process as it relates to the overall financial statement audit. Results of an IT audit over financial applications have direct bearing on the substantive testing performed by financial auditors. Substantive testing involves audit procedures necessary to examine and support the financial statements (e.g., confirming account balances, examining documentation, re-performing procedures, inquiring about or observing a transaction, etc.). These procedures provide the evidence needed to support the assertion that financial records of the organization are indeed valid, accurate, and complete.

The results or findings from an IT audit typically determine the amount of substantive tests that will be performed by financial auditors. If results are effective (i.e., IT controls are found to be in place and operating properly), the work of the financial auditor would most likely be less on that particular part of the audit. On the other hand, if there are no IT controls in place protecting the financial applications, or if existing IT controls are not operating effectively, the amount of substantive testing performed by the financial auditor will be much higher. This can have significant implications to the audit, such as the time it takes to complete the audit, increased costs to the client, etc. The remainder of this chapter is focused on IT audits conducted to support financial statement audits. The next step within the audit plan is the development of an audit schedule.

Audit Schedule

Internal auditing departments create annual audit schedules to gain agreement from the board on audit areas, communicate audit areas with the functional departments, and create a project/resource plan for the year. The audit schedule should be linked to current business objectives and risks based on their relative cost in terms of potential loss of goodwill, loss of revenue, or noncompliance with laws and regulations.

Annual schedule creation is the process of determining the total audit hours available, then assigning universe items (audit areas) to fill the available time. As mentioned previously, to maximize the risk assessment process, "high risk" universe items should be given top audit priority. Schedule creation should be performed in conjunction with the annual risk assessment process; this will enable internal audit departments to account for the changes in risk rankings and make any necessary additions or deletions to the audit universe. Of course, the audit schedule will also need to be agreed with the audit committee as part of the overall audit planning process. Once the available audit hours are determined, audit management can continue preparing the audit plan.

Planning and scheduling are ongoing tasks as risks, priorities, available resources, and timelines change. When these changes take place, it is important to communicate them to the audit committee, board, and all other impacted functional departments.

Audit Budget and Scoping

Ideally, the audit budget should be created after the audit schedule is determined. However, most organizations have budget and resource constraints. An alternative approach may be necessary when building the audit schedule. After determining the audit priorities, audit management will determine the number of available hours to decide how many audits they can complete in a year. For a particular IT audit, available hours are listed per area, staff personnel, etc. Exhibit 3.4 illustrates an example of a budget in an IT audit.

The scope of an audit defines the area(s) (e.g., relevant financial applications, databases, operating systems, networks, etc.) to be reviewed. The names of the financial applications and databases should also be described along with their hosting information (e.g., server location, etc.). The scope should clearly identify the critical business process supported by the selected financial application. This association typically justifies the relevance of the application and, hence, its inclusion as part of the audit. The scope should further state the general control areas, control objectives, and control activities that would undergo review. Exhibit 3.5a, b shows examples of scoping for applications and control objectives, respectively, in an IT audit.

Audit Team, Tasks, and Deadlines

The audit plan must include a section listing the members of the audit, their titles and positions, and the general tasks they will have. For instance, a typical audit involves staff members, seniors, managers, or senior managers, and a partner, principal, or director (PPD) who will be overseeing the entire audit. At a staff level (usually those auditors with less than 3 years of experience), most of the field work is performed, including gathering documentation, meeting with personnel, and creating audit **work papers**, among others. Senior-level auditors not only supervise the work of staff auditors, but guide them in performing the work (e.g., accompany staff auditors to meet with users, assist the staff in selecting what specific information should be gathered, how to document such information in the working papers, etc.). Next are the managers or senior managers (senior

Exhibit 3.4 Example of a Budget for an IT Audit

Company Name					
IT Budget					
Fiscal Year 20XX					
	Audit Professional				
Audit Area	*Staff/ Senior*	*Manager*	*Partner*	*Total Hours*	
Planning					
Review work papers from the prior year, if applicable; prepare IT budget; conduct planning meetings; prepare planning memo; prepare initial request of information and send to company personnel, etc.	3.0	1.0	0.0	4.0	
First year. Gather and document an understanding of the organization and its IT environment, including how the organization utilizes computer system and which applications impact critical business/financial processes, among others. *Subsequent years.* Review and update the understanding of the organization and its IT environment obtained from the prior year.	3.0	1.0	0.0	4.0	
Conduct planning meeting with company personnel.	1.0	1.0	1.0	3.0	
Subtotal	7.0	3.0	1.0	11.0	11%
Fieldwork					
Document/update understanding of the organization's IT environment and perform tests of IT controls *(per General Control IT area).*					
Information Systems Operations	16.0	0.0	0.0	16.0	
Information Security	17.0	0.0	0.0	17.0	
Change Control Management	20.0	0.0	0.0	20.0	
Subtotal	53.0	0.0	0.0	53.0	53%

(Continued)

Exhibit 3.4 (*Continued*) Example of a Budget for an IT Audit

Company Name					
IT Budget					
Fiscal Year 20XX					
		Audit Professional			
Audit Area	*Staff/ Senior*	*Manager*	*Partner*	*Total Hours*	
Review, Reporting, and Conclusion					
Review and document action(s) taken by company's Management to correct last year's IT audit findings/ deficiencies.	2.0	0.0	0.0	2.0	
Document IT audit findings/ deficiencies and opportunities to improve existing controls.	3.0	0.0	0.0	3.0	
Assess and classify identified IT audit findings/deficiencies.	1.0	0.0	0.0	1.0	
Draft IT Management letter listing all IT audit findings/deficiencies and opportunities to improve existing controls. Forward letter to IT Management for review.	0.0	1.0	1.0	2.0	
Conduct status meetings, internally or with IT personnel.	1.0	0.0	0.0	1.0	
Review work papers evidencing IT audit work performed.	0.0	9.0	4.0	13.0	
Exit meeting with IT personnel to discuss audit and results.	0.0	1.0	0.0	1.0	
Address and clear review notes from audit management (Manager and Partner) and conclude audit.	11.0	2.0	0.0	13.0	
Subtotal	18.0	13.0	5.0	36.0	36%
Grand Total	78.0	16.0	6.0	100.0	100%
			Staff/Senior	78.0	78%
			Manager	16.0	16%
			Partner	6.0	6%

Exhibit 3.5a   Example of Scoping for Financial Applications

Company Name

Financial Applications and their Association with Business Processes

Fiscal Year 20XX

Purpose: To identify relevant applications by mapping them to their corresponding business process(es). An "X" in the table on the right identifies the business process supported by the application. For example, the *SAP* application is used by (or supports) the *Financial Reporting, Expenditures, Inventory Management,* and *Revenue* business processes. This association typically justifies the relevance of the application and, hence, its inclusion as part of the audit.

# Application	Brief Description	Processing Environment (Operating System Where the Application Is Installed On)	Database Management Software	Physical Hosting Location—Application and Database	Business Process Supported						
					Financial Reporting	Expenditures	Payroll & Personnel	Inventory Management	Investment	Revenue	Fixed Assets
1 SAP	Includes the general ledger, expenditures, inventory management, and revenue accounting modules.	UNIX	Oracle	Local Data Center, Second Floor; Company's Headquarters [location]	X	X		X		X	
2 Infinium	Manages the payroll.	AS/400	Oracle	Local Data Center, Second Floor; Company's Headquarters [location]			X				

(Continued)

Exhibit 3.5a (*Continued*) Example of Scoping for Financial Applications

Company Name

Financial Applications and their Association with Business Processes

Fiscal Year 20XX

Purpose: To identify relevant applications by mapping them to their corresponding business process(es). An "X" in the table on the right identifies the business process supported by (or supports) the application. For example, the *SAP* application is used by (or supports) the *Financial Reporting, Expenditures, Inventory Management,* and *Revenue* business processes. This association typically justifies the relevance of the application and, hence, its inclusion as part of the audit.

#	Application	Brief Description	Processing Environment (Operating System Where the Application Is Installed On)	Database Management Software	Physical Hosting Location—Application and Database	Business Process Supported						
						Financial Reporting	Expenditures	Payroll & Personnel	Inventory Management	Investment	Revenue	Fixed Assets
3	APS/2	Manages investments.	Windows	Oracle	Local Data Center, Second Floor; Company's Headquarters [location]					X		
4	Timberline	Manages long term and fixed assets.	Windows	Oracle	Local Data Center, Second Floor; Company's Headquarters [location]							X

Exhibit 3.5b   Example of Scoping for General Computer Control Objectives and Activities

Company Name			
General Computer Controls Objectives and Activities Selected			
Fiscal Year 20XX			
#	*IT Area*	*Control Objective*	*Control Activity*
1	Information Systems Operations	ISO 1.00 - IT operations support adequate scheduling, execution, monitoring, and continuity of systems, programs, and processes to ensure the complete, accurate, and valid processing and recording of financial transactions.	ISO 1.01 - Batch and/or online processing is defined, timely executed, and monitored for successful completion. ISO 1.02 - Exceptions identified on batch and/or online processing are timely reviewed and corrected to ensure accurate, complete, and authorized processing of financial information.
2	Information Systems Operations	ISO 2.00 - The storage of financial information is appropriately managed, accurate, and complete.	ISO 2.02 - Automated backup tools have been implemented to manage retention data plans and schedules. ISO 2.04 - Tests for the readability of backups are performed on a periodic basis. Results support timely and successful restoration of backed up data.
3	Information Systems Operations	ISO 3.00 - Physical access is appropriately managed to safeguard relevant components of the IT infrastructure and the integrity of financial information.	ISO 3.02 - Physical access is authorized, monitored, and restricted to individuals who require such access to perform their job duties. Entry of unauthorized personnel is supervised and logged. The log is maintained and regularly reviewed by IT management.
4	Information Security	ISEC 1.00 - Security configuration of applications, databases, networks, and operating systems is adequately managed to protect against unauthorized changes to programs and data that may result in incomplete, inaccurate, or invalid processing or recording of financial information.	ISEC 1.02 - Formal policies and procedures define the organization's information security objectives and the responsibilities of employees with respect to the protection and disclosure of informational resources. Management monitors compliance with security policies and procedures, and agreement to these are evidenced by the signature of employees. ISEC 1.06 - Consistent with information security policies and procedures, local and remote users are required to authenticate to applications, databases, networks, and operating systems via passwords to enhance computer security.

(Continued)

Exhibit 3.5b (*Continued*) Example of Scoping for General Computer Control Objectives and Activities

Company Name			
General Computer Controls Objectives and Activities Selected			
Fiscal Year 20XX			
#	*IT Area*	*Control Objective*	*Control Activity*
5	Information Security	ISEC 2.00 - Adequate security is implemented to protect against unauthorized access and modifications of systems and information, which may result in the processing or recording of incomplete, inaccurate, or invalid financial information.	ISEC 2.02 - System owners authorize user accounts and the nature and extent of their access privileges. ISEC 2.04 - Users who have changed roles or tasks within the organization, or that have been transferred, or terminated are immediately informed to the security department for user account access revision in order to reflect the new and/or revised status. ISEC 2.05 - Transmission of sensitive information is encrypted consistent with security policies and procedures to protect its confidentiality.
6	Change Control Management	CCM 1.00 - Changes implemented in applications, databases, networks, and operating systems (altogether referred to as "system changes") are assessed for risk, authorized, and thoroughly documented to ensure desired results are adequate.	CCM 1.03 - Documentation related to the change implementation is adequate and complete. CCM 1.05 - Documentation related to the change implementation has been released and communicated to system users.
7	Change Control Management	CCM 2.00 - Changes implemented in applications, databases, networks, and operating systems (altogether referred to as "system changes") are appropriately tested. Tests are performed by a group other than the group responsible for the system (e.g., operating systems changes are implemented by someone other than the systems programmer, etc.).	CCM 2.01 - System changes are tested before implementation into the production environment consistent with test plans and cases. CCM 2.02 - Test plans and cases involving complete and representative test data (instead of production data) are approved by application owners and development management.

(*Continued*)

Exhibit 3.5b (*Continued*)   Example of Scoping for General Computer Control Objectives and Activities

Company Name			
General Computer Controls Objectives and Activities Selected			
Fiscal Year 20XX			
#	*IT Area*	*Control Objective*	*Control Activity*
8	Change Control Management	CCM 3.00 - Changes implemented in applications, databases, networks, and operating systems (altogether referred to as "system changes") are appropriately managed to reduce disruptions, unauthorized alterations, and errors which impact the accuracy, completeness, and valid processing and recording of financial information.	CCM 3.01 - Problems and errors encountered during the testing of system changes are identified, corrected, retested, followed up for correction, and documented.
9	Change Control Management	CCM 4.00 - Changes implemented in applications, databases, networks, and operating systems (altogether referred to as "system changes") are formally approved to support accurate, complete, and valid processing and recording of financial information.	CCM 4.04 - An overall review is performed by management after system changes have been implemented in the live or production environment to determine whether the objectives for implementing system changes were met.

managers are typically involved as part of large audits) that supervise the audit work prepared by the staff and reviewed by the senior. Managers perform detailed reviews of the work papers and ensure the audit objectives have been achieved. Managers meet frequently with audit clients, and provide them with audit status, preliminary findings identified, hours incurred and left to finish, etc. Managers also provide frequent status of the audit work to the PPD assigned, to which they report directly. Lastly, the PPD performs a high-level review of the work (as provided by managers), focusing on high-risk areas, controls in place that are not adequately designed nor operating effectively, findings identified and their impact to the overall audit, etc. PPDs tend to rely on the detailed reviews performed by managers or senior managers, and also ensure the overall objectives of the audit have been achieved.

Deadlines are a critical component of an audit plan. They should be reviewed and agreed with the client organization from the start of the audit so that they comply with requirements established by third parties (e.g., banks, financial institutions, etc.) and regulators (e.g., government, private organizations, etc.). Deadlines should be well-thought of taking into account the information and resources that must be available to perform the audit work within the established requirements.

An audit planning memo ("planning memo") is part of the auditor working papers and documents the sections just described. The planning memo is typically prepared by the audit engagement senior, and reviewed by the manager before submitting it to the PPD for approval. Appendix 1 shows the format of a typical IT planning memo, including the procedures which may be performed by an IT auditor in connection with an audit engagement. The planning memo may be tailored for the specific facts and circumstances of the audit engagement. This includes removing sections which are not applicable. The memo in Appendix 1 includes some wording in italics that is either enclosed within brackets or parentheses. This format is used to indicate information to be replaced as applicable, or that guides the completion of the memo.

Audit Process

Statement on Auditing Standards (SAS No. 1) has the effect of mandating a uniform, process-oriented approach to audit engagements. The approach depicted is a true process technique. That is, audits follow a series of logical, orderly steps, each designed to accomplish specific end results. This is also the case for an IT audit. The difference in an IT audit is the specialized approach to the audit work and the skills needed to understand technology and the IT control environment. The phases of auditing activities typically overlap and involve some reassessment and retracing of procedures performed earlier. Common phases of an audit engagement are shown in Exhibit 3.6. The first two phases, *Risk Assessment* and *Audit Plan*, have been explained above. Following are explanations of the remaining phases related to an IT audit.

Preliminary Review

In this phase, the auditor should obtain and review summary-level information and evaluate it in relation to the audit objectives. The purpose of the preliminary review phase of an IT audit engagement is to gather an understanding of the IT environment, including the controls in place

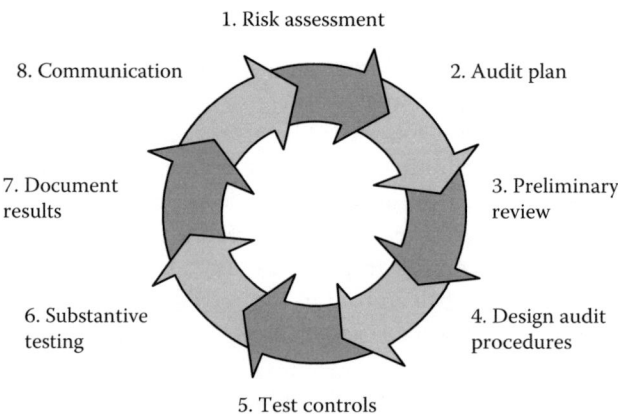

Exhibit 3.6   Phases of an audit.

that are essential to meet the overall audit objectives. The IT auditor conducts this preliminary review at a general level, without examining details of individual applications and the processes involved. Instead, the IT auditor interviews key personnel to determine policies and practices, and prepares supplemental audit information as required. Preliminary review information serves as a basis for supporting the information included in the IT audit plan.

General Information about IT Environment

As previously discussed, IT is defined as the hardware, software, communication, and other facilities used to input, store, process, transmit, and output data in whatever form. The IT environment refers to the policies, procedures, and practices implemented by organizations to program, test, deliver, monitor, control, and support their IT infrastructure (e.g., hardware, software, networks, etc.). The IT environment also includes the applications and programs used by organizations to support critical business operations (i.e., financial operations) and achieve business strategies.

The IT auditor begins the examination process by becoming acquainted, generally, with the company, its line of business, and the IT environment, including its financial application systems. Typically, an IT auditor would tour the client company's facilities and observe general business operations that bear upon customer service as well as on strictly financial functions.

Given this familiarity, the next level of general data gathering would include the preparation of organizational charts, particularly those for the accounting and IT functions. If organizational charts are unavailable, the IT auditor should develop them. Once drawn, the charts should be reviewed and verified with appropriate personnel (i.e., key executives in the accounting and IT areas) to secure an agreement that they represent the actual organization structure. During these interviews, the IT auditor would also secure copies of the company's chart of accounts and an accounting standards manual, if available.

IT auditors must gain a deep understanding of the IT environment, particularly how the organization responds to risks arising from IT, and whether the IT controls in place have been adequately designed and operate effectively to address those risks. From a financial standpoint, knowledge about the IT environment is crucial for IT auditors in order to understand how financial transactions are initiated, authorized, recorded, processed, and reported in the financial statements.

For application systems which the organization uses computers to process significant financial data, the IT auditor would gather a number of specific items of evidential matter, such as:

- Policies and procedures that the organization implements and the IT infrastructure and application software that it uses to support business operations and achieve business strategies.
- Narratives or overview flowcharts of the financial applications, including server names, make and model, supporting operating systems, databases, and physical locations, among others.
- Whether the financial applications are in-house developed, purchased with little or no customization, purchased with significant customization, or proprietary provided by a service organization.
- Whether service organizations host financial applications and if so, what are these applications and which relevant services they perform.
- Controls in place supporting the area of information systems operations, such as those supporting job scheduling, data and restoration, backups, and offsite storage.

- Controls in place supporting the area of information security, such as those supporting authentication techniques (i.e., passwords), new access or termination procedures, use of firewalls and how are they configured, physical security, etc.
- Controls in place supporting the area of change control management, such as those supporting the implementation of changes into applications, operating systems, and databases; testing whether access of programmers is adequate; etc.

Methods applied in gathering these data include reviewing computer information systems and human interface practices, procedures, documents, narratives, flowcharts, and record layouts. Other audit procedures implemented to gather data include: observing, interviewing, inspecting existing documentation, and flowcharting, among others. Physical inspection techniques are used both to gather data and to validate existing documents or representations made during the interviews. For example, a single visit to the computer/data center can provide both data gathering and validation opportunities for determining equipment configurations, library procedures, operating procedures, physical security controls, existing environmental controls, and other data control procedures.

Many of these procedures are substantially the same regardless of whether the accounting system is computerized or not. Differences associated with the audit of computerized systems center around changes in controls, documentation, audit techniques, and technical qualifications required by audit staff members. Appendix 2 shows an example of the types of questions and information that should be documented when gathering an understanding of an IT environment.

Design Audit Procedures

In this phase, the IT auditor must prepare an audit program for the areas being audited, select control objectives applicable to each area, and identify procedures or activities to assess such objectives. An audit program differs from an internal control questionnaire (ICQ) in that an ICQ involves questions to evaluate the design of the internal control system. Particularly, ICQs check whether controls are implemented to detect, prevent, or correct a **material misstatement**. Controls not in place would represent a deviation or deficiency in the internal control structure. An audit program, on the other hand, contains specific procedures to test the responses received from the questions asked, thus substantiating that the controls identified are in place and work as expected by management.

An audit program is a formal plan for reviewing and testing each significant audit subject area disclosed during fact gathering. The auditor should select subject areas for testing that have a significant impact on the control of the application and those that are within the scope defined by the audit objectives. IT audit areas are very specific to the type of audit. For IT, COBIT is an excellent starting point as it lists risks, objectives, and key controls per IT audit area. This information then has to be customized to the particular organization objectives, processes, and technology. Appendix 3 illustrates examples of IT Audit Programs for the general control IT areas.

Identifying Financial Applications

With the help of management, the IT auditor must decide what application systems will have to be examined at a more detailed level (i.e., scoping). As a basis for preparation of the audit plan, the IT auditor must also determine, in general, how much time will be required, what types of people and skills will be needed to conduct the examination; and, roughly, what the schedule will be.

The identification of financial applications can be accomplished with the auditor gaining familiarity with the organization's accounting procedures and processes. The importance of determining the significant financial applications has to be derived through preliminary analysis. The assessment of the sophistication of the application, its complexity, the business process they support, and extent of use are factors that come into play in deciding whether to select such application and how one might evaluate it. As stated before, the preliminary review phase is a critical step in the audit process that examines an organization's financial systems and provides the auditor with a basis for selecting audit areas for more detailed analysis and evaluation whether they are manual or computerized.

Auditors involved in reviewing financial applications should focus their concerns on the application's control aspects. This requires their involvement from the time a transaction is initiated until it is posted into the organization's general ledger. Specifically, auditors must ensure that provisions are made for:

- An adequate audit trail so that transactions can be traced forward and backward through the financial application
- The documentation and existence of controls over the accounting for all data (e.g., transactions, etc.) entered into the application and controls to ensure the integrity of those transactions throughout the computerized segment of the application
- Handling exceptions to, and rejections from, the financial application
- Unit and integrated testing, with controls in place to determine whether the applications perform as stated
- Controls over changes to the application to determine whether the proper authorization has been given and documented
- Authorization procedures for application system overrides and documentation of those processes
- Determining whether organization and government policies and procedures are adhered to in system implementation
- Training user personnel in the operation of the financial application
- Developing detailed evaluation criteria so that it is possible to determine whether the implemented application has met predetermined specifications
- Adequate controls between interconnected application systems
- Adequate security procedures to protect the user's data
- Backup and recovery procedures for the operation of the application and assurance of business continuity
- Ensuring technology provided by different vendors (i.e., operational platforms) is compatible and controlled
- Adequately designed and controlled databases to ensure that common definitions of data are used throughout the organization, redundancy is eliminated or controlled, and data existing in multiple databases is updated concurrently

This list affirms that the IT auditor is primarily concerned with adequate controls to safeguard the organization's assets.

Test Controls

The IT auditor executes several procedures in order to test controls, processes, and apparent exposures. These audit procedures may include examining documentary evidence, as well as performing corroborating interviews, inspections, and personal observations.

Documentary evidence may consist of a variety of forms of documentation on the application system under review. Examples include notes from meetings on subject system, programmer notes, systems documentation, screenshots, user manuals, and change control documentation from any system or operation changes since inception, and a copy of the contract if third parties involved. Examining such documentary evidence may require the IT auditor to ask questions of the user, developer and managers to help him or her establish the appropriate test criteria to be used. It also helps in identifying the critical application and processes to be tested.

Corroborating interviews are also part of the testing process, and may include procedures such as:

- Asking different personnel the same question and comparing their answers
- Asking the same question in different ways at different times
- Comparing answers to supporting documentation, work papers, programs, tests, or other verifiable results
- Comparing answers to observations and actual system results

An example would involve interviewing a programmer for an application under review. The programmer states that the application has undergone recent changes not reflected in the current documentation. It is very important to identify what those changes were if those areas of the application were to be selected for control testing.

For inspection of documentation, the IT auditor can obtain the logical settings (i.e., passwords) currently configured at the organization's network, operating system, and financial application levels. Of particular importance is to obtain and assess the network's configured logical settings as this is the first level of authentication before users can gain access to the financial applications. The settings received are then compared against the organization's password policy to determine whether they are or not in compliance with such policies. In the absence of a password policy, the organization's logical settings configured are compared against industry standards or best practices. Documentation supporting the above settings is usually first obtained through interviewing information security personnel.

Another common audit procedure to test and validate information would be to observe actual procedures taking place. In the example above, the IT auditor would observe the settings configured in the financial application and request organization personnel to print out a screenshot for documentation in the audit working papers. Exhibit 3.7a shows an example of common documentation obtained supporting the password settings configured. In this case, settings such as enforced password history, minimum (or maximum) password age, minimum password length, password complexity, account lockout duration and threshold, and whether passwords have been stored using reversible encryption are some of the setting that are typically gathered. An IT auditor working paper documenting testing of some of these settings would look like the one in Exhibit 3.7b. Notice on the table the actual password settings configured documented at the network (or the first authentication level), operating system, and financial applications levels. Also notice notes and tickmarks (explanations) about the information therein and, most importantly, the assessment of whether the client password settings comply with either the existing company policy, or industry standards and best practices. When settings do not comply with the policy or industry standards or best practices, audit exceptions (findings) are written up and listed in a separate working paper. This working paper will eventually assist when writing up the findings/deficiency section of the **Management Letter**. A second example of observation as a test procedure would involve an IT auditor examining a disaster recovery exercise. Here, the IT auditor could determine whether personnel followed appropriate procedures and processes. Through personal

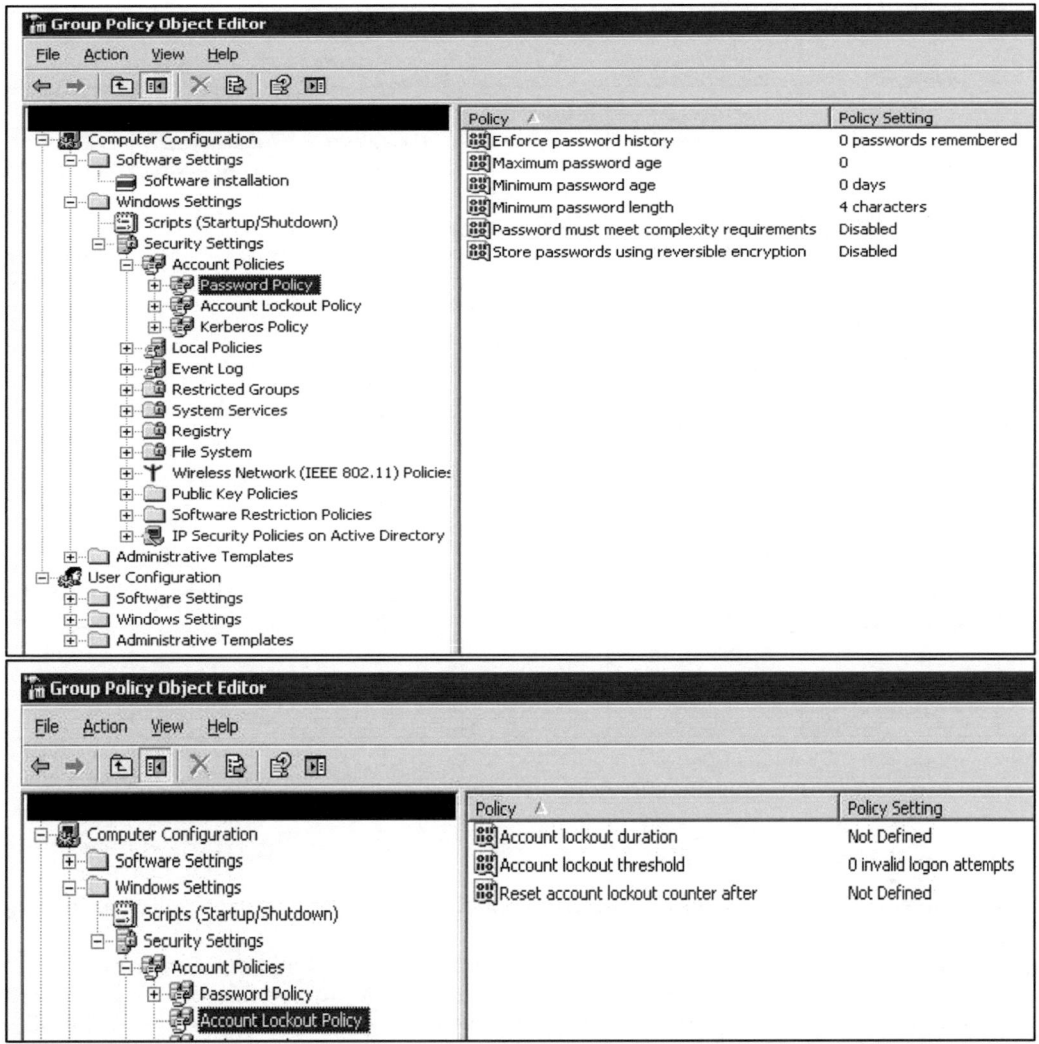

Exhibit 3.7a Example of evidence supporting logical security (password) settings currently implemented.

observations, the auditor can assess and determine whether personnel is following operating procedures and plans, and is adequately prepared for the disaster simulated.

Substantive Testing

Where controls are determined not to be effective, substantive testing may be required to determine whether there is a material issue with the resulting financial information. In an IT audit, substantive testing is used to determine the accuracy and completeness of information being generated by a process or application. Contrary to compliance testing where the auditor's goal is to confirm whether the organization is adhering to applicable policies, procedures, rules, and regulations. An example of a compliance test procedure would be verifying that a change or upgrade in a financial application was adequately tested, approved, and documented prior to its implementation.

Exhibit 3.7b   Example Supporting a Logical Security Settings Test

#	Network / System / Financial Application	Logical Setting				
		Enforce Password History	Minimum Password Age	Minimum Password Length	Password Complexity	Account Lockout
	Per Company Policy [*working paper (w/p) ##*] {1}	5 passwords remembered	90 days	6 characters	Enabled	3 invalid login attempts
Actual Testing Performed						
	Local Area Network (LAN) / Windows	0 passwords remembered {a}	0 days {a}	4 characters {a}	Disabled {a}	0 invalid login attempts {a}
1	Financial Application X	{b}	{b}	{b}	{b}	{b}
2	Financial Application Y	Option not available— Application limitation {d}	90 days {c}	6 characters {c}	Enabled {c}	3 invalid login attempts {c}

Note: The password values above were obtained through observation, and with the assistance of [*name of Information Security Administrator*].

Tickmarks (explanations):

{1}—Password settings obtained from company policy. Copy of the company policy supporting these settings is documented in w/p [##].

{a}—The Enforce Password History, Minimum Password Age, Minimum Password Length, Password Complexity, and Account Lockout settings are **not** configured consistent with company policy, and therefore, do not promote an acceptable level of security. The value configured for Password Complexity has also been set to "Disabled." Password complexity requirements establish minimum password parameters not easily compromised that users must follow in establishing their passwords, particularly at the LAN/Windows level, which serves as the first layer of authentication. Exceptions noted. Refer to w/p [##], where these exceptions have been listed.

{b}—Password security settings are controlled through the Windows operating system. Therefore, the configuration of the LAN/Windows password settings covers this application. Refer to the LAN/Windows row above.

{c}—Password security settings such as Minimum Password Age, Minimum Password Length, Password Complexity, and Account Lockout have been configured consistent with the company policy, promoting an adequate level of security. No exceptions noted.

{d}—Application functionality limitations do not allow the enforcement of password history. Exceptions noted. Refer to w/p [##], where this exception has been listed.

Substantive audit tests are designed and conducted to verify the functional accuracy, efficiency, and control of the audit subject. During the audit of a financial application, for example, the IT auditor would build and process test data to verify the processing steps of such an application.

Auditing-through-the-computer is a term that involves steps in addition to those mentioned previously. Programs are executed on the computer to test and authenticate application programs that are run in normal processing. Usually, the financial audit team will select one of the many Generalized Audit Software packages such as SAS, SPSS, Computer-Assisted Audit Techniques (CAATs), or CA-Easytrieve(T) and determine what changes are necessary to run the software at the installation. Financial auditors use this specific software to do sampling, data extraction, exception reporting, summarize and foot totals, and other tasks. They also use packages such as Microsoft Access, Excel, IDEA, or ACL because of their in-depth analyses and reporting capabilities.

CAATs, for example, use auditor-supplied specifications to generate a program that performs audit functions, such as evaluating application controls, selecting and analyzing computerized data for substantive audit tests, etc. In essence, CAATs automate and simplify the audit process, and this is why audit teams (external and internal) are increasingly using them. In fact, many organizations have Generalized Audit Software already installed for their internal auditors to allow them to gather information and conduct the planned audit tests. The appropriate selection and effective use of these audit tools are essential not only to perform adequate audit testing but also to document results.

Document Results

The next phase of an audit involves documenting results of the work performed, as well as reporting on the findings. Audit results should include a description of audit findings, conclusions, and recommendations.

Audit Findings

The terms finding, exception, deficiency, deviation, problem, and issue are basically synonymous in the audit world, and mean the auditor identified a situation where controls, procedures, or efficiencies can be improved. Findings identify and describe inaccurate, inefficient, or inadequately controlled audit subjects. An example of an IT audit finding would be a change implemented into a financial application that did not include proper management authorization. Another example would include the IT auditor discovering that the organization's procedures manual does not require management's permission before implementing changes into applications.

Audit findings should be individually documented and should at least include the following:

- Name of the IT environment (operating system hosting the relevant financial application(s)) evaluated
- IT area affected (IS operations, information security, change control management)
- Working paper test reference where the finding was identified
- General control objective(s) and activity(ies) that failed
- Brief description of the finding
- Where is the finding formally communicated to management (this should reference the Management Letter within the **Auditor Report**)

- The individual classification of the finding per audit standard AU 325, *Communications About Control Deficiencies in an Audit of Financial Statements*, as either a deficiency, significant deficiency, or a material weakness[*]
- Evaluation of the finding, specifically whether it was identified at the design level (i.e., there is no general control in place) or at the operational level (i.e., the general control was in place, but did not test effectively)
- Whether the finding represents or not a pervasive or **entity-level risk**
- Whether the finding can be mitigated by other compensating general controls, and if so, include reference to where these controls have been tested successfully

An audit finding form (e.g., General Computer Controls Findings Form, etc.) can be used to review the control issues identified with the responsible IT manager in order to agree on corrective action. This information can then be used to prepare the formal Management Letter that will accompany the Audit Report and the corrective action follow-ups. Taking corrective action could result in enhanced productivity; the deterrence of fraud; or the prevention of monetary loss, personal injury, or environmental damage. Exhibit 3.8 shows an example of a worksheet that may be used to summarize the individual findings identified during an IT audit.

Conclusions and Recommendations

Conclusions are auditor opinions, based on documented evidence, that determine whether an audit subject area meets the audit objective. All conclusions must be based on factual data obtained and documented by the auditor as a result of audit activity. The degree to which the conclusions are supported by the evidence is a function of the amount of evidence secured by the auditor. Conclusions are documented in the audit working papers and should support the audit procedures performed. Working papers are the formal collection of pertinent writings, documents, flowcharts, correspondence, results of observations, plans and results of tests, the audit plan, minutes of meetings, computerized records, data files or application results, and evaluations that document the auditor activity for the entire audit period. A complete, well-organized, cross-referenced, and legible set of working papers is essential to support the findings, conclusions, and recommendations as stated in the Audit Report. Typically, a copy of the final Audit Report is filed in the working papers.

Recommendations are formal statements that describe a course of action that should be implemented by the company's management to restore or provide accuracy, efficiency, or adequate control of audit subjects. A recommendation should be provided by the auditor for each audit finding for the report to be useful to management.

Communication

The value of an audit depends, in large part, on how efficiently and effectively its results are communicated. At the conclusion of audit tests, it is best to discuss the identified findings with IT management to gain their agreement and begin any necessary corrective action. Findings, risks as a result of those findings, and audit recommendations are usually documented on the Management Letter (in a separate section of the Audit Report). Refer to Exhibit 3.9 for an example of the format of a Management Letter from an IT audit.

---

[*] http://pcaobus.org/Standards/Auditing/Pages/AU325.aspx.

Exhibit 3.8 Example Supporting Documentation of the General Computer Control Findings Identified

Company Name

GCC Findings

Fiscal Year 20XX

No./IT Environment—IT Area/W/P Reference # Where Finding Was Identified	Control Objective	Failed Control Activity	Brief Description of Finding Communicated to Management in [W/P Reference # of IT Management Letter]	Classification of Finding as a Deficiency (Design or Operating), Significant Deficiency, or Material Weakness	Finding Mitigated By Other Compensating General Controls? (If So, List Control.)
1/Windows—Information Security/W/P Reference #	ISEC 2.00 - Adequate security is implemented to protect against unauthorized access and modifications of systems and information, which may result in the processing or recording of incomplete, inaccurate, or invalid financial information.	ISEC 2.04 - Users who have changed roles or tasks within the organization, or that have been transferred, or terminated are immediately informed to the security department for user account access revision in order to reflect the new and/or revised status.	We noted that the notification for termination from Human Resources (HR) for two user accounts selected for testing was not received by IT personnel within seven business days. We also noted that the user account of one terminated employee remained active in the [financial application name]. Moreover, for five out of six terminated employees tested, notification of the employee's termination was not sent immediately or within seven business days from the employee's supervisor or HR to IT personnel.	Operating Deficiency. Deficiency does not represent a material weakness (i.e., will not prevent, or detect and correct material misstatements in the financial statements). The deficiency is also not severe enough to merit the attention of those charged with governance (i.e., significant deficiency). Simply, the operation of the existing control does not allow management or employees, in the normal course of performing their assigned functions, to prevent, detect, and/or correct misstatements on a timely basis.	[List the general controls identified (and successfully tested) that can mitigate or compensate the identified finding.]

(Continued)

Exhibit 3.8 (*Continued*) Example Supporting Documentation of the General Computer Control Findings Identified

Company Name				
GCC Findings				
Fiscal Year 20XX				
No./IT Environment—IT Area/W/P Reference # Where Finding Was Identified	*Control Objective*	*Failed Control Activity*	*Brief Description of Finding Communicated to Management in [W/P Reference # of IT Management Letter]*	*Classification of Finding as a Deficiency (Design or Operating), Significant Deficiency, or Material Weakness*	*Finding Mitigated By Other Compensating General Controls? (If So, List Control.)*
2/UNIX— Information Security/W/P Reference #	ISEC 2.00 - Adequate security is implemented to protect against unauthorized access and modifications of systems and information, resulting in the processing or recording of incomplete, inaccurate, or invalid information.	ISEC 2.03 - User account access privileges are periodically reviewed by systems and application owners to determine whether they are reasonable and/ or remain appropriate.	We noted no formal access reviews were performed by IT personnel and/or business/application owners for the relevant financial applications in scope.	Same as above.	[List the general controls identified (and successfully tested) that can mitigate or compensate the finding.]

(Continued)

Exhibit 3.8 (Continued) Example Supporting Documentation of the General Computer Control Findings Identified

Company Name

GCC Findings

Fiscal Year 20XX

No./IT Environment—IT Area/W/P Reference # Where Finding Was Identified	Control Objective	Failed Control Activity	Brief Description of Finding Communicated to Management in [W/P Reference # of IT Management Letter]	Classification of Finding as a Deficiency (Design or Operating), Significant Deficiency, or Material Weakness	Finding Mitigated By Other Compensating General Controls? (If So, List Control.)
3/Linux—Information Security/W/P Reference #	ISEC 1.00 - Security configuration of applications, databases, networks, and operating systems is adequately managed to protect against unauthorized changes to programs and data that may result in incomplete, inaccurate, or invalid processing or recording of financial information.	ISEC 1.07 - Passwords must promote acceptable levels of security (consistent with policies and/or best industry practices) by enforcing confidentiality and a strong password format.	Even though access to [Financial Application 1, 2, etc.] requires users to first authenticate at the network level, there were application-level logical security settings identified which were not in accordance with the company's local password policy, and may therefore not promote optimal security.	Same as above.	[List the general controls identified (and successfully tested) that can mitigate or compensate the finding.]

Exhibit 3.9   Example of a Management Letter from an IT Audit

---

Company Name
Management Letter—IT Audit
Year Ended December 31, 20XX

The findings below have been prioritized in order of significance and discussed with [*name and title of company personnel responsible for IT*], on [*date when meeting took place*]. Findings marked with an asterisk (*) are repeated from prior years.

[*Name of General Control IT Area (i.e., Information Systems Operations, Information Security, or Change Control Management)—Short Description of the Failed Control Activity*]

FINDING

[*Detailed description of the finding.*]

[EXAMPLE: *During the fiscal year ended June 30, 20XX, the Company converted its core financial application from [Application #1] to [Application #2]. We noted that the Company had no established or documented formal policies and procedures regarding the change management process as it related to conversion of data from old to new systems, applications, and databases.*]

IT RISK

[*Description of the IT risk related to the finding above.*]

[EXAMPLE: *Failure to implement appropriate general controls related to the conversion of data can result in operational disruptions, degraded system performance, or compromised security.*]

RECOMMENDATION

[*Auditor recommendation is documented here.*]

[EXAMPLE: *The Company should formally document a change control policy to establish procedures over each change's life cycle, including controls on data conversions. Newly developed policies should also be formally approved, communicated, and distributed to end users.*]

MANAGEMENT RESPONSE

[*The management's response should address responsibility and accountability for implementation of a corrective action, as well as include a target implementation date for correction.*]

[EXAMPLE: *Management acknowledges and accepts the IT auditor recommendation. A plan to implement appropriate data conversion controls will be put in place and submitted to our CEO and CFO for their review and approval within the next month. General controls related to the conversion of data are expected to be designed and fully operational by the next year's IT audit.*]

On receipt of the Management Letter, IT management and affected staff should review the document immediately. Those items not already completed should be handled and followed-up. Within a relatively short time, the fact that all discrepancies have been corrected should be transmitted to the audit staff in a formal manner. These actions are noted in the audit files, and such cooperation reflects favorably in future audits.

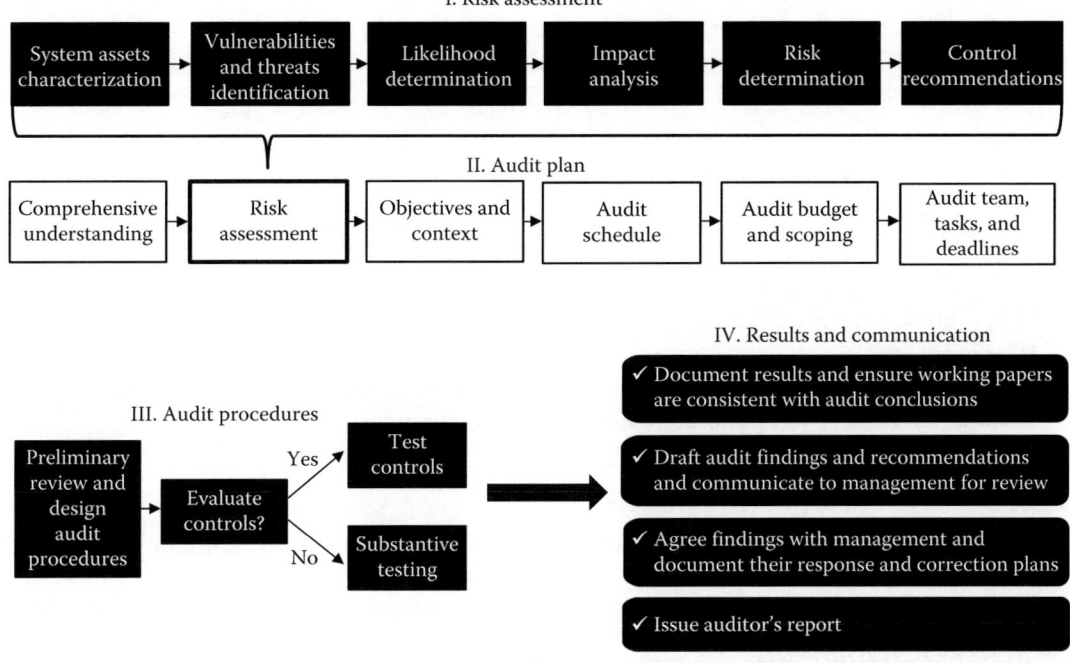

Exhibit 3.10 Summary of the audit process.

It is important to track corrective action to verify that findings have been remediated. This requires a formal process to track corrective actions, target dates, and status for reporting to IT management, the audit committee, and the board.

At the close of the audit, a draft Audit Report is issued for review by all impacted parties. The review process will go much faster if findings have already been agreed with management during the testing and conclusion phase. After the Audit Report has been finalized, it is a good practice to schedule an exit meeting involving both, IT and financial sides. Typically, invitations to the exit meeting are sent to the CIO and the Chief Financial Officer (CFO) (or Controller if the CFO is not available) to discuss the audit, as well as to review the audit objectives and ask for feedback on the performance of the audit team. This meeting will provide valuable information into the performance of the audit staff and lessons learned for improving future engagements.

To summarize the audit process explained in this chapter, refer to Exhibit 3.10.

Other Types of IT Audits

Besides supporting financial statement audits, there are other highly-demanded audit areas conducted in IT. These are briefly described next.

Enterprise Architecture

IT management must develop organizational procedures to ensure a controlled and efficient architecture for information processing. These procedures should also specify the computers and peripheral equipment required to support all functions in an economic and timely manner. With enterprise systems being very critical to medium-size and large businesses today, the need to monitor and validate operational integrity of an enterprise resource planning system is an important process. IT audit plays an important role in maintaining, validating, and monitoring the enterprise architecture.

Computerized Systems and Applications

A computerized systems and applications type of audit verifies that the organization's systems and applications (operational and non-financial in nature) are:

- appropriate to the users' needs,
- efficient, and
- adequately controlled to ensure valid, reliable, timely, and secure input, processing, and output at current and projected levels of system activity.

Information Processing Facilities

An audit of the information processing facility ensures timely, accurate, and efficient processing of applications under normal and potentially disruptive conditions.

Systems Development

An IT audit related to systems development would make certain that applications and systems under development meet the objectives of the organization, satisfy user requirements, and provide efficient, accurate, and cost-effective applications and systems. This type of audit ensures that applications and systems are written, tested, and installed in accordance with generally accepted standards for systems development.

Business Continuity Planning/Disaster Recovery Planning

According to the SysAdmin, Audit, Network, Security (SANS) Institute, a business continuity (or resiliency) plan (BCP) incorporates activities and procedures to recover all business operations (no just IT) from interruptions or adverse events.[*] A disaster recovery plan (DRP) incorporates a set of procedures to recover and protect the organization's IT infrastructure in the event of an emergency or disaster. Both plans should be formally documented, and kept updated within the organization.

A BCP audit evaluates how an organization's continuity processes are being managed. This type of audit defines the risks or threats to the success of the plan, and assesses the controls in place to determine whether those risks or threats are acceptable and in line with the organization's objectives.[†] This audit also quantifies the impact of weaknesses of the plan and offers recommendations for business continuity plan improvements.

DRP audits help ensure that the IT infrastructure and all related equipment used to develop, test, operate, monitor, manage, and/or support IT services (e.g., hardware, software, networks, data centers, etc.) are adequately maintained and protected to ensure their continued availability consistent with organizational objectives. A DRP audit considers factors such as alternate site designation, training of personnel, and insurance issues, among others.

---

[*] www.sans.org/reading-room/whitepapers/recovery/introduction-business-continuity-planning-559.
[†] http://searchdisasterrecovery.techtarget.com/definition/business-continuity-plan-audit.

Conclusion

Over decades, the computer has been used to support daily operations in business environments. Most companies find that they must use computer technology effectively to remain competitive. The nature of technology, however, continues to change rapidly. As a result, companies continue to integrate their accounting/financial systems and operations. The audit profession has made these adjustments as well. Worldwide, professional organizations have issued useful guidance and instruction to assist managers and the audit professionals.

Whether the IT audit reviews information systems operations, information security, or applications, the controls applied in those areas must be verified. The IT auditor's function (whether internal or external) provides reasonable assurance that system assets are safeguarded, information is timely and reliable, and errors and deficiencies are discovered and corrected promptly. Equally important objectives of this function are effective controls, complete audit trails, and compliance with organizational policies.

The nature of auditing will undoubtedly continue to undergo substantial change as the level of technology improves. Full automation from project initiation to the final reporting stage will enable auditors to make more efficient use of available resources and enhance the credibility of the audit performed. Effective use of computer technology can also empower auditors to better understand the design of the client's computer system, as well as conduct successful audits in today's highly automated environments.

Review Questions

1. What is an audit universe and what does it include?
2. What is Control Objectives for Information and Related Technology (COBIT) and why is it valuable to management and IT auditors?
3. Why are risk assessments significant to the audit function?
4. Summarize the importance of an audit plan. What are the four minimum steps an audit plan should have?
5. State the significance of an audit schedule.
6. Describe what an audit scoping should include.
7. Briefly describe the eight typical phases of an audit engagement.
8. What specific information or evidence can an IT auditor gather for a client that uses its IT environment to store and process financially significant data?
9. Explain what an audit program is.
10. Describe the procedures IT auditors perform in order to test controls, processes, and exposures.
11. Describe the procedures typically performed when conducting an IT audit related to:
    a. Systems Development
    b. Business Continuity Planning/Disaster Recovery Planning

Exercises

1. As the IT audit senior of the engagement, you are presenting to the IT manager and partner (as part of the planning meeting) the results of the risk assessment performed in Exhibit 3.3.

Based on such results (look at Exhibit 3.3, under the "Risk Rating" and "Action Priority" columns), it seems clear that the audit should focus on Financial Application #2 (FA2). Nevertheless, the IT manager and partner, based on previous relevant experience, believe that the audit should be performed on Financial Application #1 (FA1). The planning meeting is over, and you still feel doubtful on the decision just made. Your task: Prepare a two-page memo to the audit manager (copying the partner) stating your reasons why FA2 should be audited first. In order to convince the audit manager and partner, you are to think "outside the box." In other words, think of additional information not necessarily documented in the risk assessment shown in Exhibit 3.3, and document in your memo information related to:

a. Any additional vulnerabilities or weaknesses that may currently be in place affecting FA2
b. Any additional threat-sources that can trigger the vulnerabilities or weaknesses you just identified for FA2
c. Any additional risks or situations involving exposure to loss for the financial information in FA2
d. Any additional controls or procedures that should be implemented to mitigate the risks just identified

2. Use the following information to prepare an IT Planning Memo similar to the one in Appendix 1.

a. You are the IT audit senior (or IT auditor representative) assigned. Your audit firm has several branches, but you are working this particular client from the Melbourne, FL office.
b. The IT audit will support the financial statement audit of Company XYZ, with a fiscal year ending on December 31, 20XX.
c. Discussions with the financial audit Director regarding IT audit involvement have already taken place, and are documented in work paper (w/p) 1000.1. IT auditors have not been involved in previous audits for this client.
d. Your team is composed of: IT Partner P, IT Manager M, and IT Audit Staff AS. You are the IT audit Senior S.
e. The audit timing includes: Planning will be performed during the sixth month of the year under audit; Interim audit procedures will take place during 2 months before the end of the fiscal year; Year-end procedures are scheduled for January through March of the year following the end of the fiscal year; and all work papers and audit documentation will be due by and signed off on April 30th of the year following the end of the fiscal year.
f. The IT audit is estimated to take 100 hours. Hours will be charged to client code: Company XYZ-0000.
g. An understanding of Company XYZ's IT environment is documented in w/p 1540.
h. The three relevant applications for the IT audit include are:
    i. *All Accounting Application* (AAA)—used to capture and processing accounting-related transactions. AAA is installed on a UNIX platform (or operating system), and uses Oracle database. AAA can be accessed via a Windows network.
    ii. *Financial Document Generator Application* (FDGA)—used to produce all types of financial reports and documentation. FDGA is installed on a Windows operating system, and uses Oracle as its database. FDGA is accessed via a Windows network.
    iii. *Human Resources and Payroll Application* (HRPA)—used to manage the company's human resources and process payroll. This application is hosted outside of the company, at a third-party organization called HRP-For-All.

i.   The relevant application controls used to mitigate risks in this audit are listed in Exhibit 3.5b (these must be added to the IT Planning Memo). Use w/p 1000.2 for reference purposes.

j.   Deviations or findings resulting from testing the relevant application controls will be documented in w/p 2302.

k.   There will be no work of others (e.g., Internal Audit personnel, etc.) used in the IT audit.

l.   Human resources and payroll services are performed by a third-party service organization called HRP-For-All, located in Austin, Texas. Deloitte, the service auditor, just finished issuing a report assessing the controls at the service organization for the period July 1, 20XX through June 30, 20XX. Controls at HRP-For-All were found to be effective.

m.   There are no other areas identified within Company XYZ that IT auditors can assist with.

3. How is substantive testing used in an IT audit? Explain what does the term auditing-through-the-computer refers to.

4. What is an audit finding and which information should be included when documenting them?

5. You are an external IT auditor asked to perform a review of the following: The Financial Transactions Application (FTA) is causing a problem with the General Ledger Application (GLA) due to the timing of the transfer of transactions. Data were transferred late by FTA causing end-of-the-month reports to be inaccurately stated. Managers met to review prior month's activity reports, and noticed a shortfall of $50,000 in some accounts. Prepare an audit plan to conduct procedures to address this type of situation.

Further Reading

1. AICPA. Audit analytics and continuous audit—Looking toward the future, www.aicpa.org/InterestAreas/FRC/AssuranceAdvisoryServices/DownloadableDocuments/AuditAnalytics_LookingTowardFuture.pdf (accessed August 2017).

2. Benson, J. (August 2007). *The Importance of Monitoring*. Internal Auditor. Institute of Internal Auditors, Altamonte Springs, FL.

3. Berry, L. (October 2007). *A Kinder, Gentler Audit*. Internal Auditor. Institute of Internal Auditors, Altamonte Springs, FL.

4. Bodin, L., Gordon, L., and Loeb, M. (2008). Information security and risk management. *Commun. ACM*, 51(1), 64–68.

5. Casas, E. (October 2007). *Tell It Like It Is*. Internal Auditor. Institute of Internal Auditors, Altamonte Springs, FL.

6. Cavusoglu, H., Mishra, B., and Raghunathan, S. (2004). A model for evaluating IT security investments. *Commun. ACM*, 47(1), 87–92.

7. Chaney, C. and Gene, K. (August 2007). *The Integrated Auditor*. Internal Auditor. Institute of Internal Auditors, Altamonte Springs, FL.

8. Deloitte LLP. (2014). *IT Audit Planning Work Papers*. Unpublished internal document.

9. EY's ten key IT considerations for internal audit—Effective IT risk assessment and audit planning. (February 2013). Insights on governance, risk and compliance, www.ey.com/Publication/vwLUAssets/Ten_key_IT_considerations_for_internal_audit/$FILE/Ten_key_IT_considerations_for_internal_audit.pdf

10. Flipek, R. (June 2007). *IT Audit Skills Found Lacking*. Internal Auditor. Institute of Internal Auditors, Altamonte Springs, FL.

11. Gallegos, F. (2002). The audit report and follow up: Methods and techniques for communicating audit findings and recommendations. *Inf. Syst. Control J.*, 4, 17–20.

12. Gallegos, F. and Preiser-Houy, L. (2001). *Reviewing Focus Database Applications*, EDP Auditing Series, 74-10-23, Auerbach Publishers, Boca Raton, FL, pp. 1–24.

13. Hyde, G. (August 2007). *Enhanced Audit Testing*. Internal Auditor. Institute of Internal Auditors, Altamonte Springs, FL.
14. Information Systems Audit and Control Foundation. *COBIT*, 5th Edition, Information Systems Audit and Control Foundation, Rolling Meadows, IL, www.isaca.org/Knowledge-Center/COBIT/Pages/Overview.aspx (accessed June 2012).
15. IS Audit Basics. *The Process of Auditing Information Systems*, www.isaca.org/knowledge-center/itaf-is-assurance-audit-/pages/is-audit-basics.aspx (accessed July 2017).
16. Manson, D. and Gallegos, F. (September 2002). *Auditing DBMS Recovery Procedures*, EDP Auditing Series, 75-20-45, Auerbach Publishers, Boca Raton, FL, pp. 1–20.
17. McAfee Labs 2017 threats predictions, report issued on November 2016, www.mcafee.com/au/resources/reports/rp-threats-predictions-2017.pdf (accessed October 2017).
18. McAfee Labs threats report—December 2016, www.mcafee.com/ca/resources/reports/rp-quarterly-threats-dec-2016.pdf (accessed October 2017).
19. McCafferty, J. (2016). *Five Steps to Planning an Effective IT Audit Program*, MIS Training Institute, http://misti.com/internal-audit-insights/five-steps-to-planning-an-effective-it-audit-program
20. Menkus, B. and Gallegos, F. (2002). *Introduction to IT Auditing*, #71-10-10.1, Auerbach Publishers, Boca Raton, FL, pp. 1–20.
21. National Vulnerability Database. National Institute of Standards and Technology, https://nvd.nist.gov/vuln/search (accessed August 2017).
22. Otero, A. R. (2015). An information security control assessment methodology for organizations' financial information. *Int. J. Acc. Inform. Syst.*, 18(1), 26–45.
23. Otero, A. R. (2015). Impact of IT auditors' involvement in financial audits. *Int. J. Res. Bus. Technol.*, 6(3), 841–849.
24. Otero, A. R., Tejay, G., Otero, L. D., and Ruiz, A. (2012). A fuzzy logic-based information security control assessment for organizations, IEEE Conference on Open Systems, Kuala Lumpur, Malaysia.
25. Otero, A. R., Otero, C. E., and Qureshi, A. (2010). A multi-criteria evaluation of information security controls using Boolean features. *Int. J. Network Secur. Appl.*, 2(4), 1–11.
26. Pareek, M. (2006). Optimizing controls to test as part of a risk-based audit strategy. *Inf. Syst. Audit Control Assoc. J.*, 2, 39–42.
27. Romney, M. B. and Steinbart, P. J. (2015). *Accounting Information Systems*, 13th Edition, Pearson Education, Upper Saddle River, NJ.
28. Richardson, V. J., Chang, C. J., and Smith, R. (2014). *Accounting Information Systems*, McGraw Hill, New York.
29. SANS' Information Security Policy Templates, www.sans.org/security-resources/policies/general (accessed October 2016).
30. Sarbanes-Oxley-101. Section 404: Management Assessment of Internal Controls, www.sarbanes-oxley-101.com/SOX-404.htm (accessed August 2016).
31. Senft, S., Gallegos, F., and Davis, A. (2012). *Information Technology Control and Audit*, CRC Press/Taylor & Francis, Boca Raton, FL.
32. Singleton, T. (2003). The ramifications of the Sarbanes–Oxley Act. *Inf. Syst. Control J.*, 3, 11–16.
33. U.S. General Accounting Office, *Assessing the Reliability of Computer Processed Data Reliability*, https://digital.library.unt.edu/ark:/67531/metadc302511/ (accessed November 2016).
34. U.S. General Accounting Office, *Government Auditing Standards 2017 Exposure Draft*, www.gao.gov/yellowbook (accessed May 2017).
35. U.S. General Accounting Office, *Standards for Internal Control in the Federal Government*, September 2014, GAO/AIMD 00-21.3.1.

Chapter 4

Tools and Techniques Used in Auditing IT

LEARNING OBJECTIVES

1. Define auditor productivity tools and describe how they assist the audit process.
2. Describe techniques used to document application systems, such as flowcharting, and how these techniques are developed to assist the audit process.
3. Explain what Computer-Assisted Audit Techniques (CAATs) are and describe the role they play in the performance of audit work.
4. Describe how CAATs are used to define sample size and select the sample.
5. Describe the various CAATs used for reviewing applications, particularly, the audit command language (ACL) audit software.
6. Describe CAATs used when auditing application controls.
7. Describe CAATs used in operational reviews.
8. Differentiate between "Auditing Around the Computer" and "Auditing Through the Computer."
9. Describe computer forensics and sources to evaluate computer forensic tools and techniques.

Computer technology has become an integral part of most organizational functions. It is likely that many audit clients either have eliminated or will eliminate a substantial portion of their paper documents and replace them with electronic documents filed only in computerized form. An auditor who is unable to use computerized audit tools and techniques effectively will be at a disadvantage.

Today's auditor must be equipped with an understanding of alternative tools and techniques to test the operations of computerized systems and gather and analyze data contained in computerized files. Auditors can take advantage of those tools and techniques to be more efficient and effective when performing audit work. Tools and techniques used in IT audits include:

- *Audit productivity tools*—software that helps auditors reduce the amount of time spent on administrative tasks by automating the audit function and integrating information gathered as part of the audit process.

- *System documentation techniques*—methods, such as flowcharting, data flow diagram, and business process diagrams applied to document and test application systems, IT processes, and their integration within the IT environment.
- *Computer-assisted audit techniques (CAATs)*—software that helps auditors evaluate application controls, and select and analyze computerized data for substantive audit tests.

This chapter starts by defining auditor productivity tools and describing how they help the audit process. This chapter then touches upon the various techniques used to document application systems, in particular financial application systems, and how they assist the audit process. Explanations of CAATs and the role they play in the audit will follow along with descriptions of the various CAATs used when defining audit sample size, selecting samples, and reviewing applications (e.g., Audit Command Language (ACL), etc.). CAATs used in auditing application controls and in operational reviews will then be described followed by explanations of auditing around or through the computer. Lastly, computer forensic tools and techniques are discussed.

Audit Productivity Tools

The core of the audit process is assessing internal controls to determine if they are effective or need improvement. However, many of the tasks associated with performing an audit, such as planning, testing, and documenting results, although necessary, take time away from performing the actual control assessment work. This is where auditor productivity tools come into play. Auditor productivity tools assist auditors in automating the necessary audit functions and integrating information gathered as part of the audit process. Examples of audit functions that may be automated through auditor productivity tools include:

- Audit planning and tracking
- Documentation and presentations
- Communication
- Data management, electronic working papers, and groupware
- Resource management

Audit Planning and Tracking

Developing an audit universe with all of the potential audit areas within the organization, a risk assessment prioritizing these audit areas, an audit schedule, and a budget to track audit progress are some of the necessary tasks in any audit planning. Solutions such as spreadsheets, database software, and/or project management software can be used to document and plan audits, as well as track their current status. However, each of these solutions is standalone, as their integration may not even be possible. Because planning tasks are interdependent, an auditor productivity tool software that integrates these planning and tracking tasks would provide quicker update and ensure that all phases of planning are kept in sync. For example, the budget should provide sufficient costs to accomplish the audit schedule, or the audit schedule should not exceed the resources available, etc.

Documentation and Presentations

Tools, such as the Microsoft Office suite, provide features to facilitate the creation and presentation of documents. For example, spreadsheet data containing functional testing results can be incorporated into a report document with a few clicks of a mouse. These same data can then be copied to a presentation slide and also be linked, so that changes to the source documents will be reflected in any of the related documents. Software tools like these save time and ensure consistency and accuracy. Other tools include video conferencing and/or video capture software to provide presentations to collaborators worldwide and to document audit evidence, respectively.

Communication

Because the auditor operates as part of a team, the need to share data as well as to communicate with other members of the group is important. Providing immediate access to current data, electronic messaging, and online review capabilities allow audit staff to quickly communicate and gather research information for audits and special projects. In addition, auditors may occasionally need to operate from a host computer terminal, yet still have all the capability of a dedicated desktop processor. Therefore, it is necessary to have the required computer hardware, media hardware, protocol handlers, desired terminal software emulators, and high-speed wired or wireless connectivity at the audit site.

Electronic connectivity not only allows auditors to communicate but also provides access for organization management personnel or audit clients to exchange information. For instance, client's or organization's management personnel can be given access to the auditing risk universe database. This allows management to browse the database and suggest changes to current audit risk areas.

Video conferencing capabilities are also an effective way for communication. Video conferencing allows meetings to be conducted and members to participate worldwide. Some of the best video conferencing software includes Cisco WebEx Meeting Center, Citrix GoToMeeting, and Adobe Connect, among others.[*] Video conferencing software uses computer networks to transmit video, audit, and text data, smoothing the process of initiating and conducting live conferences between two or more parties regardless of their locations. Through video conferencing, participants can see a spreadsheet, a graph, or a video clip; receive live data feeds; and see responses from all parties involved.

Data Management, Electronic Working Papers, and Groupware

Establishing electronic connectivity provides audit personnel with the capability to access and input data into a central data repository or knowledge base. The central data repository (e.g., database, etc.) can archive historical risk, audit schedule, and budget data that can be accessed electronically by all authorized users throughout the audit group, regardless of physical location. Database applications can be developed to automatically consolidate data input electronically from all audit functions.

---

[*] www.pcmag.com/article2/0,2817,2388678,00.asp.

Through the use of databases, audit management can centrally monitor and have immediate access to critical activity such as audit schedule status, field audit status, fraud or shortage activity, and training and development progress. Database applications can automatically consolidate function-wide data and generate local and consolidated status and trending reports. Auditors can produce more effective products by leveraging off the knowledge of other auditors by having access to function-wide data.

A database can contain information such as risk areas, audit programs, findings, corrective action procedures, industry standards, best practices, and lessons learned. This information could be available for research whenever needed. In addition to historical data, databases provide a platform for interactive activities such as message boards or computer forums. Audit personnel (and others, if authorized) can post new information or update old information. Similarly, online storage of information allows auditors to search for specific information in voluminous documents (e.g., insurance code, etc.), research an audit area to determine prior risk areas and functional testing approaches, identify related or interrelated areas, and review local or organization-wide corrective action plans.

Electronic working papers or EWPs have also transformed the audit process in a significant way. EWPs deliver a consistent approach in creating, documenting, reviewing, sharing, and storing audit work.[*] When creating and documenting EWPs, auditors can reference their work to evidence, document audit procedures performed, and electronically sign-off their work without waiting for other team members to complete and sign-off their parts. Moreover, EWPs work with art imaging software allowing for incorporation of scanned images, emails, and digital pictures into the file as audit evidence.[†]

EWPs also provide access to audit management to navigate (remotely) through audit files and identify audit work completed, signed-off, and ready for review. Reviewers can add electronic notes, comments, and/or questions in the audit files that would need to be addressed, and forward to those in charge of working with those files. Upon receiving the audit files back, reviewers check and confirm that all notes, comments, and/or questions have been adequately addressed before completing their review and signing off.

Maintaining EWPs on a centralized audit file or database allows auditors to navigate through and share current and archived audit work with ease. Such centralized audit file or database facilitates the process for auditors to quickly access prior audit work (e.g., findings, areas of high risk, etc.) in order to coordinate current audit procedures.

Groupware or collaborative software is a specialized tool or assembly of compatible tools that enables business teams to work faster, share more information, communicate more effectively, and perform a better job of completing tasks. Groupware systems create collaborative work environments. Today, we are seeing desktop conferencing, videoconferencing, e-mail, message boards or forums, meeting support systems, workflow systems, and group and subgroup calendars as examples of groupware tools.

Groupware is "a natural" for automating the audit function. Groupware tools use database features and workflow processing that can be used to store and integrate information gathered and used in the audit process. For example, risk assessment information feeds audit planning, and audit results feed audit reporting and update the risk assessment model.

---

[*] www.wipo.int/export/sites/www/about-wipo/en/oversight/iaod/audit/pdf/annex_1.1_teammate_principles_guidelines.pdf.

[†] www.teammatesolutions.com/teamewp.aspx.

Resource Management

Another challenge for audit supervisors is to manage a remote workforce. Whether a staff auditor is working on a local audit or out in the field, managers need to be able to provide guidance and review work as the audit progresses. Audit managers need to provide feedback while the staff auditor is on location in case follow-up action is necessary.

A distributed workforce requires a very informed and responsive management team that can gather and disseminate information quickly. Important information can be rapidly gathered and disseminated function-wide through e-mail and message boards or computer forums. Supervisors can provide immediate feedback and direction on audit projects through online review of electronic work papers.

System Documentation Techniques to Understand Application Systems

Emphasis on understanding and documenting the organization's/client's information systems is particularly appropriate during the application analysis phase of an audit engagement. It is important for the auditor to understand the relationship of each application to the conduct of the organization's or client's business, and to document such understanding. For this, auditors typically request organizations or clients for an entity relationship diagrams (ERDs). If available, these ERDs are a great starting point for auditors, as they graphically represent the relationship between "entities" (or people, objects, places, concepts, events, etc.) within the information system (i.e., financial application system).

Documenting information systems, particularly financial application systems, help auditors, accountants, consultants, management, etc. in understanding what's going on financially at the entity and, most importantly, how to effectively evaluate those systems. Auditors also document financial application systems, as required by auditing standards, to understand the automated and manual internal control procedures the entity uses. In documenting financial application systems, auditors mostly use graphical representations, why? It has been said that a picture is worth a 1,000 words. Also, pictures tend to be easy to understand.

Documentation of application systems is commonly performed using narratives, diagrams, tables, data flow diagrams, business process diagrams, flowcharts, etc. Data flow diagrams or DFDs, for instance, are process-oriented and use graphics or symbols to describe data transformation and how it flows throughout the organization. Refer to Exhibit 4.1.

In Exhibit 4.1, the squares or rectangles represent data sources or destinations. Arrows indicate flows of data, and the circle symbol means that a transformation process is taking place. Business process diagrams visually show the various activities going on in a business process. These business process diagrams also show the organizational unit or process (e.g., payroll, accounts payable, cash disbursement, etc.) that is actually performing the activity. Refer to Exhibit 4.2.

In Exhibit 4.2, the rounded rectangles represent the activities or procedures occurring in a process. The circle indicates the start of a process, while the bolded circle indicates the end of the process. The arrow shows the flow of data. The dashed arrow is the annotation information or information that helps explain the business process. A third and most common example for documenting financial application systems is through flowcharting. Similar to the other system documentation techniques, flowcharts are a graphical description of a system representing how business processes are performed and how the various documents within the business process

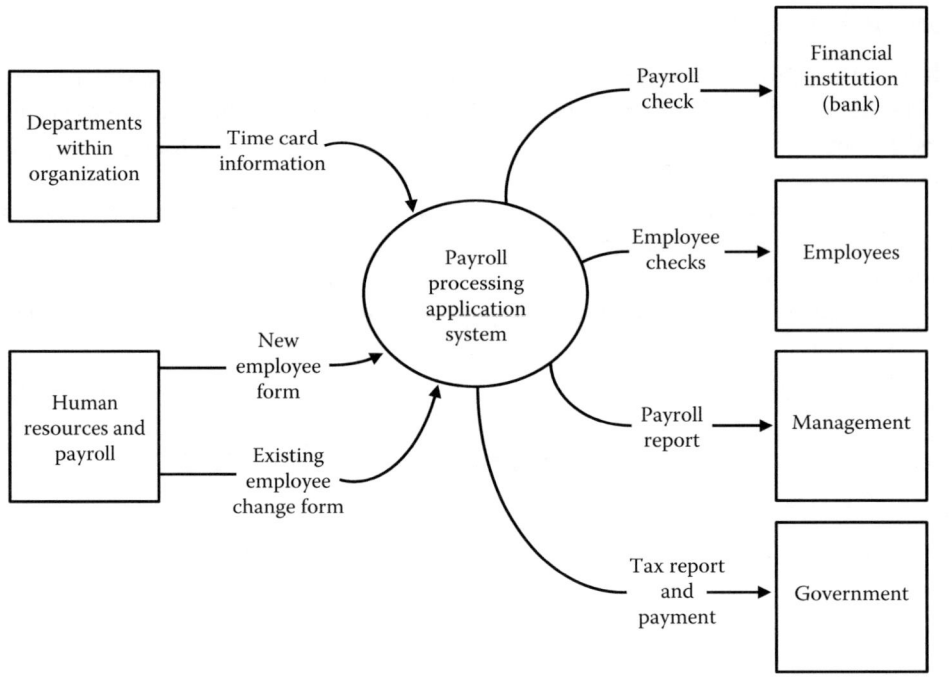

Exhibit 4.1   DFD illustrating typical payroll processing procedures.

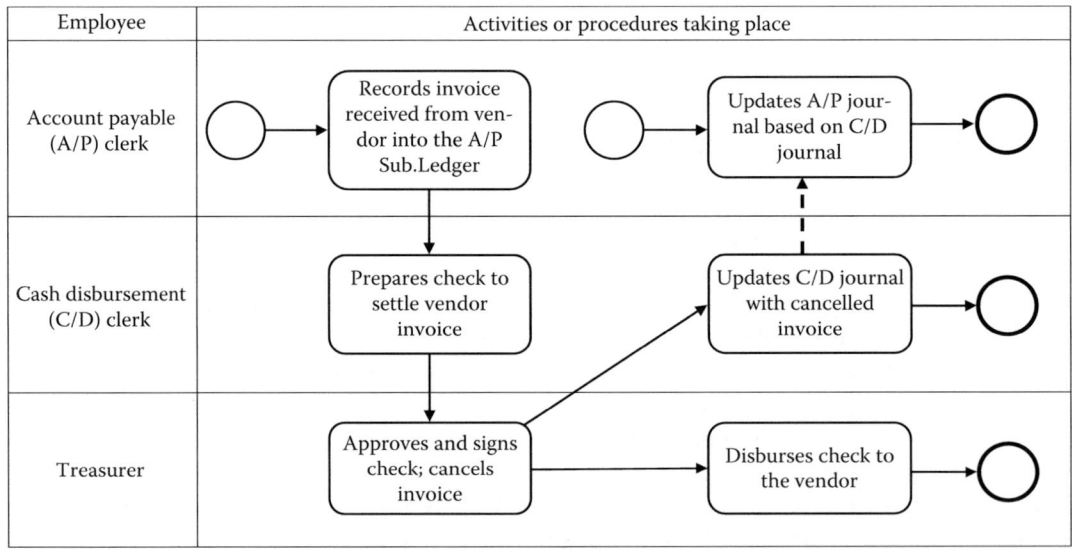

Exhibit 4.2   Example of a business process diagram for a cash disbursement process.

flow through the organization. Flowcharts use symbols to describe transaction processing and the flow of data through a system by specifically showing: inputs and outputs; information activities (processing data); data storage; data flows; and decision steps. Refer to Exhibit 4.3.

The following sections focus on the flowcharting technique to document systems.

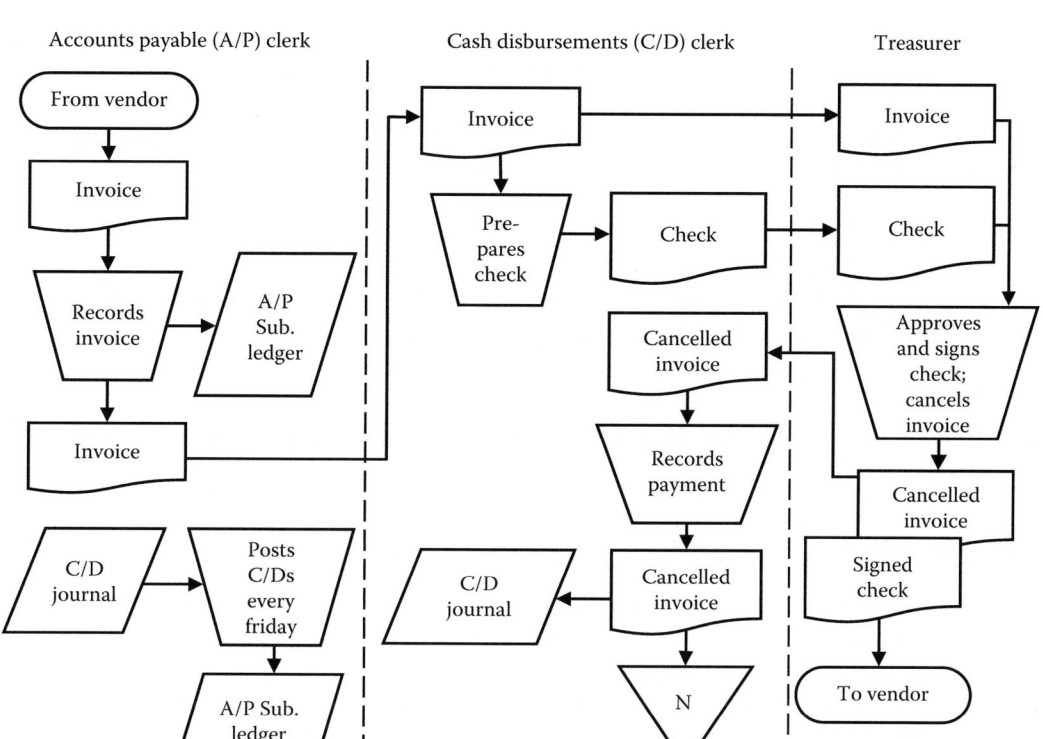

Exhibit 4.3   Example of a flowchart for a cash disbursement process.

Flowcharting as an Audit Analysis Tool

Auditors prepare flowcharts using standard symbols and techniques to represent application systems, workflows, or processes. Flowcharts developed during the application analysis phase of an audit engagement are most useful if they distinguish processing according to department, function, or company area. There are some very good application support packages for flowchart development as well as the power of the word processor to build diagrams and illustrations of the process.

For an IT auditor, flowcharts represent a method for identifying and evaluating control strengths and weaknesses within a financial application system under examination. It can be time consuming to build an understanding of strengths and weaknesses within a system to be audited. However, identification of strengths and weaknesses often is crucial because they will drive the direction of the remainder of the audit (e.g., substantiating and determining the effect of identified control weaknesses, etc.).

For example, Statement on Audit Standard (SAS) No. 109 requires the auditor gain an understanding of the entity and its environment and determine those controls relevant to the audit. The auditor must have an understanding of the nature and complexity of the systems that are part of the control environment being audited (i.e., financial application systems). One way of gaining that understanding is through any existing documentation which may provide a visual illustration of the system under review and any interaction with other systems. Any existing documentation, such as flowcharts, provides a benchmark for the auditor's review.

As a step toward building the needed understanding of control weaknesses, the audit staff should develop a flowchart diagram of all information processed. Flowcharts should encompass all information processed, from source documents to final outputs. Either automated or manual techniques can be used in preparing these flowcharts. With either approach, the process leads to the evaluation of a number of elements of a system, including the following:

- Quality of system documentation
- Adequacy of manual or automated controls over documents
- Effectiveness of processing by computer programs (i.e., whether the processing is necessary or redundant and whether the processing sequence is proper)
- Usefulness of outputs, including reports and stored files

Common flowchart symbols are described in Exhibit 4.4. Following are steps in the development of flowcharts.

Understanding How Applications Process Data

The auditor should understand how the financial application system, for example, generates its data. This understanding should encompass the entire scope of the financial system from preparation of source documents to final generation, distribution, and use of outputs. While learning how the system works, the auditor should identify potential areas for testing, using familiar audit procedures, such as:

- Reviewing corporate documentation, including system documentation files, input preparation instructions, and user manuals
- Interviewing organization personnel, including users, systems analysts, and programmers
- Inspecting, comparing, and analyzing corporate records

Identifying Documents and Their Flow through the System

To understand document flow, certain background information must be obtained through discussions with corporate officials, from previous audits or evaluations, or from system documentation files. Because this information may not be current or complete, it should be verified with the appropriate personnel (e.g., accounting, IT, etc.). A user or member of the IT department staff may already have a document flow diagram or flowchart that shows the origin of data and how it flows to and from the application. This diagram should not be confused with either a system flowchart that shows the relationship among the input, processing, and output in an IS, or a program flowchart that shows the sequence of logical operations a computer performs as it executes a program.

If not available, auditors will have to develop document flow diagrams. The document flow diagram should include:

- Sources and source document(s), by title and identification number, with copies of the forms attached
- Point of origin for each source document
- Each operating unit or office through which data are processed
- Destination of each copy of the source document(s)

Exhibit 4.4 Flowchart Common Symbols

Symbol	Description
	Manual or electronic document.
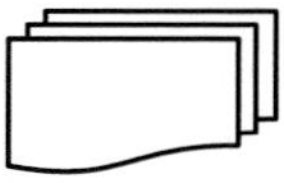	Multiple copies of manual or electronic documents.
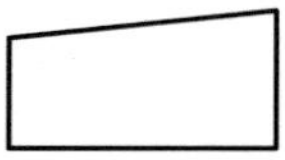	Electronic data entry device (e.g., laptop computer, mobile device, etc.)
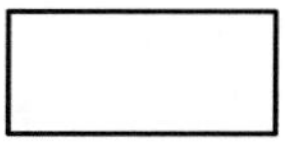	Electronic operation or processing of data taking place by the computer.
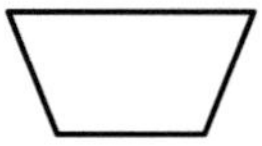	Manual operation.
	Data stored electronically in database.
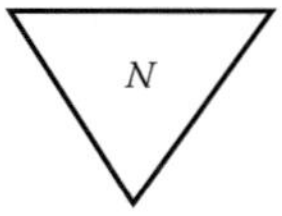 *N*	Indicates how paper documents are being filed. Typically, *N* means by number; *D* means by date; and *A* means alphabetically.
	Data stored electronically in magnetic tape (usually for backup purposes).
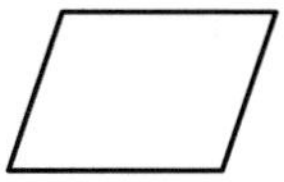	Indicates a type of manual journal or a ledger.
	Indicates the direction of a document or processing flow.
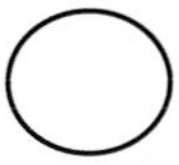	On-page connector used to link the flows of processing within the same page.
	Off-page connector to indicate the entry from or exit to another page.
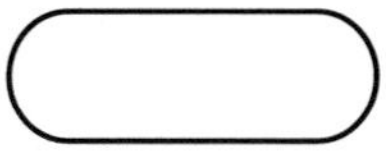	Used in a process to indicate a beginning, end, or point of interruption.
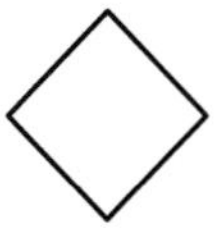	A decision is being made.

- Actions taken by each unit or office in which the data are processed (e.g., prepared, recorded, posted, filed, etc.)
- Controls over the transfer of source documents between units or offices to assure that no documents are lost, added, or changed (e.g., verifications, approvals, record counts, control totals, arithmetic totals of important data, etc.)
- Recipients of computer outputs

Defining Data Elements

The auditor must build a clear understanding of the data being recorded on the application for definition purposes. When defining individual data elements, titles can be deceptive. For example, is a cost derived from the current period or is it cumulative? Is the cost accrued or incurred? What are the components of a cost? Use descriptive names when defining data elements and action verbs for processes (e.g., update, prepare, validate, etc.). The organization's data element dictionary is a good source for such definitions. If a data dictionary is not available, a record layout may contain the needed definitions.

Developing Flowchart Diagrams

Inputs from which flowcharts are prepared should include copies of the following:

- Narrative descriptions of all major application systems
- All manually prepared source documents that affect application processing as well as corresponding coding sheets and instructions for data transcription
- Record layouts for all major computer input and output records, computer master files, and work files (such as update or file maintenance tapes and computation tapes)
- All major outputs produced by the application system
- Lists of standard codes, constants, and tables used by the application

These documents, along with the information developed in the previous tasks, should enable the audit staff to prepare a detailed and well-understood flowchart.

Evaluating the Quality of System Documentation

On the basis of user and IT staff inputs, as well as on the degree of difficulty experienced in constructing a flowchart, the auditor should be able to comment on the quality of system documentation. There are two basic questions to answer: Is the documentation accurate? Is the documentation complete?

To illustrate, if a federal auditor was examining control issues at a U.S. Navy computer facility, he or she might use the *Federal Information Systems Controls Audit Manual* (FISCAM) from the U.S. Government Accountability Office (GAO). This publication provides a basis for assessing the compliance of information system controls to federal guidelines.

Assessing Controls over Documents

Control points on the flowcharts should be identified and evaluated. By reviewing a diagram of this type, the auditor can determine whether controls have been used and if so, highlight gaps, strengths, and weaknesses within the system. Identified controls, including automated and IT dependent application controls, should be adequately designed and implemented in order to mitigate risks. They should also be assessed to determine whether they address potential misstatements, or prevent/detect unauthorized transactions that could result in a materially misstated financial statements. An example of a common control includes the three-way match verification between the vendor's invoice, purchase order, and reconciliation report that is performed by the system as confirmation before payment is released. Other examples of controls include performing verifications and approvals, as well as configuring the system to identify transactions falling outside defined tolerable ranges. If these transactions are identified, an adequate control would prevent their processing.

Determining the Effectiveness of Data Processing

The audit staff should determine how effective data processing is by identifying problem areas, such as the ones below, in the processing cycle:

- Redundant processing of data or other forms of duplication
- Bottleneck points that delay or congest processing
- Points in the operating cycle at which clerks do not have enough time to review output reports and make corrections

Upon identification, the auditor should make recommendations on how to address these problem areas.

Evaluating the Accuracy, Completeness, and Usefulness of Reports

The audit staff should review key or major outputs (e.g., edit listings, error listings, control of hour listings, etc.) of the financial application system and determine if the outputs are accurate, complete, and useful as intended. The auditor should confirm the accuracy, completeness, and usefulness of the generated reports by interviewing appropriate users. One suitable technique might be the completion of a questionnaire or survey, perhaps conducted by e-mail, on user satisfaction with output reports.

Appropriateness of Flowcharting Techniques

A distinction should be noted between the use of flowcharts in computer auditing and in the broader field of systems analysis. In recent years, systems analysts have begun to favor other methods of modeling and documentation. DFDs, for example, are often preferred over flowcharts for purposes of analysis (see Exhibit 4.1). As stated earlier, DFDs are process-oriented and emphasize logical flows and transformations of data. By contrast, flowcharts emphasize physical processing steps and controls. It is just this type of control-oriented view, however, that is the auditor's primary focus. Thus, although the use of flowcharting may be declining for systems development purposes, this modeling tool remains important for IT auditors.

Flowcharting is not necessarily always the most practical approach for the auditor. Existing documentation including DFDs, narratives, or descriptions of programs in pseudocode may be used as points of departure. Based on a review of existing documentation, the auditor can decide what additional modeling is needed to gain adequate understanding of the financial application systems under examination.

The auditor should also be aware of the increasing use of automated techniques in preparing flowcharts. Software packages are available, many of which run on mainframes and microcomputers that accept program source code as input and generate finished flowcharts. Also, microcomputer-based software packages now available can aid in documentation or verification of spreadsheets or database applications, for instance.

The technique for departmental segregation of processing in the preparation of flowcharts is important. Segregating departments (e.g., Accounts Payable, Cash Disbursements, Treasurer, Accounts Receivable, etc.) in vertical columns when creating flowcharts show processing by function or department. This representation is useful because one of the important controls the auditor evaluates is the segregation of duties within the financial accounting system. Structuring flowcharts in this way helps to discipline the auditor's thinking and identify any incompatible functions that may exist within financial applications. This segregation also aids in documenting the role of IT in the initiation, authorization, recording, processing, and reporting of transactions handled by the application.

An example of a flowchart for a cash disbursement process is shown in Exhibit 4.3. The following describes summarized steps taking place in the process and used to create the flowchart:

1. Vendor sends invoice to Company for business consulting services.
2. Invoice is directed to the Company's Accounts Payable (A/P) Clerk for recording.
3. A/P Clerk manually records invoice in the A/P subsidiary ledger.
4. Invoice is then sent to the Company's Cash Disbursement (C/D) Clerk for processing.
5. C/D Clerk prepares a check to settle the invoice, then sends both, check and invoice, to the Company's Treasurer.
6. Treasurer reviews both documents, then approves and signs check. Treasurer also marks the invoice as cancelled (there are several controls taking place here).
7. Treasurer then mails check to the Vendor, and forwards back the cancelled invoice to the C/D Clerk to record the payment or cash disbursement in the C/D journal.
8. The cancelled invoice is then filed by number (depicted as "*N*" in the flowchart).
9. Each Friday, the A/P Clerk manually posts payments from the C/D journal to the A/P subsidiary ledger.

When creating or reviewing flowcharts depicting business processes, the auditor should be accumulating notes to be considered for later inclusion as comments within a letter of recommendations to organization or client management personnel. At the conclusion of the review, the audit team briefs management personnel associated with the audit. All responsible parties should have a clear understanding of the sources and procedures depicted in the development of the flowchart, and ultimately how they reflect in the financial statements on which the audit firm will render an opinion. On completing such a review, the audit team should have built an understanding that includes:

- Establishing of sources for all financially significant accounting information
- Identifying processing steps, particularly of points within applications at which major changes in accounting information take place
- Determining and understanding processing results
- Analyzing the nature and progress of audit trails to the extent that they exist and can be followed within individual applications

Computer-Assisted Audit Techniques (CAATs)

Another type of software techniques used in IT audits is CAATs. As mentioned in an earlier chapter, the American Institute of Certified Public Accountants issued SAS No. 94, "The Effect of Information Technology on the Auditor's Consideration of Internal Control in a Financial Statement Audit." This SAS does not change the requirement to perform substantive tests on significant amounts but states, "It is not practical or possible to restrict detection risk to an acceptable level by performing only substantive tests." When assessing the effectiveness of the design and operation of IT controls, it is necessary for the auditor (IT or financial auditor) to evaluate and test these controls. The decision to evaluate and test is not related to the size of the organization but to the complexity of the IT environment.

CAATs can be used by both IT or financial auditors in a variety of ways to evaluate the integrity of an application, determine compliance with procedures, and continuously monitor processing results. IT auditors, for instance, review applications to gain an understanding of the controls in place to ensure the accuracy and completeness of the information generated. When adequate application controls are identified, the IT auditor performs tests to verify their design and effectiveness. When controls are not adequate, IT auditors perform extensive testing to verify the integrity of the data. To perform tests of applications and data, the auditor may use CAATs.

Automated techniques have proven to be better than manual techniques when confronted with large volumes of information. The auditor, by using automated techniques, can evaluate greater volumes of data and quickly perform analysis on data to gather a broader view of a process. Common CAATs like ACL and Interactive Data Extraction and Analysis (IDEA) can be used to select a sample, analyze the characteristics of a data file, identify trends in data, and evaluate data integrity. Other techniques used for analyzing data include, for example, Microsoft Access and Microsoft Excel. Microsoft Access can be used to analyze data, create reports, and query data files. Microsoft Excel also analyzes data, generates samples, creates graphs, and performs regression or trend analysis. SAP Audit Management (part of the SAP Assurance and Compliance Software that comes encapsulated with SAP GRC) also streamlines the auditing process by providing cost-effective alternatives to spreadsheets and manual tools.* SAP Audit Management facilitates the documentation of evidence, organization of working papers, and creation of audit reports. This technique also provides analytical capabilities to shift the focus of audits from basic assurance to providing insight and advice.†

A large part of the professional skills required to use CAATs lies in planning, understanding, and supervising (e.g., SAS No. 108—Planning and Supervision, etc.) these audit techniques, and conducting the appropriate audit functions and tests. The computer has a broad range of capabilities. By way of illustration, three broad categories of computer auditing functions can be identified:

- Items of audit interest
- Audit mathematics
- Data analysis

---

* www.complianceweek.com/blogs/grc-announcements/sap-delivers-new-audit-management-tool-for-internal-audit-teams#.WEhtkE0zW72.

† https://www.sap.com/products/audit-management.html.

Items of Audit Interest

The auditor can use the computer to select items of interest, such as material items, unusual items, or statistical samples of items by, for instance, stipulating specific criteria for the selection of sample items, or by stating relative criteria and let the computer do the selection.

An example of selection by specific criteria might be a specification that the computer identifies all transactions of $100,000 or more and prepares a report including such transactions for audit review. However, the auditor could take a relative approach and instruct the computer to select the largest transactions that make up 20% of the total dollar volume for a given application. This approach abridges manual audit procedures because the auditor can rely on the computer's selection of items of interest. If computers were not used, the auditor would have to validate the selection process. Under traditional approaches, for example, it would be common for an auditor to ask organization or client personnel to list all transactions of $100,000 or more. With the computer, the auditor can be satisfied that the CAAT used has looked at the total universe of accounts payable items, for example. The validation of the selection process is inherent in the auditor's developing and accepting the computer-auditing application program.

Audit Mathematics

Performing **extensions** or **footings** can be a cost-effective payoff area for the application of computers in auditing—particularly if the calculations can be performed as a by-product of another audit function. For example, suppose the computer is being used to select significant items from an accounts receivable file. In the process of looking at this file, the computer can be programmed to extend and foot all invoicing transactions. Because of the speed of the computer, these calculations can be performed on 100% of the items in a file with no significant addition of time or cost for this processing.

By contrast, extensions and footings are both tedious and costly under conventional manual examination techniques. Typically, the auditor must limit examination of any given application to extension and footing of a judgmental sample covering a few short intervals of the period under examination. Clearly, reliance can be far higher when these verification calculations are performed on complete files.

Remember, however, that the computer has limitations in this area. Although it can be programmed to make many logical comparisons and tests, the computer cannot supplant human judgment in examining items to be tested.

Data Analysis

Using the computer for analysis of data represents a major opportunity for innovation by the auditor. The computer can compare and summarize data and can represent data in graphic form. Data analysis programs use techniques such as:

- Histograms
- Modeling
- Comparative Analysis

Histograms are bar charts showing graphic relationships among strata of data. In computer-assisted auditing, histograms typically represent graphic frequency distributions of records within data files. By picturing these relationships in graphic form, histograms give the auditor an improved perspective on the analysis of financial statements. The histogram is, in effect, a snapshot showing the substance, makeup, and distribution of data within an organization's accounting or financial system. Refer to Exhibit 4.5 for an example of a histogram.

Auditors can apply their judgment in identifying and selecting appropriate testing techniques using histograms. By comparison, given a large collection of data about which such distribution data are not known, the auditor performs testing on a relatively blind basis. In such cases, the auditor cannot be sure of the significance of data until after testing is well along. With a histogram, items of significance for testing can be identified in advance because their relationship to the accounting universe is emphasized graphically.

Modeling is a technique by which the auditor can compare current data with a trend or pattern as a basis for evaluating reasonableness. Refer to Exhibit 4.6 for an illustration of a comparison model. Common modeling examples developed by auditors are based on several years of financial statements. The computer can generate a ***pro forma* financial statement** based on past revenue or cost relationships. The *pro forma* statement is compared with the actual financial statements as a test of reasonableness.

Both techniques—histograms and modeling—add new content and dimensions of information to the audit process through the use of the computer. With these methods, the auditor is no longer restricted simply to validating data provided by organization or client personnel. With these automated techniques, the auditor generates figures or snapshots of financial data to test the reasonableness of representations under examination.

Comparative analysis, another common technique used in analyzing data, is a proven, cost-effective audit examination that involves the comparison of sets of data to determine relationships that may be of audit interest. Refer to Exhibit 4.7 for an illustration of a comparative income statement analysis.

Other common data analysis examples use the computer to compare inventory files of the previous and current years. Wide variations in year-end balances could lead to reviews for possible obsolescence. A failure to match part numbers from the previous and current years might trigger testing procedures to determine whether old items have been dropped or new ones added.

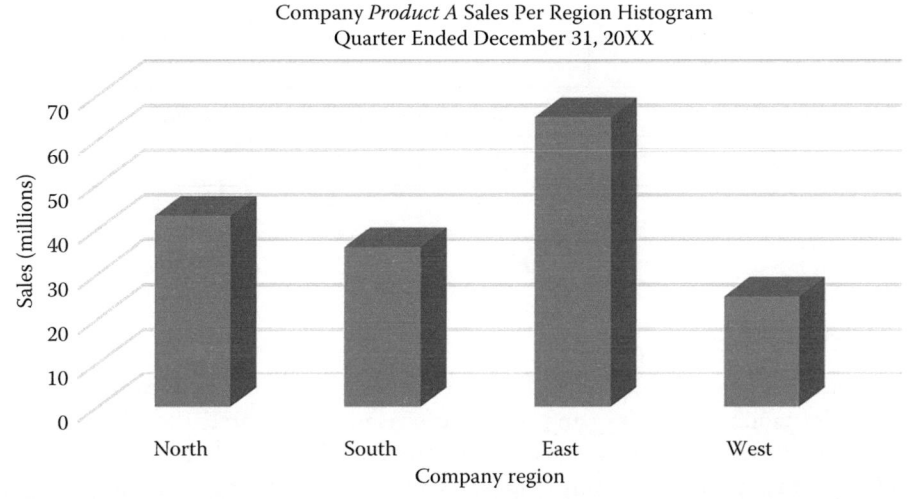

Exhibit 4.5 Histogram example.

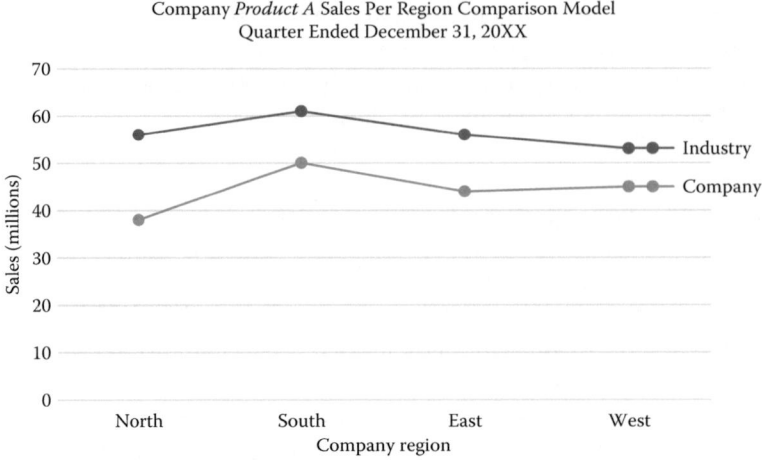

Exhibit 4.6   Example of a comparison model.

Exhibit 4.7   Example of a Comparative Income Statement Analysis

Company XYZ				
Comparative Income Statement				
For the Years Ended December 31, 20X1 and 20X2 (in thousands except percent)				
	20X1	*20X2*	*Increase (Decrease)*	
	Amount	*Amount*	*Amount*	*Percent*
Sales	$840.0	$600	$240.0	40.0
Cost of Goods Sold	724.5	525	199.5	38.0
Gross Profit	$115.5	$75	$40.5	54.0
Selling Expenses	52.5	37.5	15.0	40.0
Administrative Expenses	41.4	30	11.4	38.0
Total Operating Expenses	$93.9	$67.5	$26.4	39.1
Income Before Income Tax	21.6	7.5	14.1	188.0
Income Tax Expense	10.8	2.7	8.1	300.0
Net Income	$10.8	$4.8	$6.0	125.0

Data analysis, critical for performing audit functions and tests, must follow a thorough and complete understanding of the client's IT system (i.e., accounting information system). SAS No. 109, "Understanding the Entity and Its Environment…," reinforces the statement above, and requires auditors to understand the client's environment, which includes their accounting and financial IT systems. Auditors typically employ CAATs when understanding and examining these types of IT systems.

CAATs for Sampling

Some audit techniques assist in defining sample size and selecting the sample. For example, ACL automatically calculates the sample size and selects a sample from a population. Spreadsheet applications also generate random numbers for selecting a sample. There are two types of sampling techniques:

- *Judgmental sampling*: The sample selected is based on the auditor's knowledge and experience. The judgment may be to select a specific block of time, geographic region, or function.
- *Statistical sampling*: The sample is randomly selected and evaluated through the application of the probability theory.

Both methods allow the auditor to project to the population. However, only statistical sampling allows the auditor to quantify the risk that the sample is not representative of the population. The specific method selected for a sample will depend on the audit objectives and the characteristics of the population. The appropriateness of the method selected should be reviewed for validity purposes by statistical or actuarial staff with expertise in this area. Also, the applied sampling method should be revisited and reassessed over time to see if there is any change to the characteristics or attributes of the population under review. Two common statistical sampling methods are: Random Attribute Sampling and Variable Sampling.

Random Attribute Sampling and Variable Sampling

Random attribute sampling is a statistical technique that tests for specific, predefined attributes of transactions selected on a random basis from a file. Attributes for which such testing is done could include signatures, account distribution, documentation, and compliance with policies and procedures. To perform attribute sampling, the auditor must specify three parameters that determine sample size:

1. Estimate the "expected error rate," or estimated percentage of exception transactions, in the total population.
2. Specify the required "precision," or degree of accuracy desired, of the sample conclusion to be made.
3. Establish an acceptable "confidence level" that the conclusion drawn from the sample will be representative of the population.

The size of the sample will be determined by the combination of the expected error rate, precision, and confidence level parameters.

Variable sampling is another statistical technique that estimates the dollar value of a population or some other quantifiable characteristic. To determine the sample size using variable sampling, the auditor must specify four parameters:

1. Acceptable "confidence level" that the conclusion drawn from the sample will be representative of the population.
2. Absolute value of the "population" for the field being sampled.
3. "Materiality" or maximum amount of error allowable in the population without detection.
4. "Expected error rate" or estimated percentage of exception transactions in the total population.

The size of the sample will be determined by the combination of all four parameters listed above. Exhibit 4.8 describes other statistical sampling techniques commonly used. Again, the auditor must watch for changes and updates to guidance in the use of sampling to perform audit work. A good example is SAS No. 111 (Amendment to the Statement on Auditing Standard No. 39, Audit Sampling). SAS No. 111 addresses the concepts of establishing "tolerable deviation rates" when sampling test of controls such as matching and authorization. It also defines the appropriate use of dual-purpose sampling.

Exhibit 4.8   Statistical Sampling Techniques

Sampling Technique	Description
Random Number Sampling	Items are randomly selected from a population so that each item has an equal chance of being selected.
Systematic Sampling (Interval Sampling)	A method of random sampling that begins the sample by selecting a random starting point in a population and then selecting the remaining items at fixed intervals. This method should not be used for selection from a population that has a fixed pattern.
Stratified Sampling	A method of random sampling that separates the population into homogeneous groups before selecting a random sample. This method should be used for selection from a population with wide variances in value.
Cluster Sampling (Block Sampling)	A method of random sampling that separates the population into similar groups, and then selects a random sample from the group.
Stop-or-go Sampling (Sequential Sampling)	Minimizes the sample size by assuming a low error rate. It estimates the error rate of the population within a specified interval (e.g., plus or minus number, etc.).
Discovery Sampling	Tests for a significant error or irregularity. It should not be used where there are known deviant conditions.
Dollar-unit Sampling (Probability Proportional to Size)	This method uses the dollar as a sampling unit, which increases the probability that larger dollar values will be selected. It primarily detects overpayments.
Mean Per Unit	The mean value of a sample is calculated and multiplied by the units in the population to estimate the total value of the population.
Difference Estimation	The average difference between the audit value and book value for a sample unit is calculated. This difference is then multiplied by the population to estimate the total value.
Ratio Estimation	The sample ratio to book value is multiplied by the population book value to estimate the total value.

Source: From Senft, S., Gallegos, F., and Davis, A. (2012). *Information Technology Control and Audit.* CRC Press/Taylor & Francis, Boca Raton, FL.

CAATs for Application Reviews

There is a variety of CAATs that are useful when auditing applications and data integrity. An example of such techniques includes generalized audit software. Generalized audit software can be used to analyze spreadsheet logic and calculations for accuracy and completeness, evaluate data produced from applications (residing in databases), and produce logical data flowcharts, among others. In auditing databases, for example, techniques related to data mining can search "through large amounts of computerized data to find useful patterns or trends."* Data mining techniques help analyzing data from different perspectives and summarizing it into useful information. Another related example include data analytics (DA), or procedures to examine raw data in order to draw conclusions. DA is used in many industries to allow for better decision making, and in science to verify or disprove existing models or theories.† DA differentiates from data mining by the scope, purpose, and focus of the analysis. Data mining sorts through huge amounts of data using sophisticated software in order to identify undiscovered patterns and establish hidden relationships. DA, on the other hand, focuses on the process of deriving a conclusion (or inferring) based solely on what is already known.

Generalized audit software makes it possible to perform required functions directly on application files as it uses auditor-supplied specifications to generate a program that performs audit functions. Financial auditors, for example, use generalized audit software to:

- Analyze and compare files
- Select specific records for examination
- Conduct random samples
- Validate calculations
- Prepare confirmation letters
- Analyze aging of transaction files

IT auditors also use these software techniques for testing and/or documentation of selected processes within the IT environment in the form of flowcharts, and data flow diagrams, for instance. Generalized audit software allow IT auditors to evaluate application controls as well as query and analyze computerized data for substantive audit tests, among others. Some of the most popular software packages include Audit Analytics by Arbutus Software, TopCAATs, CaseWare Analytics IDEA Data Analysis, Easy2Analyse, TeamMate, and ACL. These are all virtually similar in regards to functionality. The ACL software package is described below as an example of what these techniques can do.

Audit Command Language (ACL)

ACL is a general audit software that reads from most formats (e.g., databases, delimited files, text files, Excel files, etc.) and provides data selection, analysis, and reporting. More specifically, ACL is a file interrogation tool designed to assist the audit of applications as it can handle and process large amounts of data. ACL functions range from: (1) identifying negative, minimum, and maximum balances; (2) performing statistical sampling and aging analyses; (3) identifying duplicates

---

* www.merriam-webster.com/dictionary/data%20mining.
† http://searchdatamanagement.techtarget.com/definition/data-analytics.

or gaps in sequence testing; and (4) performing comparative joining and matching; among others. Benefits within ACL include:

- Effective overviews of a file's format and structure
- Ability to import various types of raw data files
- Easy creation of audit samples and summaries
- Increased testing coverage and improved efficiency
- Scripts that can be executed in current and subsequent periods
- Dialog boxes for use in interactive applications
- Increased process and system understanding of complex environments
- Reduced manual procedures

Some of the common ACL features include:

- *Defining and importing data into ACL.* Defining what type of organization or client data will be used for ACL analysis purposes is one of the most important steps when using ACL. Here, the auditor needs to identify two things: first, the location of the data to be used; second, the format and structure of such data. To avoid running into issues when accessing the data or accessing protected drives by mistake, etc., it is a good practice for the auditor to ask organization or client personnel to place the data required for analysis into a separate drive (e.g., auditor's USB drive, CD, separate hard drive, etc.). ACL has the ability to read and import data from carriage return (CR) files (plain text reports where CRs mark the end of each record); databases, such as dBASE, DB2, and Open Database Connectivity (ODBC) (e.g., MS Access, MS Excel, Oracle, etc.); flat files (data arranged sequentially in rows); and delimited files (fields separated by a comma, a semicolon, etc.). Other types of files read by ACL include report files, segmented files, variable-length files, CR/Line Feed (LF) files, data sources with or without file layout, fixed-length files, LF files, mainframe databases, and multiple-record-type files.
- *Customizing a view.* With ACL, the auditor can modify the view of the original file being defined into one that better meets the data analysis in question. ACL also allows the auditor to create a new view or change existing ones without "touching" the actual data from the file. This means that changes to viewing the data are only for presentation purposes and would, therefore, not change nor delete the data.
- *Filtering data.* Filters are easily created in ACL and are useful for quick analysis and control totals once a file is imported. Filtering data through ACL allows auditors to support their tests and analyses. Filtering in ACL uses logical operators (e.g., AND, OR, NOT, etc.) as conditions to effectively generate information matching a specific criteria. For instance, auditors can search for specific accounting journal entries posted in a holiday or during off hours. ACL would allow for such condition-based information to be produced for analysis. There are two types of filters: global filters and command filters. A global filter, when activated, applies to all commands and to all views for the active table. Global filters stay in place until removed or until the table is closed. A command filter, on the other hand, applies to a single command and remains in effect only until ACL processes the command. Filters can be named, saved, and reused when required.
- *Data analysis.* Auditors use ACL to evaluate and transform data into usable information. Data can be of any size, format, and from almost any platform. Both analytical and logical reasoning are used to examine each component of the data, and distill meaning even from significant amounts of data. The ACL commands in Exhibit 4.9 are commonly used to perform data analytical tasks.

Exhibit 4.9 ACL Commands Commonly Used in Performing Data Analyses

ACL Command	Description
Extract	Selects records or fields from a file or current table, and copies them to a different file or table.
Export	Sends data to external file (e.g., database, Excel, text file, etc.) for use outside ACL.
Sorting	Sorts or organizes the active table into ascending or descending order based on specified key fields.
Verify	Checks for data validity errors in the active table. Ensures that data in a table conform to the table layout and reports on any errors encountered.
Search	Locates first record in an indexed table that meets a specified criteria/ condition.
Append	Adds command output to the end of an existing file instead of overwriting the existing file.
Count	Totals the number of records in the current table, or only those records that meet a specified criteria or test condition.
Total	Sums numeric fields or expressions in the active table.
Statistical	Used on date and numeric fields to obtain an overview of the data. The Statistical command produces (1) record counts, field totals, and average field values for positive, null, and negative field values; (2) absolute values; (3) ranges; and (4) highest and lowest field values.
Stratify	Allows the view of a distribution of records that fall into specified intervals or strata. That is, it counts the number of records falling into specified intervals of numeric field or expression values, and subtotals one or more fields for each stratum.
Classify	Counts the number of records relating to each unique value of a character field and subtotals specified numeric fields for each of these unique values.
Histogram	Provides an overview of a table's contents. Specifically, it produces a 3-D vertical bar graph of the distribution of records over the values of a field or expression.
Age	Produces aged summaries of data (e.g., classification of outstanding invoices)
Summarize	Generates a record count and numeric field value totals for each distinct value of key character fields in a sorted table.
Examine Sequence	Determines whether key fields in the active table are in sequential order, and detects gaps, duplicates, or missing numbers in the sequence.
Look for Gaps	Detects gaps in the key fields of the current table (e.g., gaps in numbers, dates, etc.)

(Continued)

Exhibit 4.9 (*Continued*)   ACL Commands Commonly Used in Performing Data Analyses

ACL Command	Description
Look for Duplicates	Detects whether key fields in the current table contain duplicates in the sequence.
Sampling	ACL offers many sampling methods for statistical analysis. Two of the most frequently used are record sampling (RS) and monetary unit sampling (MUS), both created from a population within a table. Each method allows random and interval sampling. MUS extracts sample records from a data set. MUS is typically used if the file is heavily skewed with large value items. RS, on the other hand, treats each record equally, using a nominal value of one. RS is used when records in a file are fairly evenly distributed across data, resulting in each record having an equal chance of being selected. The choice of methods will depend on the sampling overall purpose as well as the composition of the file being audited.

When planning for an ACL data analysis project, it is important for IT auditors to follow the steps below:

- Step 1: Acquiring the data
- Step 2: Accessing the data
- Step 3: Verifying the integrity of the data
- Step 4: Analyzing and testing the data
- Step 5: Reporting findings

Step 1: Acquiring the Data

The auditor must be aware of the data he or she requires to meet the goals of the specified project. For this, the auditor should gather the necessary information by meeting with various organization and/or client personnel, including, but not limited to, IT, MIS, and/or accounting or finance personnel to understand the data, its size, format, structure, and required data fields.

Step 2: Accessing the Data

The auditor must become familiarized with the data he/she is about to work with, particularly, the file where the data are stored, the file's structure, format, layout, size, data fields, number of records, etc. The auditor must assess the data included within the file in order to determine which data analysis task should be used and what platform or environment.

Step 3: Verifying the Integrity of the Data

The auditor must ensure that data are good data. In other words, the data that are to be analyzed must be valid, accurate, and complete, particularly, when working with data files that are not organized in records. ACL provides tools such as count, total, and verify to deal with these types of data files.

Step 4: Analyzing and Testing the Data

Data to be analyzed and tested can be of any size, format, and from almost any platform. Auditors use ACL to evaluate and transform such data into meaningful information that can assist their decisions-making process. There are several ACL commands that are commonly used when performing these types of data analyses and tests (refer to Exhibit 4.9). Appendix 4 also shows best practice audit procedures when using ACL to perform, for example, testing on accounting journal entries.

Step 5: Reporting Findings

Upon completing performing data analyses and tests, IT auditors must present and communicate their results and findings in an easily readable format. As part of reporting the results, auditors must maintain file layouts and ACL projects for backup purpose and to allow recreation, if necessary. Auditors must include ACL-related information in the audit work papers, including, but not limited to, copy of ACL program, ACL logs/file layouts, and data requests for future year audits.

CAATs for Auditing Application Controls

When auditing application controls, auditors examine input, processing, and output controls specific to the application. Application controls are also referred to as "automated controls." Automated input controls validate the data entered in the system, and minimize the chances for errors and omissions. Examples of input controls include checking for: characters in a field; appropriate positive/negative signs; amounts against fixed/limited values; amounts against lower and upper limits; data size; and data completeness, among others. Processing controls are those controls that prevent, detect, and/or correct errors while processing. Examples of processing controls include matching data before actions take place (e.g., matching invoice amount against purchase order and receiving report, etc.), recalculating batch totals, cross-footing data to verify accuracy of calculations, and ensuring that only the correct and most updated files are used. Output controls detect errors after processing is completed. Examples of output controls include performing report data reconciliations (e.g., general ledger with subsidiary ledgers, etc.), reviewing reports for accuracy and completeness (e.g., performing comparisons of key data field, checks for missing information, etc.), and protecting the transfer of data to ensure data are being transmitted completely and adequately (e.g., encryption, etc.).

CAATs come very handy to the auditor when evaluating application controls related to the processing of transactions. As described above, controls regarding the processing of transactions are concerned with the accuracy, completeness, validity, and authorization of the data captured, entered, processed, stored, transmitted, and reported. Auditors typically work with organization- or client-provided spreadsheets and/or databases when performing their procedures. Application controls found on spreadsheets and/or databases that are commonly tested by auditors include checking for mathematical accuracy of records, validating data input, and performing numerical sequence checks, among others. Auditors must ensure these types of controls are effectively implemented to ensure accurate results.

Spreadsheet Controls

Spreadsheets may seem to be relatively straightforward because of their widespread use. However, the risks presented are significant if the spreadsheet results are relied on for decision making.

Lack of reliability, lack of auditability, and lack of modifiability are all risks that are associated with poor spreadsheet design. Auditors use CAATs to assess client- or organization-prepared spreadsheets for analyzing their data and ultimately forming opinions. Controls should be implemented to minimize the risk of bad data and incorrect logic, particularly, if spreadsheets are reused. Some of the key controls that minimize the risks in spreadsheet development and use include:

- *Analysis.* Understanding the requirements before building the spreadsheet
- *Source of data.* Assurances that data being used are valid, reliable, and can be authenticated to originating source
- *Design review.* Reviews performed by peers or system professionals
- *Documentation.* Formulas, macro commands, and any changes to the spreadsheet should be documented externally and within the spreadsheet
- *Verification of logic.* Reasonableness checks and comparisons with known outputs
- *Extent of training.* Formal training in spreadsheet design, testing, and implementation
- *Extent of audit.* Informal design reviews or formal audit procedures
- *Support commitment.* Ongoing application maintenance and support from IT personnel

Database Controls

Department databases should be protected with controls that prevent unauthorized changes to the data. In addition, once the database is implemented, it should be kept in a separate program directory and limited to "execute only." The database can also be protected by enabling "read-only" abilities to users for data that remain static. Access rights should be assigned to specific users for specific tables (access groups). The input screens should include editing controls that limit data entry to valid options. This can be accomplished by having a table of acceptable values for the data fields. Data accuracy can also be enhanced by limiting the number of free-form fields and providing key entry codes with lookup values for the full description. Controls that auditors commonly expect to identify (and ultimately assess) within client or organization-prepared databases include:

- *Referential integrity.* Prevent deleting key values from related tables
- *Transaction integrity.* Restore value of unsuccessful transactions
- *Entity integrity.* Create unique record identification
- *Value constraints.* Limit values to a selected range
- *Concurrent update protection.* Prevent data contention
- *Backup and recovery protection.* Ability to back up critical information and applications and restore to continue
- *Testing protection.* Perform tests at the systems, application, and unit level

CAATs for Operational Reviews

Earlier, we covered a number of techniques used for performing tasks to support the audit of applications. Most of these techniques can be used to support operational reviews as well as collect information about the effectiveness of general controls over IT operations. However, the use of techniques need not be limited to specialized packages. Computer languages can be useful in performing operational tests and collecting information about the effectiveness of general controls. Even basic tools such as Access in MS Office can be used to take an imported data file of operational data (e.g., users' account information and file accesses, rights to number of file accesses, etc.), perform analysis on the file (histograms, frequencies, summaries), and then move data into MS Excel and visually portray information for management or even forecast trends with regard to workload, growth, and other IT operational areas.

The focus of an operational review is on the evaluation of effectiveness, efficiency, and goal achievement related to information systems management operations. Basic audit steps in an operational review are similar to IT audits or financial statement audits, except for the scope. Specific activities in an operational review include:

- Review operating policies and documentation
- Confirm procedures with management and operating personnel
- Observe operating functions and activities
- Examine financial and operating plans and reports
- Test accuracy of operating information
- Test operational controls

Auditing Around the Computer Versus Auditing Through the Computer

There may be situations in the IT environment where *auditing around the computer* or "black box auditing approach" may be more adequate than *auditing through the computer* when automated applications are relatively simple and straightforward. When performing *auditing around the computer*, the auditor obtains source documents that are associated with particular input transactions and reconciles them against output results. Hence, audit supporting documentation is drawn and conclusions are reached without considering how inputs are being processed to provide outputs. Unfortunately, SAS No. 94 does not eliminate the use of this technique. The major weakness of the *auditing around the computer* approach is that it does not verify or validate whether the program logic of the application being tested is correct. This is characteristic of the *auditing through the computer* approach (or the "white box auditing approach").

The *auditing through the computer* approach includes a variety of techniques to evaluate how the application and their embedded controls respond to various types of transactions (anomalies) that can contain errors. When audits involve the use of advanced technologies or complex applications, the IT auditor must draw upon techniques combined with tools to successfully test and evaluate the application. This audit approach is relevant given technology's significant increase and its impact on the audit process. The techniques most commonly used include integrated test facility, test data, parallel simulation, embedded audit module, systems control audit review file (scarf), and transaction tagging. Again, many of these techniques should be embedded into the application for use by auditors and information security personnel. These techniques provide continuous audit and evaluation of the application or systems and provide management and the audit or security personnel assurances that controls are working as planned, designed, and implemented. These are described, with their advantages and disadvantages, in Exhibit 4.10.

Exhibit 4.10   Computer-Assisted Audit Techniques for Computer Programs

Audit Technique	Description	Advantages (A) / Disadvantages (D)
Integrated Test Facility	Integrated test facilities are built-in test environments within a system. This approach is used primarily with large-scale, online systems serving multiple locations within the company or organization. The test facility is composed of a fictitious company or branch, set up in the application and file structure to accept or process test transactions as though it was an actual operating entity. Throughout the financial period, auditors can submit transactions to test the system.	Designed into the application during system development. (A) Expertise is required to design the audit modules (built-in test environments) into the application and to ensure that test transactions do not affect actual data. (D) Since the audit module is set up in the organization or client application, the risk of disrupting the data is high. Controls must be adequately designed and implemented to identify and remove the effects of test transactions. (D)
Test Data	This technique involves methods of providing test transactions to a system for processing by existing applications. Test data provide a full spectrum of transactions to test the processes within the application and system. Both valid and invalid transactions should be included in the test data as the objective is to test how the system processes both correct and erroneous transaction input. For a consumer credit card service, such transactions may be invalid account numbers, accounts that have been suspended or deleted, and others. If reliance is placed on program, application, or system testing, some form of intermittent testing is essential. Test data generators are very good tools to support this technique but should not be relied on entirely for extreme condition testing.	Minimal expertise required to run test data techniques. (A) Risk of disrupting organization or client data is minimal due to the fact that a copy of the application is used. (A) Personnel from the organization or the client provides copy of the application, however, it may be difficult to determine if the copy provided is exact, thereby reducing reliability of the test method. (D)

(Continued)

Exhibit 4.10 (Continued)   Computer-Assisted Audit Techniques for Computer Programs

Audit Technique	Description	Advantages (A) / Disadvantages (D)
Parallel Simulation	Parallel simulation involves the separate maintenance of two presumably identical sets of programs. The original set of programs is the production copy used in the application under examination. The second set could be a copy secured by auditors at the same time that the original version was placed into production. As changes or modifications are made to the production programs, the auditors make the same updates to their copies. If no unauthorized alteration has taken place, using the same inputs, comparing the results from each set of programs should yield the same results. Another way is for the auditor to develop pseudocode using higher-level languages (Vbasic, SQL, JAVA, etc.) from the base documentation following the process logic and requirements. For audit purposes, both software applications (test versus actual) would utilize same inputs and generate independent results that can be compared to validate the internal processing steps.	Risk of disrupting organization or client data is minimal. Simulation does not affect processing. (A) Auditor obtains output information directly without intervention from organization or client personnel. (A) Extent of to which expertise is required depends upon the complexity of the organization or client's processing being simulated. (A or D)
Embedded Audit Module	Programmed audit module that is added to the application under review.	The embedded module allows auditors to monitor and collect data for analysis and to assess control risks and effectiveness. (A) Level of expertise required is considered medium to high, as auditors require knowledge and skills in programming to design and implements the module. (D) Risk of disrupting client data may be high. Because all transactions would be subjected to the module's screening algorithm, it can significantly affect the speed of processing. (D)

(Continued)

Exhibit 4.10 (*Continued*)   Computer-Assisted Audit Techniques for Computer Programs

Audit Technique	Description	Advantages (A) / Disadvantages (D)
Systems Control Audit Review File (SCARF)	Systems Control Audit Review File (SCARF) is another real-time technique that can collect specific transactions or processes that violate certain predetermined conditions or patterns. This may be enhanced by decision support software that alerts designated personnel (audit, security, etc.) of unusual activity or items out of the ordinary. Computer forensic specialists can collect data to log files for further review and examination.	Allow auditors to embed audit routines into an application system and collect data on events that are of interest to them. (A) Expertise is required to embed audit routines into an application system, and ensure those routines do not affect actual data. (D) Risk of disrupting organization or client data is high. Controls must be adequately designed and implemented to identify and remove the effects of the embedded audit routines. (D)
Transaction Tagging	Follows a selected transaction through the application from input, transmission, processing, and storage to its output to verify the integrity, validity, and reliability of the application. Some applications have a trace or debug function, which can allow one to follow the transaction through the application. This may be a way to ensure that the process for handling unusual transactions is followed within the application modules and code.	Tags transactions from beginning to end. (A) Allows auditors to log all of the transactions or snapshot of activities. (A) Expertise required to add special designation (or tag) to the transaction record. (D) Risk of disrupting client data may be medium to high. Controls must be adequately designed and implemented to identify and remove the tag or special designation added to the transaction being evaluated. (D)

Source: From Senft, S., Gallegos, F., and Davis, A. (2012). *Information Technology Control and Audit.* CRC Press/Taylor & Francis, Boca Raton, FL.

Computer Forensics Tools

Computer forensics is the examination, analysis, testing, and evaluation of computer-based material conducted to provide relevant and valid information to a court of law. Computer forensics tools are increasingly used to support law enforcement, computer security, and computer audit investigations.

A good source for evaluating computer forensics tools is the Computer Forensics Tool Testing (CFTT) Project Website at www.cftt.nist.gov/. CFTT is a joint project of the NIST, the U.S. Department of Justice's National Institute of Justice (NIJ), the Federal Bureau of Investigation (FBI), the Defense Computer Forensics Laboratory (DCFL), the U.S. Customs Service, and others to develop programs for testing computer forensics tools used in the investigation of crimes involving computers.

One tool recently reviewed by the CFTT was EnCase Forensics by Guidance Software, Inc. EnCase enables "noninvasive" computer forensic investigations, allowing examiners to view relevant files including "deleted" files, file slack, and unallocated space. Other valuable resources for experience in the use of computer forensics tools would be those professional associations or organizations that support this area. Some of those would be The International High Technology Crime Investigators Association, Association of Certified Fraud Examiners, the Institute of Internal Auditors, Federal Government's Electronic Crimes Task Force, FBI Regional Computer Forensics Laboratory, and the Colloquium for Information Systems Security Education. Note that when applying computer forensics tools, one must be aware of the investigative methodology, processes, and procedures that must be followed to ensure that the evidence can be gathered successfully, documented, and not contaminated as evidential matter that could be used in court. An excellent resource here is the U.S. Department of Justice publication, *Prosecuting Computer Crimes* (2nd edition) published in 2010, as well as information provided by the High Tech Criminal Investigation Association (www.htcia.org).

Conclusion

The continued evolution of IT has placed advanced (software) features in the hands of IT auditors to apply in support of conducting, documenting, and executing the audit process. These software tools and techniques allow the auditor to apply innovative approaches to validating processes at the applications level.

Auditor productivity tools, for instance, include software to automate the audit function and integrate information gathered as part of the audit process. These tools allow auditors to reduce the amount of time spent on administrative tasks. System documentation techniques are also very common, and are mainly used to document and test application systems, IT processes, and their integration within the IT environment. Flowcharts, data flow diagram, and business process diagrams are good examples of system documentation techniques. Lastly, CAATs assist auditors when assessing application controls, and selecting and analyzing computerized data for substantive audit tests.

CAATs can be used by IT and/or financial auditors, in a variety of ways, to define sample size and select samples, determine compliance with procedures, and continuously monitor processing results. IT auditors, for instance, use CAATs to review applications in order to gain an understanding of the controls in place to ensure the accuracy and completeness of the information generated.

Auditors use generalized audit software (a type of CAAT) to evaluate the integrity of applications. Generalized audit software allows auditors to analyze and compare files, select specific records for examination, conduct random samples, validate calculations, prepare confirmation letters, and analyze aging of transaction files, among others. Some of the most popular generalized audit software include Audit Analytics by Arbutus Software, TopCAATs, CaseWare Analytics IDEA Data Analysis, Easy2Analyse, TeamMate, and ACL. These are all virtually similar in regards to functionality.

The ACL software package, described in this chapter, is a file interrogation tool designed to read data from most formats (e.g., databases, delimited files, text files, Excel files, etc.) and to provide data selection, analysis, and reporting. ACL handles and processes large amounts of data in order to identify negative, minimum, and maximum balances; perform statistical sampling and aging analyses; identify duplicates or gaps in sequence testing; and perform comparative joining and matching.

CAATs are also used when conducting operational reviews and as a computer forensic tool. An operational review focuses on the evaluation of effectiveness, efficiency, and goal achievement related to information systems management operations. As a computer forensic tool, auditors examine, analyze, test, and evaluate computer-based material in order to provide relevant and valid information to a court of law. Computer forensics tools are increasingly used to support law enforcement, computer security, and computer audit investigations.

Review Questions

1. What are audit productivity tools? How do they assist auditors?
2. What are CAATs and what benefits they provide to IT auditors?
3. Describe the following system documentation techniques commonly used to understand financial application systems:
   a. Data flow diagrams
   b. Business Process Diagrams
   c. Flowcharts
4. List the steps required in the development of flowcharts.
5. CAATs are known to assist auditors in defining sample size and selecting a sample for testing purposes. Describe two techniques used by CAATs to define sample size and select the sample.
6. What is the audit command language (ACL) audit software? List the benefits it provides.
7. Explain the four steps to follow when planning for an ACL data analysis project.
8. Spreadsheet controls are one type of application controls used by auditors. List and describe five key spreadsheet controls.
9. What is the emphasis or focus of an operational review? List specific activities when performing an operational review.
10. What is computer forensics? What do computer forensic tools support? How do you think computer forensic tools may assist the IT auditor?

Exercises

1. List and describe three broad categories of computer auditing functions IT professionals use to support the audit of an application. Explain their application.

2. You are a Senior IT auditor having a planning meeting with your two Staff members. The task at hand is an ACL data analysis project for the client. List and describe the steps you and your team should follow in order to deliver a successful project.

3. Differentiate between "auditing around the computer" and "auditing through the computer."

CASE—CHANGE CONTROL MANAGEMENT PROCESS

OVERVIEW: A change control management process is a method that formally defines, evaluates, and approves application changes prior to their implementation into live or production environments. The process includes several control procedures to ensure that implemented changes will cause minimal impact to the objectives of the organization. These procedures involve submission of change requests, determination of feasibility, approval, and implementation. The following describes typical roles and procedures undertaken in a change control management process.

ROLE: CHANGE REQUESTER

The Change Requester identifies a requirement for change to the application (e.g., upgrades to new editions, etc.). The Requester then prepares a Change Request Form (CRF), including description of the change, cost and benefits analyses, impact, approvals, and any other supporting documentation deemed necessary. He or she then submits the CRF to the Project Manager for further review.

ROLE: PROJECT MANAGER

Upon receipt, the Project Manager reviews the CRF and determines whether or not additional information is required for the Change Control Board to assess the full impact of the change in terms of time, scope, and cost (i.e., feasibility). The decision is based among others on factors, such as:

- Number of change options presented
- Feasibility and benefits of the change
- Risks and impact to the organization
- Complexity and/or difficulty of the change options requested
- Scale of the change solutions proposed

If the Project Manager determines the change is feasible, he/she will log the CRF in the change log by number, and track its status. The Project Manager then submits the CRF to the Change Control Board. On the other hand, if the CRF is not deemed feasible, the Project Manager will close the CRF.

ROLE: CHANGE CONTROL BOARD

Upon receipt, the Change Control Board reviews the CRF and any supporting documentation provided by the Project Manager. The Change Control Board represents an authorized body who is ultimately responsible for approving or rejecting CRFs based on relevant analyses (i.e., feasibility).

After a formal review, the Change Control Board may:

■ Reject the change (the reasons for the rejection are notified back to the Change Requester)
■ Request more information related to the change
■ Approve the change as requested or subject to specified conditions

Once approved, the Change Control Board forwards the change and any related supporting documentation to the Implementation Team.

ROLE: IMPLEMENTATION TEAM

The Implementation Team schedules and tests the approved change. If test results are not successful, the change and all related supporting documentation are sent back for re-testing. If results are successful, the Implementation Team formally implements the change, and notifies the Change Requester.

TASK: Prepare a flowchart depicting the change control management process just described. Make sure you segregate the roles (i.e., Change Requester, Project Manager, Change Control Board, and Implementation Team) in vertical columns when creating the flowchart to illustrate the procedures performed in the process. This representation is useful for auditors to evaluate segregation of duties and identify incompatible functions within the process.

Further Reading

1. AICPA. Audit analytics and continuous audit—Looking toward the future, www.aicpa.org/InterestAreas/FRC/AssuranceAdvisoryServices/DownloadableDocuments/AuditAnalytics_LookingTowardFuture.pdf (accessed August 2017).
2. AICPA. (2007). *Information Technology Considerations in Risk-Based Auditing: A Strategic Overview*, Top Technology Initiatives. New York: American Institute of Certified Public Accountants (AICPA).
3. Barbin, D. and Patzakis, J. (2002). Cybercrime and forensics. *IS Control J.*, 3, 25–27.
4. Bates, T. J. (2000). Computer evidence—Recent issues. *Inf. Sec. Tech. Rep.*, 5(2), 15–22.
5. Braun, R. L. and Davis, H. E. (2003). Computer-assisted audit tools and techniques: Analysis and perspectives.*Manage.Audit.J.*,18(9),725–731.www.emeraldinsight.com/doi/full/10.1108/02686900310500488
6. Cerullo, V. M. and Cerullo, M. J. (2003). Impact of SAS No. 94 on computer audit techniques. *IS Control J.*, 1, 53–57.
7. Computer Forensic Tool Testing (CFTT) Project Website, National Institute of Standards and Technology, www.cftt.nist.gov/ (accessed March 2017).
8. Deloitte, LLP (2014). *ACL for Auditors*. Unpublished internal document.
9. Deloitte, LLP (2014). *IT Audit Planning Work Papers*. Unpublished internal document.
10. EY's ten key IT considerations for internal audit—Effective IT risk assessment and audit planning. (February 2013). Insights on Governance, Risk and Compliance, www.ey.com/Publication/vwLUAssets/Ten_key_IT_considerations_for_internal_audit/$FILE/Ten_key_IT_considerations_for_internal_audit.pdf
11. Gallegos, F. (2001). *WebMetrics: Computer-Assisted Audit Tools*, EDP Auditing Series, #73-20-50, Auerbach Publishers, Boca Raton, FL, pp. 1–16.
12. Gallegos, F. (2002). *Personal Computers in IT Auditing*, EDP Auditing, #73-20-05, Auerbach Publishers, Boca Raton, FL, pp. 1–7.

13. Guidance Software, Inc., EnCase Enterprise, Pasadena, CA, www.guidancesoftware.com (accessed September 2016).
14. Heiser, J. and Kruse, W. (2002). *Computer Forensics—Incident Response Essentials*, Addison-Wesley, Reading, MA.
15. IS Audit Basics. *The Process of Auditing Information Systems*, www.isaca.org/knowledge-center/ itaf-is-assurance-audit-/pages/is-audit-basics.aspx (accessed July 2017).
16. James, H. (2011). *Information Technology Auditing*, 3rd Edition, South-Western Cengage Learning, Nashville, TN.
17. Kaplin, J. (June 2007). *Leverage the Internet*. Internal Auditor. Institute of Internal Auditors, Lake Mary, FL.
18. Kneer, D. C. (2003). Continuous assurance: We are way overdue. *IS Control J.*, 1, 30–34.
19. Laudon, K. C. and Laudon, J. P. (2014). *Management Information Systems—Managing the Digital Firm*, 13th Edition, Pearson, Upper Saddle River, NJ.
20. McCafferty, J. (2016). *Five Steps to Planning an Effective IT Audit Program*. MIS Training Institute, http://misti.com/internal-audit-insights/five-steps-to-planning-an-effective-it-audit-program
21. Otero, A. R. (2015). An information security control assessment methodology for organizations' financial information. *Int. J. Acc. Inform. Syst.*, 18(1), 26–45.
22. Otero, A. R. (2015). Impact of IT auditors' involvement in financial audits. *Int. J. Res. Bus. Technol.*, 6(3), 841–849.
23. Otero, A. R., Tejay, G., Otero, L. D., and Ruiz, A. (2012). A fuzzy logic-based information security control assessment for organizations, IEEE Conference on Open Systems, Kuala Lumpur, Malaysia.
24. Otero, A. R., Otero, C. E., and Qureshi, A. (2010). A multi-criteria evaluation of information security controls using Boolean features. *Int. J. Network Secur. Appl.*, 2(4), 1–11.
25. Richardson, V. J., Chang, C. J., and Smith, R. (2014). *Accounting Information Systems*, McGraw Hill, New York.
26. Romney, M. B. and Steinbart, P. J. (2015). *Accounting Information Systems*, 13th Edition, Pearson Education, Upper Saddle River, NJ.
27. Sarva, S. (2006). Continuous auditing through leveraging technology. *Inf. Syst. Audit Control Assoc. J.*, 2, 1–20.
28. Sayana, S. A. (2003). Using CAATs to support IS audit. *IS Control J.*, 1, 21–23.
29. Senft, S., Gallegos, F., and Davis, A. (2012). *Information Technology Control and Audit*, CRC Press/ Taylor & Francis, Boca Raton, FL.
30. Singleton, T. (2006). Generalized audit software: Effective and efficient tool for today's IT audits. *Inf. Syst. Audit Control Assoc. J.*, 2, 1–3.
31. U.S. General Accounting Office, *Assessing the Reliability of Computer Processed Data Reliability*, https:// digital.library.unt.edu/ark:/67531/metadc302511/ (accessed November 2016).

PLANNING AND ORGANIZATION

Chapter 5

IT Governance and Strategy

LEARNING OBJECTIVES

1. Describe IT governance and explain the significance of aligning IT with business objectives.
2. Describe relevant IT governance frameworks.
3. Explain the importance of implementing IT performance metrics within the organization, particularly, the IT Balanced Scorecard. Describe the steps in building an IT Balanced Scorecard, and illustrate supporting example.
4. Discuss the importance of regulatory compliance and internal controls in organizations.
5. Define IT strategy and discuss the IT strategic plan, and its significance in aligning business objectives with IT.
6. Explain what an IT Steering Committee is, and describe its tasks in an organization.
7. Discuss the importance of effective communication of the IT strategy to members of the organization.
8. Describe the operational governance processes and how they control delivery of IT projects, while aligning with business objectives.

IT governance has taken on greater importance in many organizations. Based on the fifth version of Control Objectives for Information and Related Technologies (COBIT), governance "ensures that stakeholder needs, conditions, and options are evaluated to determine balanced, agreed-on enterprise objectives to be achieved; setting direction through prioritization and decision making; and monitoring performance and compliance against agreed-on direction and objectives."* With the globalization of many industries and financial markets, developed and developing economies are recognizing the importance of effective governance and controls to the success of organizations. The Sarbanes–Oxley Act of 2002 (SOX) and the Committee of Sponsoring Organizations of the Treadway Commission (COSO) (both from the United States), as well as the Combined Code on Governance in the United Kingdom and the Organization for Economic Co-Operation and Development Principles of Corporate Governance in Europe, all, have set the bar for corporate governance. For IT, COBIT has become the global standard for governance and controls. COBIT provides a framework for implementing IT controls to comply with SOX

---

* www.isaca.org/cobit/pages/default.aspx.

and other global governance standards. Organizations around the world are using the principles defined in COBIT to improve IT performance. A strategy is a formal vision to guide in the acquisition, allocation, and management of resources to fulfill the organization's objectives. An IT strategy, for instance, is part of the overall corporate strategy for IT and includes the future direction of technology in fulfilling the organization's objectives. IT governance provides the structure and direction to achieve the alignment of the IT strategy with the business strategy. Close alignment of the IT strategy with the business strategy is essential to the success of a well-functioning partnership.

IT Governance—Alignment of IT with Business Objectives

In a survey conducted by the IT Governance Institute, 94% of participating organizations considered IT to be very important to the overall organization strategy. The same survey noted that the higher the level of IT governance maturity, the higher the return on IT investment. To achieve IT governance maturity and a higher return on IT investment requires a close partnership between IT and business management. Close alignment of the IT strategy with the business strategy is essential to the success of a well-functioning partnership. It is important for the organization to understand the business it supports and for the business to understand the technology it uses. For this to happen, the organization must have a seat at the table with the Chief Executive Officer (CEO) and other business leaders.

Communicating with senior management is not an easy task as IT is only a small portion of the issues faced by organizations today. IT leaders must be seen as valuable members of the team, and not just as service providers. For this to happen, the Chief Information Officer (CIO) and IT management must first seek to understand the business issues and offer proactive solutions to the organization's needs. IT management must also have a clear understanding of their current strengths and weaknesses and be able to communicate this information to the business management.

IT governance provides the structure to achieve alignment of the IT activities and processes with business objectives, incorporate IT into the enterprise risk management program, manage the performance of IT, ensure the delivery of IT value, and make certain of regulatory compliance and adequate implementation of internal controls.

Effectively managing an organization requires a solid foundation of governance and control over IT resources. Governance guides the decision rights, accountability, and behaviors of an organization. This is controlled through a series of processes and procedures that identify who can make decisions, what decisions can be made, how decisions are made, how investments are managed, and how results are measured. Implemented effectively, IT governance allows IT activities and processes to be in alignment with the direction set by the governance body to achieve the enterprise objectives.

Delivering value from IT is a joint effort between business and IT to develop the right requirements and work together for successful delivery of the promised benefits. To be effective, the Board of Directors (Board), an organization's governing body including the audit committee to whom the chief audit executive may functionally report, must understand the current state of IT and actively participate in establishing the future direction of IT.

Effectively communicating with the Board about IT is not always easy. IT is a very complex environment, which is difficult to explain to non-IT professionals. In addition, many members of the Board or senior management will have their own issues and a vested interest in certain IT projects and services that may influence the decision-making process. Getting agreement up front on the measures of IT performance will go a long way toward focusing senior management on the key issues in managing IT. Measuring both business and IT performance will also help hold both parties accountable for the success of IT projects and service delivery.

IT Governance Frameworks

Three widely recognized and best practice IT-related frameworks include: IT Infrastructure Library (ITIL), COBIT, and the British Standard International Organization for Standardization (ISO)/International Electrotechnical Commission 27002 (ISO/IEC 27002). These three frameworks provide organizations with the means to address different angles within the IT arena.

ITIL

ITIL was developed by the United Kingdom's Cabinet Office of Government Commerce (OGC) as a library of best practice processes for IT service management. Widely adopted around the world, ITIL provides guidelines for best practices in the IT services management field. Specifically, an ITIL's service management environment effectively and efficiently delivers business services to end-users and customers by adhering to five core guidelines related to:

- *Strategy*—guidelines or best practice processes to map the IT strategy with overall business goals and objectives.
- *Design*—best practice processes (or requirements) implemented to guide toward a solution designed to meet business needs.
- *Transition*—aims at managing change, risk, and quality assurance during the deployment of an IT service.
- *Operation*—guidelines or best practice processes put in place to maintain adequate and effective IT services once implemented into the production environment.
- *Continuous Improvement*—constantly looks for ways to improve the overall process and service provision.

The ITIL framework should be chosen when the goal of the organization is to improve the quality of the IT management services. The ITIL framework assists organizations in creating IT services that can effectively help to manage the daily tasks, particularly when the focus is on either customer or end-user.

COBIT

COBIT is an IT governance framework that helps organizations meet today's business challenges in the areas of regulatory compliance, risk management, and alignment of the IT strategy with organizational goals. COBIT is an authoritative, international set of generally accepted IT practices or control objectives, designed to help employees, managers, executives, and auditors in: understanding IT systems, discharging fiduciary responsibilities, and deciding adequate levels of security and controls.

COBIT supports the need to research, develop, publicize, and promote up-to-date internationally accepted IT control objectives. The primary emphasis of the COBIT framework is to ensure that technology provides businesses with relevant, timely, and quality information for decision-making purposes.

The COBIT framework, now on its fifth edition (COBIT 5), allows management to benchmark its environment and compare it to other organizations. IT auditors can also use COBIT to substantiate their internal control assessments and opinions. Because the framework is comprehensive, it provides assurances that IT security and controls exist.

COBIT 5 helps organizations create optimal value from IT by maintaining a balance between realizing benefits and optimizing risk levels and resource use. COBIT 5 is based on five principles (see Exhibit 3.2). COBIT 5 considers the IT needs of internal and external stakeholders (Principle 1), while fully covering the organization's governance and management of information and related technology (Principle 2). COBIT 5 provides an integrated framework that aligns and integrates easily with other frameworks (e.g., Committee of Sponsoring Organizations of the Treadway Commission-Enterprise Risk Management (COSO-ERM), etc.), standards, and best practices used (Principle 3). COBIT 5 enables IT to be governed and managed in a holistic manner for the entire organization (Principle 4). Lastly, COBIT 5 assists organizations in adequately separating governance from management objectives (Principle 5).

The framework is valuable for all size types organizations, including commercial, not-for-profit, or in the public sector. The comprehensive framework provides a set of control objectives that not only helps IT management and governance professionals manage their IT operations but also IT auditors in their quests for examining those objectives.

Selection of COBIT may be appropriate when the goal of the organization is not only to understand and align IT and business objectives but also to address the areas of regulatory compliance and risk management.

ISO/IEC 27002

The ISO/IEC 27002 framework is a global standard (used together with the **ISO/IEC 27001** framework) that provides best practice recommendations related to the management of information security. The standard applies to those in charge of initiating, implementing, and/or maintaining information security management systems. This framework also assists in implementing commonly accepted information security controls and procedures.

The ISO/IEC 27000 family of standards includes techniques that help organizations secure their information assets. Some standards, in addition to the one mentioned above, involve IT security techniques related to:

- Requirements for establishing, implementing, maintaining, assessing, and continually improving an information security management system within the context of the organization. These requirements are generic and are intended to be applicable to all organizations, regardless of type, size, or nature. (ISO/IEC 27001:2013)
- Guidance for information security management system implementation. (ISO/IEC DIS 27003)
- Guidelines for implementing information security management (i.e., initiating, implementing, maintaining, and improving information security) for inter-sector and inter-organizational communications. (ISO/IEC 27010:2015)
- ISO/IEC 27013:2015. Guidance on the integrated implementation of an information security management system, as specified in ISO/IEC 27001, and a service management system, as specified in ISO/IEC 20000-1.

Using the family of standards above will assist organizations to manage the security of assets, including, but not limited to, financial information, intellectual property, employee details or information entrusted by third parties.

The purpose of the ISO/IEC 27002 framework is to help organizations select proper security measures by utilizing available domains of security controls. Each domain specifies control objectives that provides further guidance on how organizations may attempt to implement the framework.

The ISO/IEC 27002 framework should be chosen when IT senior management (i.e., CIO) targets an information security architecture that provides generic security measures to comply with federal laws and regulations.

A Joint Framework

As seen, ITIL, COBIT, and the ISO/IEC 27002 are all best-practice IT-related frameworks to regulatory and corporate governance compliance. A challenge for many organizations, however, is to implement an integrated framework that draws on these three standards. The Joint Framework, put together by the IT Governance Institute (ITGI) and the OGC, is a significant step leading into such direction.

Aligning ITIL, COBIT, and ISO/IEC 27002 not only formalizes the relationship between them but, most importantly, allows organizations to:

- implement a single, integrated, compliance method that delivers corporate governance general control objectives;
- meet the regulatory requirements of data and privacy-related regulation; and
- get ready for external certification to ISO 27001 and ISO 20000, both of which demonstrate compliance.

Implementing a joint framework leads organizations toward effective regulatory compliance and improves their competitiveness. Implementation of the frameworks just discussed is paramount in addressing relevant areas within the IT field. Of equal importance is the establishment of metrics to measure IT performance. These metrics should not only be in place but also regularly assessed for consistency with the goals and objectives of the organization.

IT Performance Metrics

Developing a measurement process takes time and resources to implement. To be successful, both the organization and IT management must be in full support. They should also be consulted as to the types of measurements they believe will be most beneficial. The areas to be measured should be closely aligned to the objectives of the organization. It makes no sense to measure something that no one cares about. Management will be most supportive when it sees the metrics applied to the areas that are most in need of improvement. Typically, the areas that are measured have a tendency to attract focus and improve over time. A critical metric set—the few key metrics that are critical to the successful management of the function—should be identified and applied to the environment.

Once the critical metric set has been identified, personnel in the areas that are to be measured should be consulted, and a set of measurements that will provide meaningful data should be devised. Personnel responsible for doing the work should select the best means to measure the quality and productivity of their work. Metrics that are developed should only be applied to data that are both measurable and meaningful. It is useless to waste time on developing measures on areas that do not fall within the critical metric set, as these measures will not satisfy the needs of management.

After initial implementation of the first measurements, it is important to show the results. Data should be compiled over a predefined period, and results should be provided to management on a regular basis. As the metrics database grows, the reliability of the data will increase and the usefulness of the reports to management will also increase.

Although it is quite easy to get management to support metrics (if they are informed as to what metrics are and the impact they can have), it is also difficult to get management support if they are skeptical or have not been educated on the matter. In this situation, a different task should be taken. First, management must be made to realize that it is next to impossible to manage what cannot be or is not measured. The easiest way to strengthen this argument is to back it up with some sample metrics.

Second, survey data from other organizations can be compiled and presented to encourage adoption of a metrics frame of mind. For sample metrics, identify several areas that can be measured and provide reports on these areas. Again, it is important to provide short-term payback to show results and continue to produce reports showing progress in those areas.

Once all metric data are gathered, it must be presented in a format that is easy for the reader to understand. A combination of graphics and text is important to illustrate the context and performance trends. The reports must stress the progress in the areas selected for measurement. This is a key point in that it shows short-term results in the long-term measurement process. Areas of improvement must be stressed to show that the process is working.

When management has accepted the concept of metrics, it is time to begin implementing some measurements in critical areas. During this step of the measurement process, it is important to be sensitive to the resistance to change. The implementation of metrics causes uneasiness and fear in the ranks.

The most important rule to remember in the design and implementation of a metric is that in all cases, the area that is to be measured must help in the development of the metrics. This will create a sense of ownership over the measurements and will ease the resistance to their implementation. The group should be informed as to the needs of management and should be empowered to develop the metrics to meet the need. This will result in more relevant data being produced and an increase in quality in that area.

The second important rule to remember in the design and implementation of a metric is that it is absolutely vital that the measures are applied to events and processes, and never to individuals. If people get the idea that their performance is being measured, they will be less likely to comply with the metrics process. It must be explicitly stated that the results of the metrics will not be used to measure the productivity or effectiveness of individuals, but of the processes used by the individuals to create their products or services.

Keeping these two rules in mind, the next step is to identify the attributes of an effective measure. An effective measure must be able to pass tests of reliability and validity. Reliability defines consistency of a measure, and validity determines the degree to which it actually measures what it was intended to measure. The measure must be meaningful and provide useful data to management. An example to measure IT performance is through implementing a balanced scorecard.

IT Balanced Scorecard

As the implementation of IT systems continues to grow rapidly in organizations, questions such as the following are being asked (and assessed) more frequently than ever before: Is our current IT investment plan consistent with the organization's strategic goals and objectives? Was the IT application just developed a success? Was it implemented effectively and efficiently? Is our IT department adding value to the organization? Should our current IT services be outsourced to third parties?

These types of questions are not uncommon, and support the continuous needs organizations have to measure the value of IT and evaluate its performance. This is essentially what an IT Balanced Scorecard (IBS) does. An IBS provides an overall picture of IT performance aligned to the objectives of the organization. It specifically measures and evaluates IT-related activities (e.g., IT application projects, functions performed by the IT department, etc.) from various perspectives, such as IT-generated business value, future orientation, operational efficiency and effectiveness, and end-user service satisfaction. These perspectives are then translated into corresponding metrics that reconcile with the organization's mission and strategic objectives. Results from the metrics are assessed for adequacy against target values and/or organization initiatives. The four perspectives, described next, should be periodically revised for adequacy by management personnel.

IT-Generated Business Value

Measuring IT performance is dependent on the strategy and objectives of the organization. However, it comes down to the business value IT is delivering to the organization. In general, IT provides value through delivering successful projects and keeping operations running. If an organization is looking for reduced costs, it may measure the cost of IT and the business function cost before and after automation. If an organization is focused on growing new markets, it may measure the time to market for new products. IT adds value to an organization through project and service delivery.

IT projects deliver business value by automating business processes. As these projects are enabled by technology, IT is adding value to the organization. Measuring the amount of benefit delivered from these projects is one way of representing the value of IT. Automating business processes typically results in higher IT costs and lower business costs (or higher revenue). An original application development project's business case made certain assumptions about the cost and benefit of the new application. Although the project's business case will be validated as part of the post-implementation review, it is important to continue measuring the ongoing costs over time. There may be a perception that IT costs are growing without the recognition that business costs should be dropping or revenue growing by a greater **margin**. It is important to keep this information in front of the Board and senior management as a reminder of the value of IT. Delivering the promised value is the responsibility of both IT and the business functions. Reporting on the actual results holds both parties accountable for the expected results. Another measure of value is how quickly the organization can respond to new business opportunities. If IT has been successful at implementing flexible infrastructure, applications, and processes, it will be able to respond to business needs.

IT services deliver value by being available for the organization as needed. Organizations rely heavily on automated systems to function on a day-to-day basis. The failure of these systems results in loss of revenue or increased expense to the organization. A more positive perspective is the amount of revenue or productivity generated by these systems. As part of the strategic and operational planning process, an organization must decide the level of service required of IT. The service levels will depend on the type of organization, application portfolio, services provided by IT, and the objectives of the organization. An online auction house that depends on 24/7 service availability for its existence will have a different need than a brick and mortar grocery store.

Metrics to measure business value may address the functions of the IT department, value generated by IT projects, management of IT investments, and sales made to outsiders or third parties. These metrics may include: percentages of resources devoted to strategic projects; perceived relationship between IT management and senior-level management; computation of traditional financial evaluation methods, such as **return on investment (ROI)** and **pay-back period**; actual versus budgeted expenses; percentages over/under overall IT budget; and revenues from IT-related services and/or products; among others.

Future Orientation

Future orientation is concerned with positioning IT for the future by focusing on the following objectives: (1) training and educating IT personnel for future IT challenges; (2) improving service capabilities; (3) staffing management effectiveness; (4) enhancing enterprise architecture; and (5) researching for emerging technologies and their potential value to the organization.

A sample mission for this perspective would be to deliver continuous improvement and preparing for future challenges. Sample metrics within this perspective would address the following:

- Continuously improving IT skills through education, training, and development.
- Delivering internal projects consistent to plan.
- Staffing metrics by function (e.g., using utilization/billable ratios, voluntary turnover by performance level, etc.).
- Developing and approving an enterprise architecture plan, and adherence to its standards.
- Conducting relevant research on newly-emerging technologies and their suitability for the organization.

Operational Efficiency and Effectiveness

The operational efficiency and effectiveness perspective focuses on the internal processes in place to deliver IT products and services in an efficient and effective manner. Internal operations may be assessed by measuring and evaluating IT processes in areas, such as quality, responsiveness, security, and safety, among others. Other processes to be considered may include hardware and software supply and support, problem management, management of IT personnel, and the effectiveness and efficiency of current communication channels.

Measurements of the operational efficiency and effectiveness perspective may result in useful data about the productivity of different internal processes as well as resources. Metrics here can yield productivity information about the performance of technologies and of specific personnel.

End-User Service Satisfaction

End-user satisfaction should play an important role in the overall evaluation of the IT department or function. The end-user, for IT purposes, may be internal personnel or external (e.g., users accessing inter-organizational IT systems or services, etc.). From an end-user's perspective, the value of IT will be based on whether their jobs are completed timely and accurately. For instance, managers rely on IT-generated reports to make critical decisions related to their organization. These reports must not only by timely made but they should be accurate and involve relevant information for them to make well-informed and necessary business decisions.

A mission for this perspective would be to deliver value-adding products and services to end-users. Related objectives would include maintaining acceptable levels of customer satisfaction, partnerships between IT and business, application development performance, and service-level performance. Metrics used to measure the aforementioned objectives should focus on three areas:

- being the preferred supplier for applications and operations
- establishing and maintaining relationship with the user community
- satisfying end-user needs

It would be necessary for IT personnel to establish and maintain positive relationships with the user community in order to understand and anticipate their needs. Such a relationship is critical for building up and/or improving the credibility of the IT department among the end-users.

Steps in Building an IT Balanced Scorecard

Having an understanding of both corporate-level and IT strategies, as well as the specific goals related to each type of strategy, is crucial before developing an IBS. The following steps are recommended when building a company-specific IBS:

1. Have both senior management and IT management on board from the start; make them aware with the concept of the IBS.
2. Coordinate the collection and analysis of data related to:
   - corporate strategy and objectives (e.g., business strategy, IT strategy, company mission, company specific goals, etc.);
   - traditional business evaluation metrics and methods (e.g., ROI, payback period, etc.) currently implemented for IT performance measurement; and
   - potential metrics applicable to the four IBS perspectives.
3. Define the company-specific objectives and goals of the IT department or functional area from each of the four perspectives.
4. Develop a preliminary IBS based on the defined objectives and goals of the organization and the data outlined in the previous steps.
5. Request revisions, comments, and feedback from management after revising the IBS.
6. Have the IBS formally approved and ready to be used by the organization.
7. Communicate the IBS development process and its underlying rationale to all stakeholders.

The IBS provides value to the business when it addresses IT management processes, including, individual and team IT goal-setting, performance appraisal and rewards for IT personnel, resource allocation, and feedback-based learning, among others. Having a systematic framework like the IBS that is based on goals and measures that have been agreed upon in advance will likely benefit management of both IT people and projects.

All metrics included in the IBS should be quantifiable, easy to understand, and ones for which data can be collected and analyzed in a cost-effective manner. A sample IBS is illustrated in Exhibit 5.1.

Exhibit 5.1   Example of an IT Balanced Scorecard

Mission	Objectives	Metric to Measure	Target Values/ Initiatives
To contribute to the value of the business	**IT-GENERATED BUSINESS VALUE**		
	Business value and strategic contribution of IT department	– Completion of strategic initiatives – Percentage of resources devoted to strategic projects – Perceived relationship between IT management and senior-level management	
	Business value of IT projects	– Business evaluation based on financial measures (ROI, payback period, etc.)	
	Management of IT investment	– Actual versus budgeted expenses – Percentage over/under overall IT budget	
	Sales to outsiders or third parties	– Revenues from IT-related services and/or products	
To deliver continuous improvement and prepare for future challenges	**FUTURE ORIENTATION**		
	Knowledge management	– Completion of education, training, and development courses – Percentage of positions with qualified backup personnel – Expertise with specific technologies	
	Service capability improvement	Deliver internal projects to plan: – Internal process improvement – Organization development – Technology renewal – Professional development	

(Continued)

Exhibit 5.1 (*Continued*)   Example of an IT Balanced Scorecard

Mission	Objectives	Metric to Measure	*Target Values/ Initiatives*
	Staff management effectiveness	Staff metrics by function: – Utilization/billable ratios – Voluntary turnover by performance level – Percent of staff with completed professional performance plans	
	Enterprise architecture evolution	– Development/approval of enterprise architecture plan (EAP) – Systems adherence to EAP and IT standards	
	Emerging technologies research	– Percent of IT budget allocated to research of new and updated technologies	
To deliver IT products and services that are efficient and effective	**OPERATIONAL EFFICIENCY AND EFFECTIVENESS**		
	Process excellence	– Process maturity rating and performance (i.e., quality, cost, and speed)	
	Responsiveness	– Process cycle time and cycle time to market	
	Backlog management and aging	– Staff days of budgeted work in backlog status – Days outstanding of oldest budgeted work	
	Internal cost of quality	– Time/cost of process improvement and quality assurance initiatives per IT employee	
	Security and safety	– Absence of major issues in audit reports and unrecoverable failures or security breaches	

(Continued)

Exhibit 5.1 (*Continued*)   Example of an IT Balanced Scorecard

Mission	Objectives	Metric to Measure	Target Values/ Initiatives
To deliver products and services that add value to end-users	**END-USER SERVICE SATISFACTION**		
	End-user satisfaction	– Score on end-user satisfaction survey	
	IT/business partnership	– Frequency of IT Business Group meetings – Index of both, user and IT involvement in generating new strategic application systems	
	Application development performance	– Delivery to end-users expectations: quality (user acceptance); cost (budget); and speed (schedule)	
	Service-level performance	– Weighted percent of applications and operations services meeting service-level targets for availability and performance	

Source: Adapted from: Senft, S., Gallegos, F., and Davis, A. (2012). *Information Technology Control and Audit.* CRC Press/Taylor & Francis, Boca Raton, FL; Adapted from: Martinsons, M., Davison, R., and Tse, D. (1999). The balanced scorecard: A foundation for the strategic management of information systems, *Decis. Support Syst.*, 25(1), 71–88.

Measuring and assessing IT activities from multiple points of view or perspectives, say through an IBS for instance, help in evaluating the efficiency, effectiveness, and potential of those activities. Such scorecard permits managers to assess the impact of IT systems, applications, and activities on the factors considered important to the organization.

Regulatory Compliance and Internal Controls

One of the key processes that organizations need to manage is compliance with laws and regulations. The sheer number of laws and regulations applicable to a global organization can be overwhelming (refer to Chapter 2). It can take a dedicated team to sift through all the financial, security, privacy, and industry-specific regulatory requirements to determine the impact on processes and information systems. Fortunately, many of the IT requirements are satisfied with the implementation of the controls outlined in COBIT.

There are tools that can help an organization identify laws and regulations and track the control processes implemented to address them.* There are also tools that can help with mapping controls to regulatory requirements (e.g., SOX of 2002 etc.). These tools provide key information for auditors, regulators, and user groups to determine where controls are effective for testing, and which are the gaps that will need to be filled. IT should work together with the organization's compliance officer to ensure that it is aware of new requirements and report on the resolution of existing requirements.

As mentioned earlier, the implementation of SOX created greater awareness and focus on IT controls. Although there is some debate on the value of SOX to enterprises, there is no doubt that it has increased investment in IT general controls and application controls in many organizations. SOX compliance has forced many organizations to review existing applications that process financial transactions with an eye to controlling these processes. Business and IT professionals now need to work together in developing control requirements that can be incorporated into the development of applications. Having more IT controls implemented in application systems translates into more opportunities for IT auditors to perform controls assessment work!

Cases such as the above have prompted organizations to review and revise their existing game plan or IT strategy so that they not only comply with regulatory agency bodies like SOX but also meet the constantly-changing requirements of their business environments.

IT Strategy

IT has become the critical ingredient in business strategies as both enabler and enhancer of the organization's goals and objectives. Organizations must be positioned to take best advantage of emerging opportunities while also responding to the global requirements of the twenty-first century.

A strategy is an important first step toward meeting the challenging and changing business environment. A strategy is a formal vision to guide in the acquisition, allocation, and management of resources to fulfill the organization's objectives. An IT strategy, for example, should be developed with the involvement of the business users to address the future direction of technology. The IT strategy or IT strategic plan formally guides the acquisition, allocation, and management of IT resources consistent with goals and objectives of the organization. It should be part of an overall corporate strategy for IT and should align to the business strategy it supports. The technology strategy needs to be in lockstep with the business strategy to ensure that resources are not wasted on projects or processes that do not contribute to achieving the organization's overall objectives. This alignment should occur at all levels of the planning process to provide continued assurance that the operational plans continue to support the business objectives. Supporting the strategy, architectural standards and technology planning ensure that investments in IT lead to efficient maintenance and a secure environment.

IT governance (discussed early in the chapter) provides the structure and direction to achieve the alignment of the IT strategy with the business strategy. Close alignment of the IT strategy with the business strategy is essential to the success of a well-functioning partnership.

The most effective strategy will be determined by the combination of the environment, culture, and technology used by an organization. IT management involves combining technology, people, and processes to provide solutions to organizational problems. IT must take the lead in gathering information to incorporate organizational needs with technological feasibility to create an overall strategy.

An IT strategic plan provides a roadmap for operating plans and a framework for evaluating technology investments. The IT strategy supports the business strategy to ensure that technology

---

* Filipek, R., Compliance automation, *Internal Auditor*, February 2007, pp. 27–29.

resources are applied to meeting business objectives while minimizing ongoing support costs. This task sounds fairly simple, but according to a Gartner Group report, "95% of enterprises lack a well-defined business strategy." In most cases, the business strategy has to be assumed based on conversations with business executives. The first step in defining an IT strategic plan is to understand the business objectives, whether stated or implied. These objectives guide management in evaluating investments, assessing risk, and implementing controls.

So, why should IT have a strategic plan if the organization has none? The main risk of not having an IT strategic plan is the increased cost of technology. If there is no roadmap, organizations run the risk of investing in technology that increases costs but adds no business value. According to the IT Governance Institute, aligning IT investments with business strategies is the biggest single issue organizations face.

As IT exists to support and enable business, the ultimate responsibility for setting and implementing the IT strategy should rest with the organization's senior management. However, business leaders need IT management to take the lead in identifying ways IT can support the transformation of an organization to meet its long-term goals. A strong business and IT partnership in the strategic planning process provides the best foundation for success. One way to achieve alignment is to involve business leaders in the development of the IT strategy by establishing an IT Steering Committee.

IT Steering Committee

An IT Steering Committee is composed of decision makers from the various constituencies in the organization to resolve conflicting priorities. Even when business objectives are clearly stated, conflicts will arise with the interpretation of the actions necessary to fulfill those objectives. The IT Steering Committee is responsible for determining the overall IT investment strategy, ensuring that IT investments are aligned with business priorities and that IT and business resources are available to enable IT to deliver upon its expectations.

An IT Steering Committee can help ensure integration of the business and the IT strategic plan. This committee facilitates the integration of business and technology strategies, plans, and operations by employing the principles of joint ownership, teamwork, accountability, and understanding of major projects. The committee should be composed of members of senior management and the CIO. The CIO, according to Gartner, "oversees the people, processes and technologies within a company's IT organization to ensure they deliver outcomes that support the goals of the business."[*] In other words, the CIO is key when identifying critical strategic, technical, and management initiatives that can be implemented to mitigate risks and threats, as well as drive business growth. Essentials functions of the CIO role, as described by the Society for Human Resource Management,[†] include:

1. Create, maintain, and implement written policies and procedures regarding all computer operations in the Management Information Systems or IT Department and throughout the organization.
2. Communicate new or revised information systems policies and procedures formally to all users within the organization.
3. Review and assess the productivity of the department, including the quality of the output and cost of service. Implement methods and procedures to continually improve results.

---

[*] www.gartner.com/it-glossary/cio-chief-information-officer/.
[†] https://www.shrm.org/resourcesandtools/tools-and-samples/job-descriptions/pages/default.aspx.

4. Employ necessary functions to manage department personnel.
5. Develop department annual budgets, segregating per activity/personnel, and administer funds according to budget approval.
6. Maintain security of all data proprietary to the organization, and provide for the complete backup of all computer systems in case of system failure or disaster.
7. Procure, install, and maintain all computer equipment (hardware and software) and all other products and supplies necessary to keep computer systems operable and to fulfill managements requests for computer support.
8. Act as liaison between hardware/software suppliers and organization management for informational updates and problem resolution.
9. Provide employees with top quality, consistently available computer service, support training and maintenance of all computer systems used throughout the organization.
10. Assess new equipment, software, and processes continuously, recommend changes as appropriate and supervise their installation.

As part of the IT Steering Committee, the CIO oversees the IT strategy and the computer systems required to support the objectives and goals of the organization. The IT Steering Committee helps ensure integration of the business objectives and goals with the IT strategy. To attain this, the IT Steering Committee tasks may involve:

- Reviewing business and technology strategies and plans.
- Prioritizing major development projects.
- Developing communication strategies.
- Reviewing development and implementation plans for all major projects.
- Providing business decisions on major design issues for all major projects.
- Monitoring status, schedule, and milestones for all major projects.
- Reviewing and approving major change requests for all major projects.
- Reviewing project budgets and ROIs.
- Resolving conflicts between business and technology groups.
- Monitoring business benefits during and after implementation of major projects.

Once an IT strategy has been established by the IT Steering Committee, it must be communicated to all levels of management and to the users to ensure alignment and reduce conflict.

Communication

Effective communication is critical to coordinate the efforts of internal and external resources to accomplish the organization's goals. Communication should occur at multiple levels, starting by having internal weekly staff meetings. This should cover the employees within the department. Communication should also takes place via town hall meetings, which are typically attended by (and addressed to) all employees in the organization. Communication between IT and the organization, particularly of matters such as IT strategy, goals, etc., should be timely and consistent. Communication should also include all (external) business partners and customers related to the organization.

After the completion of the strategic planning process, the business and IT goals must be translated into actionable goals for the coming year. This is performed through a process called operational planning.

Operational Planning

Once there is an understanding of the organization's objectives and IT strategy, that strategy needs to be translated into operating plans (also called operationalization). The annual operating planning process includes setting the top priorities for the overall IT function as well as for individual IT departments, including developing their annual budget, creating resource and capacity plans, and preparing individual performance plans for all IT staff.

Operating plans will also identify and schedule the IT projects that will be initiated and the IT service levels expected. Delivery of these plans should be controlled by a series of governance processes. These governance processes, listed in Exhibit 5.2, are needed to ensure the effective use of resources and delivery of IT projects, as well as proper alignment with business objectives. This includes processes to: manage project demands, initiate projects, perform technical reviews, procure products and manage vendors, and control financial investments. These processes are explained next.

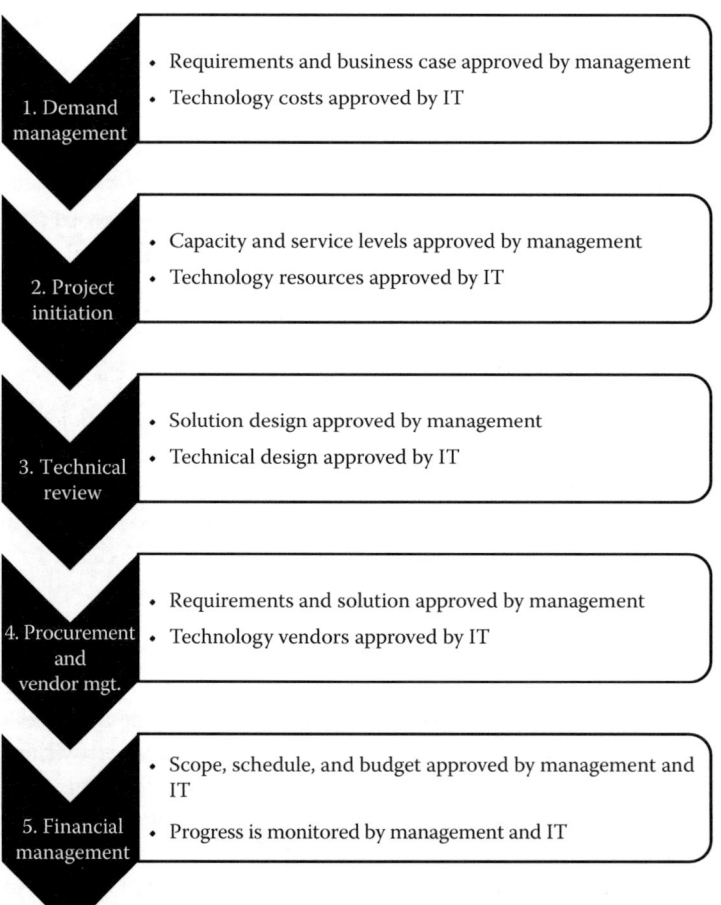

Exhibit 5.2   Governance processes.

Demand Management

Projects need to be reviewed at the beginning of their life cycles to make sure they have a strong business case, as well as senior management support. Researching technology solutions takes time and consumes resources that could be devoted to providing business value. A demand management process can help ensure that resources are devoted to projects that have a strong business case and also approved by senior management. The demand management process helps ensure that senior management is on board, and has provided conceptual approval to the project to proceed through the initial requirements definition and conceptual design phases of the development life cycle. All projects should have an appropriate sponsor from senior management before evaluating the costs of implementing a solution. This is highly advisable in order to avoid wasted effort on a project that will not get approved.

A demand management process ensures that a project has business justification, a business and IT sponsor, and a consistent approach for approving projects. A demand management process also ensures alignment of application and infrastructure groups; that all project costs are identified to improve decision making; that there are means to "weed out" nonessential projects; and that means are identified to control IT capacity and spending.

Project Initiation

Once a project with a strong business case has been approved, it should undergo an initiation process that determines its total cost and benefit. This is usually done by defining high-level business requirements and a conceptual solution. Building a project estimate takes time and resources. It takes time from business users to develop requirements and a business case. It also takes time from software developers to develop a solution and cost estimates. After a project has conceptual approval, business users and software programmers can work together to develop detailed requirements and project estimates that will be used in the final business case and form the basis for the project budget.

Technical Review

The technical solution needs to be evaluated before moving forward to ensure compliance with technology standards. A technical review process helps ensure that the right solution is selected, that it integrates effectively with other components of technology (e.g., network, etc.), and that it can be supported with minimal investments in infrastructure. One way to control technology solutions is to implement a Technical Steering Committee (not to be confused with an IT Steering Committee) with representatives from the various technical disciplines and enterprise architects. A Technical Steering Committee provides a control mechanism for evaluating and approving new technology solutions. A formal technology solution evaluation process includes the assessments of:

- Technical feasibility
- Alternative technologies
- Architecture
- In-house skill compatibility
- Existing environments/replacements
- Implementation, licensing, and cost considerations
- Research and analyst views
- Vendor company profile and financial viability

Procurement and Vendor Management

Processes and procedures should be in place to define how the procurement of IT resources, including people, hardware, software, and other services will be performed. IT procurement involves strategic and administrative tasks, such as defining requirements and specifications; performing the actual IT service or resource acquisition (only after assessing and selecting the appropriate vendor); and fulfilling contract requirements. Vendor selection usually involves the evaluation of three to five vendors. The IT Steering Committee regularly evaluates IT vendors and suppliers and makes the ultimate decision of which vendors or suppliers to bring on board.

Financial Management

In the financial management governance process, potential investments, services, and asset portfolios are evaluated so that they get incorporated in cost/benefit analyses and ultimately within the budget. IT budgeting, for instance, considers existing IT products, resources, and services in order to assist the planning of IT operations. Budgeting is a strategic planning tool (typically expressed in quantitative terms) which aids in the monitoring of specific activities and events. Budgeting also provides forecasts and projections of income and expenses which are used strategically for measuring financial activities and events. Budgets are useful to management when determining whether specific revenues/costs activities are being controlled (i.e., revenues being higher than budget estimates or costs being lower than estimated budget amounts). Budgets lead how organizations might perform financially, operationally, etc. should certain strategies and/or events take place.

Conclusion

IT governance establishes a fundamental basis for managing IT to deliver value to the organization. Effective governance aligns IT to the organization and establishes controls to measure meeting this objective. Three effective and best practice IT-related frameworks commonly used by organizations are ITIL, COBIT, and ISO/IEC 27002. These three frameworks provide organizations with value and the means to address different angles within the IT arena. Realizing the value of IT requires a partnership between management and IT. This partnership should include managing enterprise risk, as well as establishing measuring performance assessments consistent with existing strategies and goals. These performance measures should be aligned to the objectives of the organization, result in accurate and timely data, and report needs in a format that is easy to understand.

An example of a common tool to measure IT performance is the IBS. An IBS provides an overall picture of IT performance aligned to the objectives of the organization. It specifically measures and evaluates IT-related activities, such as IT projects and functions performed by the IT department from perspectives like IT-generated business value, future orientation, operational efficiency and effectiveness, and end-user service satisfaction.

Establishing effective controls in IT and ensuring regulatory compliance is also a joint effort. Well-controlled technology is the result of an organization that considers controls a priority. Organizations need to include controls in system requirements to make this happen. Internal and external auditors can add tremendous value to an organization by providing independent assurance that controls are working as intended. With the implementation of SOX, the knowledge and skills of auditors is a valuable resource to any organization. IT auditors can assist the organization in documenting and evaluating internal control structures to comply with SOX or other governance models.

A strategy is an important first step toward meeting the challenging and changing business environment. An IT strategic plan is a formal vision to guide in the acquisition, allocation, and management of IT resources to fulfill the organization's objectives. One way to achieve alignment is to involve business leaders in the development of the IT strategic plan via establishing an IT Steering Committee. The committee helps ensure integration of the business and IT strategic plan.

To ensure the effective use of resources and delivery of IT projects, as well as proper alignment with business objectives, organizations employ governance processes within their annual operating plan. These processes address how to manage project demands, initiate projects, perform technical reviews, procure products and manage vendors, and control financial investments.

Review Questions

1. How does COBIT define governance?
2. In regards to delivering IT value, why is it so important for the business and its IT department to join efforts?
3. Describe the three widely recognized best practice IT-related frameworks, and state when each framework should be used.
4. Discuss why should organizations consider implementing a joint framework between ITIL, COBIT, and ISO/IEC 27002.
5. Explain what an IT balanced scorecard is.
6. The chapter mentioned three ways that IT can deliver value to the organization, through:
    a. Implementing successful projects and keeping operations running
    b. Automating business processes
    c. Being available for the organization as needed
    Explain in your own words how these three actually provide value to organizations. Provide examples supporting each.
7. What is a strategy? What is an IT strategic plan and why is it significant in aligning business objectives with IT?
8. What is an IT Steering Committee? Summarize the various activities included as part of its scope.
9. Operationalization translates the understanding of both, organization and IT objectives, into operating plans. Operating plans identify and schedule the IT projects that will be initiated and the IT service levels that will be expected. Delivery of these operating plans should be controlled by a series of governance processes. List and describe these processes.
10. What is a Technical Steering Committee and what does it assess related to a technology solution?

Exercises

1. Choose one of the three widely recognized and best practice IT-related frameworks discussed in the chapter. Perform research, outside of the chapter, and provide the following:
    a. Summary of the framework, including advantages, disadvantages, and under which circumstances it should be adopted by organizations.
    b. Provide two or three examples of the framework being used, as appropriate.
    Be ready to present your work to the class.

2. Summarize the steps in building an IT Balanced Scorecard.
3. Describe the essentials functions of Chief Information Officers in organizations.

CASE—SIGNIFICANCE OF IT

INSTRUCTIONS: Read the Harvard Business Review article "IT Doesn't Matter" by Nicholas G. Carr.

TASK: Summarize the article. Then, state whether IT should or should not matter, and why. Support your reasons and justifications with IT literature and/or any other valid external source. Include examples as appropriate to evidence your case point. Submit a word file with a cover page, responses to the tasks above, and a reference section at the end. The submitted file should be between 8 and 10 pages long (double line spacing), including cover page and references. Be ready to present your work to the class.

Further Reading

1. Anzola, L. (2005). IT governance regulation—A Latin American perspective. *Inf. Syst. Control J.*, 2, 21.
2. Bagranoff, N. and Hendry, L. (2005). Choosing and using Sarbanes–Oxley software. *Inf. Syst. Control J.*, 2, 49–51.
3. Basel Committee on Banking Supervision. (2010). Sound practices for the management and supervision of operational risk, Consultative Document, www.bis.org/publ/bcbs183.pdf
4. Brancheau, J., Janz, B., and Wetherbe, J. (1996). Key issues in information systems management: 1994–95 SIM delphi results. *MIS Q.*, 20(2), 225–242.
5. Burg, W. and Singleton, T. (2005). Assessing the value of IT: Understanding and measuring the link between IT and strategy. *Inf. Syst. Control J.*, 3, 44.
6. Carr, N. (2003). IT doesn't matter, *Harvard Business Review*, Harvard Business School Publications, Boston, MA.
7. Dietrich, R. (2005). After year one—Automating IT controls for Sarbanes–Oxley compliance. *Inf. Syst. Control J.*, 3, 53–55.
8. Global Technology Audit Guide (GTAG) 17 Auditing IT Governance. (July 2012). https://na.theiia.org/standards-guidance/recommended-guidance/practice-guides/Pages/GTAG17.aspx
9. Ho Chi, J. (2005). IT governance regulation—An Asian perspective. *Inf. Syst. Control J.*, 2, 21–22.
10. Information Systems Audit and Control Foundation. *COBIT*, 5th Edition, Information Systems Audit and Control Foundation, Rolling Meadows, IL, www.isaca.org/Knowledge-Center/COBIT/Pages/Overview.aspx (accessed June 2017).
11. ISACA. (2012) COBIT 5: A business framework for the governance and management of enterprise IT, 94.
12. ISO/IEC. 27001-Information Security Management, https://www.iso.org/isoiec-27001-information-security.html (January 2017).
13. IT governance defined, www.itgovernance.co.uk/it_governance (accessed 2017).
14. IT Governance Institute. (2008). *COBIT Mapping of ITIL V3 with COBIT 4.1, ISACA*, Rolling Meadows, IL. Digital.
15. IT Governance Institute. (2006). *COBIT Mapping of ISO/IEC 17799–2005 with COBIT 4.0*, ISACA, Rolling Meadows, IL. Digital.
16. IT Governance Institute. (2007). *COBIT Mapping of NIST SP800–53 Rev 1 with COBIT 4.1*, ISACA, Rolling Meadows, IL. Digital.

17. IT Governance Institute. Global status report on the governance of enterprise IT (GEIT)—2011, http://www.isaca.org/Knowledge-Center/Research/Documents/Global-Status-Report-GEIT-2011_ res_Eng_0111.pdf

18. Schroeder, J. (September 6, 2015). Framework comparison, *NeverSys*, http://neversys.com/wp-content/uploads/2015/09/Framework-Comparison.pdf

19. Jones, W. (2005). IT governance regulation—An Australian perspective. *Inf. Syst. Control J.*, 2, 20.

20. Kendall, K. (2007). Streamlining Sarbanes–Oxley compliance. *Internal Auditor*, pp. 39–44.

21. KPMG. (2006). *Leveraging IT to Reduce Costs and Improve Responsiveness*, KPMG International, New York.

22. Leung, L. (2007). ISACA introduces IT governance certification. *Network World*. www.networkworld.com/newsletters/edu/2007/0910ed1.html

23. Mack, R. and Frey, N. (December 11, 2002). *Six Building Blocks for Creating Real IT Strategies*, R-17-63607, Gartner Group, Stamford, CT.

24. Martinsons, M., Davison, R., and Tse, D. (1999). The balanced scorecard: A foundation for the strategic management of information systems. *Decis. Support Syst.*, 25(1), 71–88.

25. Parkinson, M. and Baker, N. (2005). IT and enterprise governance. *Inf. Syst. Control J.*, 3.

26. Pohlman, M. B. (2008). *Compliance Frameworks. Oracle Identity Management: Governance, Risk, and Compliance Architecture*, Third Edition, Auerbach Publications, New York. www.infosectoday.com/Articles/Compliance_Frameworks.htm

27. Senft, S., Gallegos, F., and Davis, A. (2012). *Information Technology Control and Audit*, CRC Press/Taylor & Francis, Boca Raton, FL.

28. Van Grembergen, W. and De Haes, S. (2005). Measuring and improving IT governance through the balanced scorecard. *Inf. Syst. Control J.*, 2, 35.

29. Van Grembergen, W. (2000). The balanced scorecard and IT governance. Challenges of information technology management in the 21st century, 2000 Information Resources Management Association International Conference, Anchorage, AL, May 21–24, https://www.isaca.org/Certification/CGEIT-Certified-in-the-Governance-of-Enterprise-IT/Prepare-for-the-Exam/Study-Materials/Documents/The-Balanced-Scorecard-and-IT-Governance.pdf

30. Williams, P. (2005). *IT Alignment: Who Is in Charge?* IT Governance Institute, Rolling Meadows, IL.

Chapter 6

Risk Management

LEARNING OBJECTIVES

1. Discuss the risk management process, and how it plays an important role in protecting organizations' information from IT threats.
2. Describe the Enterprise Risk Management—Integrated Framework, as well as its eight risk and control components, and how they apply to objectives set by management.
3. Explain what risk assessment is in the context of an organization.
4. Summarize professional standards that provide guidance to auditors and managers about risk assessments.
5. Support the need of insurance coverage as part of the risk assessment process for IT operations.

This chapter discusses the process of managing and evaluating risks in an IT environment. Risk management should be integrated into strategic planning, operational planning, project management, resource allocation, and daily operations. Risk management enables organizations to focus on areas that have the highest impact. Risk assessments, on the other hand, occur at multiple levels of the organization with a focus on different areas. Executive management may focus on business risks, while the technology officer focuses on technology risks, the security officer focuses on security risks, and auditors focus on control risks. But, what are the characteristics and components of a risk management process? What are the professional standards of practice for risk assessments? What are examples of risk assessment practices used in varied environments? Why is insurance coverage so important when dealing with risk assessments? What are the risks typically insured? These are some of the questions that will be answered within the chapter.

Risk Management

Mismanagement of risk can carry an enormous cost. In recent years, businesses have experienced numerous risk-associated reversals that have resulted in considerable financial loss, decrease in shareholder value, damage to the organization reputation, dismissals of senior management, and, in some cases, dissolution of the business. This increasingly risky environment prompts management to adopt a more proactive perspective on risk management.

Risk management ensures that losses do not prevent organizations' management from seeking its goals of conserving assets and realizing the expected value from investments. The National Institute of Standards and Technology (NIST) Special Publication (SP) 800-30[*] defines risk management as the process of identifying and assessing risk, followed by implementing the necessary procedures to reduce such risk to acceptable levels. Risk management plays an important role in protecting organizations' information from IT threats. For instance, IT risk management focuses on risks resulting from IT systems with threats such as fraud, erroneous decisions, loss of productive time, data inaccuracy, unauthorized data disclosure, and loss of public confidence that can put organizations at risk. A well-designed IT risk management process is essential for developing a successful security program to protect the IT assets of an organization. When used effectively, a well-structured risk management methodology will assist organizations' management in identifying adequate controls for supporting their IT systems.

Historically, risk management in even the most successful businesses has tended to be in "silos"—the insurance risk, the technology risk, the financial risk, and the environmental risk—all managed independently in separate compartments. Coordination of risk management has usually been nonexistent, and identification of emerging risks has been sluggish.

The COSO Enterprise Risk Management (ERM) Framework defines enterprise risk management as follows:

> A process, effected by an entity's board of directors, management and other personnel, applied in strategy setting and across the enterprise, designed to identify potential events that may affect the entity, and manage risks to be within its risk appetite, to provide reasonable assurance regarding the achievement of entity objectives.[†]

At first glimpse, there is much similarity between ERM and other classes of risk (e.g., **credit**, **market**, **liquidity**, **operational risk**, etc.) and the tools and techniques applied to them. In fact, the principles applied are nearly identical. ERM must identify, measure, mitigate, and monitor risk. However, at a more detailed level, there are numerous differences, ranging from the risk classes themselves to the skills needed to work with operational risk.

ERM has become more widely accepted as a means of managing organizations. In a survey conducted by the Professional Risk Managers' International Association, over 90% of respondents believed that ERM is or will be part of their business processes. Should organizations be able to develop successful ERM programs, the next step will be for these organizations to integrate ERM with all other classes of risks into truly enterprise-wide risk management frameworks.

Senior managers need to encourage the development of integrated systems that aggregate credit, market, liquidity, operational, and other risks generated by business units in a consistent framework across the organization. Consistency may become a necessary condition to regulatory approval of internal risk management models. An environment where each business unit calculates its risk separately with different rules will not provide a meaningful oversight of enterprise-wide risk. The increasing complexity of products, linkages between markets, and potential benefits offered by overall portfolio effects are pushing organizations toward standardizing and integrating risk management.

---

[*] http://nvlpubs.nist.gov/nistpubs/Legacy/SP/nistspecialpublication800-30r1.pdf.
[†] Enterprise Risk Management—Integrated Framework. Committee of Sponsoring Organizations of the Treadway Commission. September 2004, p. 2. www.coso.org/documents/coso_erm_-executivesummary.pdf.

Enterprise Risk Management—Integrated Framework

The strongest defense against operational risk and losses resides and flows from the highest level of the organization—the Board and senior management. The Board, the management team they hire, and the policies they develop, all set the tone for an organization. As guardians of shareholder value, the Board must be acutely attuned to market reaction to negative news. In fact, they can find themselves castigated by the public if the reaction is severe enough. As representatives of the shareholders, Board members are responsible for policy matters relative to corporate governance, including, but not limited to, setting the stage for the framework and foundation for ERM.

In 2010, the Basel Risk Management Committee issued updated guidance on managing operational risk that further highlights the importance of enterprise risk management. Meanwhile, shareholders are aware of operational risks that can add up to billions of dollars every year and include frequent, low-level losses and also infrequent but catastrophic losses that have actually wiped out businesses. Regulators and shareholders have already signaled that they will hold the Board and executives accountable for managing operational risk.

The ERM—Integrated Framework, developed by COSO, is an effective tool for senior management and the Board to set goals and strategies; identify, evaluate, and manage risk areas; select and implement controls to mitigate or address the risk areas; and ensure that the company ultimately achieves its objectives and goals.

The ERM—Integrated Framework model is illustrated in Exhibit 6.1. The top four columns are the objectives management typically establishes in order to achieve the company goals. The right side of the model shows the four units that a company may be composed of. The model also shows the eight specific interrelated components of the ERM. These eight risk and control components apply to each of the four management objectives, as well as to the company units on the right side of the model.

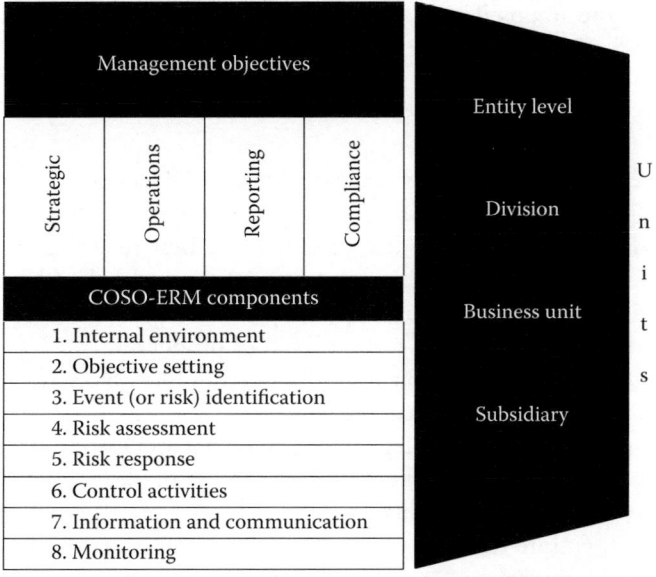

Exhibit 6.1   COSO-ERM model.

The ERM—Integrated Framework takes a risk-based rather than a controls-based approach when evaluating internal controls, as is the case with the widely adopted, SOX-required COSO's Internal Control—Integrated Framework ("IC framework"). The ERM risk-based approach resulted from the addition of four elements to the previous IC framework: Objective Setting, Event (or Risk) Identification, Risk Assessment, and Risk Response. These additional elements to the IC framework make the ERM—Integrated Framework a more comprehensive one to assist companies with not only setting goals and evaluating risks, but also identifying and implementing procedures to control risk (i.e., accept, avoid, diversify, share, or transfer risk). The eight components of the framework are described below.

Internal Environment

The internal environment of a company is everything. It refers to its culture, its behaviors, its actions, its policies, its procedures, its tone, its heart. The internal environment is crucial in setting the company's goals, strategies, and objectives; establishing procedures to assess or mitigate risk business areas; and identifying and implementing adequate controls to respond to those risk areas. A strong internal environment often prevents a company from breakdowns in risk management and control. The internal environment is the base and infrastructure for all other seven ERM components, and consists of:

- Management's beliefs, attitudes, operating style, and risk appetite.
- Management's commitment to integrity, ethical values, and competence.
- Management's oversight over the company's internal control and structure.
- Methods of assigning authority and responsibility through the establishment of formal policies and procedures that are consistent with goals and objectives.
- Human resource policies, procedures, and practices overseeing existing working conditions, job incentives, promotion, and career advancement.
- Procedures in place to comply with industry external requirements, as well as regulatory laws, such as those imposed by banks, utilities, insurance companies, the SEC and the PCAOB, among others.

Objective Setting

Objectives refer to the goals the company wants to achieve. Objectives are established at various levels within a company. That is, companies may set objectives at the top/management level, say to guide their direction or strategy (e.g., become the best seller in the market, acquire a separate business, merge with a competitor, etc.); or at lower levels, like improving existing operations (e.g., hiring quality personnel, improving current processes, implementing controls to address additional risks, maintaining certain levels of production, etc.). Companies may also set goals for reporting and compliance purposes. Reporting-like objectives are set, for instance, to ensure reliability, completeness, and accuracy of reports (e.g., financial statements, etc.). These objectives are achieved via adequately safeguarding financial application systems, as well as performing timely and thorough management reviews, for example. Compliance objectives, on the other hand, ensure all applicable industry-specific, local, state, and federal laws are properly followed and observed. Failure to comply with these can result in serious consequences, leaving the company vulnerable to lawsuits, on-demand audits, and sanctions that can ultimate lead to dissolution.

Event (or Risk) Identification

Events impact companies internally or externally. For instance, events could occur outside the company (e.g., natural disasters, enactment of new laws and regulations, etc.) that can significantly affect its goals, objectives, and/or strategy. Identification of these events or risks can result from responding to management questions, such as: (1) What could go wrong? (2) How can it go wrong? (3) What is the potential harm? and (4) What can be done about it? An example would be an office desk manufacturer that relies on sourcing the wood necessary to build the desks from specific regions in the Caribbean. The manufacturer's organizational objective is to keep up with production demand levels. So, here are the management questions from above with hypothetical responses to identify internal or external events:

1. *What could go wrong?* Shipment of wood may fail or may not be received on time resulting in not having enough supplied wood to meet customer demands and/or required production levels.
2. *How can it go wrong?* Weather conditions (e.g., hurricanes, flooding, etc.) may affect safe conditions to cut trees and prepare the necessary wood; or prevent timely shipment of the wood to the manufacturing site.
3. *What is the potential harm?* The lack of or limited supply may prompt the manufacturer higher costs which could translate into higher costs and prices to customers.
4. *What can be done about it?* Solutions may include identifying at least one or two additional suppliers (outside of the Caribbean), and/or having higher amounts of wood inventory on hand. These will help in preventing or mitigating the issues just identified, and ensure that minimum production levels are kept consistent with organizational objectives.

The key is to identify potential events or risks that can significantly impact business operations and revenues. Risks are classified as either inherent (they exist before plans are made to control them) or residual (risks left over after being controlled), and can be identified through:

■ Audits or inspections by managers, workers, or independent parties of the company's operational sites or practices
■ Operations or process flowcharts of the company's operations
■ Risk analysis questionnaires where information can be captured about the company's operations and ongoing activities
■ Financial statement analyses to depict trends in revenue and cost areas, identifying asset exposure analysis
■ Insurance policy checklists

Risk Assessment

In view of the increased reliance on IT and automated systems, special emphasis must be placed in the review and analysis of risk in these areas. IT facilities and hardware are often included in the company's overall plant and property review; however, automated systems require a separate analysis, especially when these systems are the sole source of critical information to the company as in today's e-business environments. There are many risks that affect today's IT environment. Companies face loss from traditional events, such as natural disasters, accidents, vandalism, and theft, and also from similar events in electronic form. These can result from computer viruses, theft of information or data, electronic sabotage, and so on. Some examples of resources to assist in the identification and evaluation of these IT-related risks include:

- *NIST.gov.* The NIST has been a leader in providing tools and techniques to support IT. It has a number of support tools that can be used by private small-to-large organizations for risk assessment purposes.
- *GAO.gov.* The U.S. Government Accountability Office (GAO) has provided a number of audit, control, and security resources as well as identification of best practices in managing and reviewing IT risk in many areas.
- *Expected loss approach.* A method developed by IBM that assesses the probable loss and the frequency of occurrence for all unacceptable events for each automated system or data file. Unacceptable events are categorized as either: accidental or deliberate disclosure; accidental or deliberate modification; or accidental or deliberate destruction.
- *Scoring approach.* Identifies and weighs various characteristics of IT systems. The approach uses the final score to compare and rank their importance.

Once identified, risks are assessed, meaning that the probability of their potential losses is quantified and ranked. Risks are assessed from two perspectives: Likelihood and Impact. Likelihood refers to the probability that the event will occur. Impact, on the other hand, is the estimated potential loss should such particular event occurs. Risks are categorized as follows:

- *Critical*—exposures would result in bankruptcy, for instance.
- *Important*—possible losses would not lead to bankruptcy, but require the company to take out loans to continue operations.
- *Unimportant*—exposures that could be accommodated by existing assets or current income without imposing undue financial strain.

Assigning identified risks to one of the above categories gives them a level of significance and helps determine the proper means for treating such risks. Assessment of risks is discussed in more detail in a later section.

Risk Response

After assessing risks, the next step is to put an action plan together and determine the applicable technique(s) to respond to the identified risks. Typically, the risk response process starts with companies evaluating their inherent risks, then selecting the appropriate response technique, and finally assessing the residual risk. Management can react or respond to identified risks in one of the following four ways: Avoid, Prevent, Reduce, or Transfer.

- *Avoid* or completely eliminate the risk. For example, a new feature included within the next application software release is estimated to downgrade application performance by slowing down some critical processing. To avoid the risk, the software feature is eliminated from the next release.
- *Prevent* a risk through implementing IT controls, such as (1) performing validity checks upon inputting data; (2) cleaning disk drives and properly storing magnetic and optical media to reduce the risk of hardware and software failures; (3) configuring logical setting security controls (i.e., passwords) in the application system.

Exhibit 6.2 Risk Response Techniques

Sample Questions before Choosing a Technique	Risk Response Technique
• Is the risk impossible to avoid? • Is the risk impractical to avoid? • Is the risk too expensive to avoid? • Is the risk too time consuming to avoid?	Avoidance
• Are there controls in place to prevent the risk from occurring? • If so, are these controls cost effective? (i.e., the benefits of implementing the controls outweigh their costs?)	Prevention
• Are there effective controls in place that reduce the risk? (i.e., the benefits of implementing the controls outweigh their costs?) • Will other risks be reduced as well from the implemented controls?	Reduction
• Can the risk be transferred to a third party through purchasing insurance? • Can the risk be partially reduced and partially transferred?	Transfer

▪ *Reduce* the risk through taking mitigation actions, such as having controls detecting errors after data are complete. Examples of these include implementing user access reviews, conducting reconciliations, and performing data transmission controls, among others.

▪ *Transfer* all or part of the risk to a third party. Common methods of risk transfer include acquiring insurance or outsourcing (subcontracting) services. As an example, a company that needs to update its financial application system may choose to outsource or subcontract such a project (along with all of its risks) to an outsider.

A last option would involve management assuming or retaining the risk. That is, after assessing the risk, management feels comfortable knowing about the risk and decides to go forward with it. An example here would be an investor assuming the risk that a company he/she is buying ownership (stock) from will likely go bankrupt. More than one technique can be applied to a given risk (e.g., risk can be reduced, then transferred, etc.). The IT risk management objectives should be used as a guide in choosing a technique. Exhibit 6.2 shows common key questions IT and management personnel ask before selecting from the four response techniques mentioned above.

Once the appropriate technique has been chosen, it must be implemented. The techniques implemented must be evaluated and reviewed on a frequent basis. This is important because variables that went in the selection of a previous technique may change. Techniques that were appropriate last year may not be so this year, and mistakes may occur. The application of the wrong technique(s) must be detected early and corrected.

Control Activities

COBIT defines control activities as the "policies, procedures, practices, and organization structures designed to provide reasonable assurance that business objectives will be achieved and that undesired events will be prevented or detected and corrected." In other words, control activities (or controls) are procedures management implement to safeguard assets, keep accurate and complete information, as well as achieve established business goals and objectives. Implementing controls is an effective way to: (1) reduce identified risks to acceptable levels; (2) comply with company policies, procedures, laws, and regulations; and (3) enhance efficiency of existing operations. Once in place, controls must be monitored for effective implementation. They should also be assessed to determine whether they do operate effectively and as expected when originally designed.

There are three types of controls: Preventive, Detective, and Corrective. Management should identify and implement controls from the three types above in order to protect the company from undesired events. Preventive controls, for instance, deter problems from occurring and are usually superior than detective controls. Examples of preventive controls include hiring qualified personnel, segregating employee duties, and controlling physical access. The second type of controls, detective controls, are intended to discover problems that cannot be prevented. Examples of a detective control include performing reconciliations of bank accounts, trial balances, etc. Detective controls are designed to trigger when preventive controls fail. Corrective controls, the third type of controls, are designed to identify, correct, and recover from the problems identified. Similar to detective controls, corrective controls "react to what just happened." Examples include maintaining backup copies of files and correcting data entry errors. An effective internal control system should implement all three types of controls. Areas where controls can be implemented include, among others, duties segregation; approval and authorization of transactions; change management; assets, records, and data protection; and systems performance checks and monitoring.

Information and Communication

To describe the seventh component of the ERM—Integrated Framework model, information and communication, it is crucial to explain what information is and what communication refers to. Companies need information to carry out their internal control responsibilities and ultimately to support the achievement of their business goals and objectives. Information is data organized and processed to provide meaning and, thus, improve decision making. Management needs that such information, generated from either internal or external sources, be useful (i.e., quality information) in order to make effective and efficient business decisions, as well as to adequately support the functioning of its internal control system. Information is useful when it is:

1. *Relevant*: information is pertinent and applicable to make a decision (e.g., the decision to extend customer credit would need relevant information on customer balance from an Accounts Receivable aging report, etc.).
2. *Reliable*: information is free from bias, dependable, trusted.
3. *Complete*: information does not omit important aspects of events or activities.
4. *Timely*: information needs to be provided in time to make the decision.
5. *Understandable*: information must be presented in a meaningful manner.
6. *Verifiable*: two or more independent people can produce the same conclusion.
7. *Accessible*: information is available when needed.

Communication, on the other hand, refers to the process of providing, sharing, and obtaining necessary information in a continuing and frequent basis. Communication of information could occur internally within the company (e.g., message from the CEO or CIO to all company employees, etc.) or externally (e.g., information received from regulators, information submitted for audit purposes, etc.).

An information and communication system, such as an accounting information system (AIS), should be implemented to allow for capturing and exchanging the information needed, as well as conducting, managing, and controlling the company's operations. AISs should gather, record, process, store, summarize, and communicate information about an organization. This includes understanding how transactions are initiated, data are captured, files are accessed and updated, data are processed, and information is reported. AISs also include understanding accounting records and procedures, supporting documents, and financial statements.

Monitoring

Monitoring activities, either on a continuing or separate basis, must occur to ensure that the information and communication system (i.e., AIS) is implemented effectively and, most importantly, operates as designed. Continuous monitoring assessments that have been incorporated into existing business processes at various levels, for instance, provide timely and relevant information supporting whether the AIS is or not working as expected. Monitoring assessments that are performed separately vary in scope and frequency, and are conducted depending on how effective they are, the results from risk assessments, and specific management goals and objectives.

Examples of monitoring activities may include having internal audits or internal control evaluations; assessing for effective supervision; monitoring against established and approved budgets; tracking purchased software and mobile devices; conducting periodic external, internal, and/or network security audits; bringing on board a **Chief Information Security Officer** and **forensic specialists**; installing **fraud detection software**; and implementing a **fraud hotline**, among others.

Deficiencies, if any, that result from monitoring activities and evaluations against criteria established by regulators and standard-setting bodies, as well as policies and procedures established by management, are to be documented, evaluated, and communicated. Deficiencies are communicated to management and to the Board as appropriate.

Risk Assessment

Risk assessment forms the first step in the risk management methodology. Risk assessments, based on NIST, are used by organizations to determine the extent of potential threats and evaluate the risks associated with IT systems. The results of the above assist management in identifying and implementing appropriate IT controls for reducing and/or eliminating those threats and risks. Risk assessments provide a framework for allocating resources to achieve maximum benefits. Given the significant number of IT areas, but limited amount of resources, it is important to focus on the right areas. Risk assessments are both a tool and a technique that can be used to self-evaluate the level of risk of a given process or function, such as IT. They represent a way of applying objective measurement to a process that is really subjective.

A **chief risk officer (CRO)**, in collaboration with the Board of Directors (Board), should determine risk limits the organization is willing to take on. These risk limits should not be static but should be subject to change—a working document. These risk limits should be published and available to the business units, as each business manager will be held accountable for assessing the line of business' risks, creating a risk action plan, and determining if their risks fall within or outside of the established tolerances.

As part of the strategic planning process each year, business managers should be required to complete a risk assessment of his or her area. Included in that is a risk assessment of the business risks of each application or system that the line of business owns. COBIT or similar standards like NIST, the International Organization for Standardization/International Electro Technical

Commission (ISO/IEC), and others should be agreed upon as the guideline to be used. This will place all IT risk assessments on similar terms and make them somewhat standardized as to the types of risks identified.

Risk assessments should be completed by the line of business with assistance from the IT risk management coordinator or internal audit. The IT risk management coordinator can give insight and information to the line of business regarding the specific risks faced by the application or system. The business manager would be able to assess these in light of the overall risk facing the line of business. The IT department should perform the risk assessments of enterprise-wide applications and systems such as the network or the enterprise-wide e-mail software. The IT department, headed by the **chief technology officer (CTO)**, would be evaluating, managing, and accepting the risks associated with this type of enterprise-wide technology.

In some ways, the CRO and the CTO's staff will serve as facilitators of this process. They will determine if the risk assessment is not adequate or lacks information. They will create tools to assist the line of business in identifying risks and possible controls, deciding which controls to implement, and monitoring and measuring those controls for effectiveness.

After the risk assessment is filled out and all risks the particular line of business is facing are fully identified, the business manager, with the assistance of the CRO's staff, should review the risks and associated controls. These should be compared to applicable regulatory requirements and Board-approved limits to risk taking. If any risks fall outside of either regulatory or Board limits, the CRO and the business management work together to find solutions to lower the risks to acceptable levels. This could include implementing more controls—for example, requiring two management signatures before processing a master file change. It could include purchasing insurance to transfer some of the risk to a third party, such as hazard insurance for the data center if a natural disaster were to strike. Or, it could mean deciding not to offer a particular service, such as opening accounts online, due to an unacceptably high risk of fraud. All of these possible solutions result in the risk being lowered, and the goal is to reduce the risk to a level acceptable to both the organization's regulatory agencies and its Board.

Risk assessments should be reviewed and reconsidered each year. This review should include adding any new risks to the business unit due to new products or services, or perhaps a new technology that has just been implemented. The review should also assess whether the ratings for each risk were warranted or may need to be adjusted. The organization may decide to require review of the risk assessment more frequently in the beginning of the implementation until satisfied that all potential risks have been identified and included in the risk management process. The CRO should also implement a scorecard and metrics, such as a maturity model, against which the line of business risk management can be measured. Lines of business with good risk management practices should be rewarded.

Internal audit will independently evaluate risk assessments each time they audit a function, area, or application. If the audit feels the risk assessments are not adequate or that all the potential risks have not been identified or adequately controlled, it would be an issue for both the business and the CRO. Periodic audits by external auditors and regulatory bodies are also a necessary part of IT risk management program.

Available Guidance

There are several professional standards that provide guidance to auditors and managers involved in the risk assessment process. These standards come from widely recognized organizations like

COBIT and the ISO/IEC. Other standards for risk assessments are available from NIST, GAO, American Institute of Certified Public Accountants, ISACA, Institute of Internal Auditors, and the Committee of Sponsoring Organizations of the Treadway Commission.

COBIT

As stated previously, COBIT is a well-known IT governance framework that helps organizations in the areas of regulatory compliance and alignment of IT strategy and organizational goals. COBIT is also crucial to organizations in the area of risk management. Specifically, COBIT's international set of generally accepted IT practices or control objectives help employees, managers, executives, and auditors in: understanding IT systems, discharging fiduciary responsibilities, managing and assessing IT risks, and deciding adequate levels of security and controls.

COBIT helps organizations create optimal value from IT by maintaining a balance between realizing benefits and optimizing risk levels and resource use. The framework is valuable for all size types organizations, including commercial, not-for-profit, or in the public sector. The comprehensive framework provides a set of control objectives that not only helps IT management and governance professionals manage their IT operations, but also IT auditors in their quests for examining those objectives. Selection of COBIT may be appropriate when the goal of the organization is not only to understand and align IT and business objectives, but also to address the areas of regulatory compliance and risk management.

ISO/IEC

The ISO/IEC 27000 family of standards includes techniques that help organizations secure their information assets. The ISO/IEC 27005:2011 Information Technology—Security Techniques—Information Security Risk Management, for example, provides guidelines for the satisfactory management of information security risks. It supports the general concepts specified in ISO/IEC 27001, and applies to organizations within most types of industries (e.g., commercial/private, government, non-for-profit, etc.). The ISO/IEC 27005:2011 as well as the rest of the family of ISO/IEC standards all assists organizations manage the security of assets, including, but not limited to, financial information, intellectual property, employee details, or information entrusted by third parties.

The ISO/IEC 27005:2011 standard does not specify nor recommend any specific risk management method, but does suggest a process consisting of a structured sequence of continuous activities, which include:

■ Establishing the risk management context, including the scope, compliance objectives, approaches/methods to be used, and relevant policies and criteria (e.g., organization's risk tolerance, risk appetite, etc.).

■ Assessing quantitatively or qualitatively relevant information risks considering information assets, threats, vulnerabilities, and existing controls. This assessment will be helpful in determining the probability of incidents or incident scenarios, and the predicted business consequences if they were to occur (i.e., risk level).

■ Determining, based on the risk level, how will management react or respond to identified risks (i.e., whether management will completely avoid, reduce, transfer to a third party, or finally accept the risk).

■ Maintaining stakeholders aware and informed throughout the information security risk management process.
■ Monitoring and reviewing risks, risk treatments, risk objectives, obligations, and criteria continuously.
■ Identifying and responding appropriately to significant changes.

National Institute of Standards and Technology (NIST)

A major focus of NIST activities in IT is providing measurement criteria to support the development of pivotal, forward-looking technology. NIST standards and guidelines are issued as **Federal Information Processing Standards (FIPS)** for government-wide use. NIST develops FIPS when there are compelling federal government requirements for IT standards related to security and interoperability, and there are no acceptable industry standards or solutions.

One of the first of several federal standards issued by NIST in 1974 was FIPS 31, "Guidelines for Automatic Data Processing Physical Security and Risk Management." This standard provided the initial guidance to federal organizations in developing physical security and risk management programs for information system (IS) facilities. Then, in March 2006, NIST issued FIPS 200 "Minimum Security Requirements for Federal Information and Information Systems," where federal agencies were responsible for including within their information "policies and procedures that ensure compliance with minimally acceptable system configuration requirements, as determined by the agency."

Managing system configurations is also a minimum security requirement identified in FIPS 200, and NIST SP 800-53, "Security and Privacy Controls for Federal Information Systems and Organizations," came to define security and privacy controls that supported this requirement. In August 2011, NIST issued SP 800-128, "Guide for Security-Focused Configuration Management of IS." Configuration management concepts and principles described in this special publication provided supporting information for NIST SP 800-53, and complied with the Risk Management Framework (RMF) that is discussed in NIST SP 800-37, "Guide for Applying the Risk Management Framework to Federal Information Systems: A Security Life Cycle Approach," as amended. More specific guidelines on the implementation of the monitor step of the RMF are provided in Draft NIST SP 800-137, "Information Security Continuous Monitoring for Federal IS and Organizations." The purpose of the NIST SP 800-137 in the RMF is to continuously monitor the effectiveness of all security controls selected, implemented, and authorized for protecting organizational information and IS, which includes the configuration management security controls identified in SP 800-53. These documents are a very good starting point for understanding the basis and many approaches one can use in assessing risk in IT today.

When assessing risks related to IT, particular attention should be provided to NIST SP 800-30 guide, "Guide for Conducting Risk Assessments."* The NIST SP 800-30 guide provides a common foundation for organizations' personnel with or without experience, who either use or support the risk management process for their IT systems. Organizations' personnel include: senior management, IT security managers, technical support personnel, IT consultants, and IT auditors, among others. The NIST SP 800-30's risk assessment standard can be implemented in single or multiple interrelated systems, from small-to-large organizations.

---

* http://nvlpubs.nist.gov/nistpubs/Legacy/SP/nistspecialpublication800-30r1.pdf.

NIST guidelines, including the SP 800-30, have assisted federal agencies and organizations in significantly improving their overall IT security quality by:

- providing a standard framework for managing and assessing organizations' IS risks, while supporting organizational missions and business functions;
- allowing for making risk-based determinations, while ensuring cost-effective implementations;
- describing a more flexible and dynamic approach that can be used for monitoring the information security status of organizations' IS;
- supporting a bottom-up approach in regards to information security, centering on individual IS that support the organization; and
- promoting a top-down approach related to information security, focusing on specific IT-related issues from a corporate perspective.

Organizations within the private sector (small, mid-sized, and large) are significantly using NIST guidelines to promote secured critical business functions, including customers' confidence in organizations' abilities to protect their personal and sensitive information. Furthermore, the flexibility of implementing NIST guidelines provides organization's appropriate tools to demonstrate compliance with regulations.

The NIST SP 800-30 standard is used frequently when performing risk assessments because of its flexibility and ease of use in: (1) identifying potential risks associated with IS, as well as (2) determining the probability of occurrence, impact, and additional safeguards for mitigation. The guideline has proved to conduct accurate and thorough assessments of risks and vulnerabilities in regards to information's confidentiality, integrity, and availability. Appendix 5 shows an example of an IT risk assessment performed for an organization using NIST SP 800-30.

Government Accountability Office (GAO)

The GAO is a nonpartisan agency within the legislative branch of the government. The GAO conducts audits, surveys, investigations, and evaluations of federal programs. This may include audits of federal agencies and state, county, and city governments, and extend to private industry, where federal funds are spent. Often, the GAO's work is done at the request of congressional committees or members, or to fulfill specifically mandated or basic legislative requirements. The GAO's findings and recommendations are published as reports to congressional members or delivered as testimony to congressional committees. The GAO has issued numerous reports on computer security, IT vulnerabilities, and risk assessments.

The U.S. federal government has invested an extraordinary amount of resources in examining risk dating back to the early 1960s. Examples of these include Government Accounting Standards (GAS) and GAO's Information Management and Technology (IMTEC) reports. GAS 4.29, "Safeguarding Controls," for instance, is used to help auditors recognize risk factors involving computer processing. IMTEC 8.1.4, "Information Technology: An Audit Guide for Assessing Acquisition Risk," is used in planning and conducting risk assessments of computer hardware and software, telecommunications, and system development acquisitions.

American Institute of Certified Public Accountants (AICPA)

Statements on Auditing Standards (SAS) are issued by the Auditing Standards Board of the AICPA and are recognized as interpretations of the 10 generally accepted auditing standards. As mentioned in earlier chapters, the AICPA has played a major role in the issuance of guidance to the accounting and control profession. An example in applying audit risk and materiality concepts comes with the issuance of SAS 47, "Audit Risk and Materiality in Conducting an Audit," which relates to risk assessment. In SAS 47, control risk is defined as the possibility of a misstatement occurring in an account balance or class of transactions that (1) could be material when aggregated with misstatements in other balances or classes and (2) will not be prevented or detected on a timely basis by the system of internal control.

SAS 65, "The Auditor's Consideration of the Internal Audit Function in an Audit of Financial Statements," requires that, in all engagements, the auditor develops some understanding of the internal audit function (IT audit, if available) and determine whether that function is relevant to the assessment of control risk. Thus, if there is an internal audit function, it must be evaluated. The evaluation is not optional. In 1996, the AICPA issued SAS 80, which amended SAS 31, "Evidential Matter." SAS 80 was directly aimed at improving auditing in the IT environment. This SAS made a profound impact on the auditing profession. An excerpt from SAS 80 states: "In entities where significant information is transmitted, processed, maintained, or accessed electronically, the auditor may determine that it is not practical or possible to reduce detection risk to an acceptable level by performing only substantive tests for one or more financial statement assertions. For example, the potential for improper initiation or alteration of information to occur and not be detected may be greater if information is produced, maintained, or accessed only in electronic form. In such circumstances, the auditor should perform tests of controls to gather evidential matter to use in assessing control risk, or consider the effect on his or her report."

SAS 94, "The Effect of Information Technology on the Auditor's Consideration of Internal Control in a Financial Statement Audit," was adopted in 2001 and provided guidance to auditors about the effect of IT on internal control and on the auditor's understanding of internal control and assessment of control risk. SAS 109, "Understanding the Entity and Its Environment and Assessing the Risks of Material Misstatement," was adopted in 2006 and also emphasized the auditor's understanding of the entity to validate and verify how IT contributes to the risk of material misstatement, and whether controls exist to prevent or detect errors or fraud.

ISACA

ISACA (formerly known as the Information Systems Audit and Control Association) is a worldwide not-for-profit association of more than 28,000 practitioners dedicated to IT audit, control, and security in over 100 countries. The Information Systems Audit and Control Foundation is an associated not-for-profit foundation committed to expanding the knowledge base of the profession through a commitment to research. The ISACA standards board has updated and issued several Information System Audit Guidelines that have been recognized as system auditing standards.

The ISACA's guideline titled "Use of Risk Assessment in Audit Planning" specifies the level of audit work required to meet a specific audit objective; it is a subjective decision made by the IT auditor. The risk of reaching an incorrect conclusion based on the audit findings (audit risk) is one aspect of this decision. The other is the risk of errors occurring in the area being audited (error risk). Recommended practices for risk assessment in carrying out financial audits are well documented in auditing standards for financial auditors, but guidance is required on how to apply such techniques to IT audits.

Management also bases its decisions on how much control is appropriate upon assessment of the level of risk exposure it is prepared to accept. For example, the inability to process computer applications for a period of time is an exposure that could result from unexpected and undesirable events (e.g., data center fire, flooding, etc.). Exposures can be reduced by the implementation of appropriately designed controls. These controls are ordinarily based on the estimation of the occurrence of adverse events and are intended to decrease such probability. For example, a fire alarm does not prevent fires but is intended to reduce the extent of fire damage.

The ISACA guideline provides guidance in applying IT auditing standards. The IT auditor should consider such guidance in determining how to achieve implementation of the preceding standards, use professional judgment in its application, and be prepared to justify any departure.

Institute of Internal Auditors (IIA)

Established in 1941, the IIA serves more than 85,000 members in internal auditing, governance and internal control, IT audits education, and security in more than 120 countries.

The IIA has in place Performance Standard 2110 titled "Risk Management," which specifies that the internal audit activity should assist the organization by identifying and evaluating significant exposures to risk and contributing to the improvement of risk management and control systems. It provides additional guidance in the form of Implementation Standard 2110.A1 (Assurance Engagements) with which the internal audit activity should monitor and evaluate the effectiveness of the organization's risk management system. Implementation Standard 2110.A2 (Assurance Engagements) stipulates that the internal audit activity should evaluate risk exposures relating to the organization's governance, operations, and IS regarding:

- Reliability and integrity of financial and operational information
- Effectiveness and efficiency of operations
- Safeguarding of assets
- Compliance with laws, regulations, and contracts

The last performance standard addresses consulting engagements in Implementation Standard 2110.C1 (Consulting Engagements). The IIA recommends that during consulting engagements, internal auditors should address risk consistent with the engagement's objectives and be alert to the existence of other significant risks.

The IIA has also developed a series of publications that aid in the assessment of internal controls over financial reporting, particularly IT controls. These are referred to as the Guides to the Assessment of IT Risk, or GAIT. The GAIT for "IT General Controls Deficiency Assessment" is a top-down and risk-based approach to assessing IT general controls. Such GAIT provides an approach for evaluating IT general controls deficiencies identified during a financial audit or Sarbanes–Oxley control assessment. The GAIT for "Business and IT Risk," or GAIT-R, is a risk-based audit methodology to align IT audits to business risks.

Committee of Sponsoring Organizations of the Treadway Commission (COSO)

The COSO was formed in 1985 as an independent, voluntary, private-sector organization dedicated to improving the quality of financial reporting through business ethics, effective internal controls, and corporate governance. COSO consists of representatives from industry, public accounting agencies, investment firms, and the New York Stock Exchange. The first chairman of COSO was James C. Treadway, Jr., executive vice president and general counsel for Paine Webber Inc. at the time, and a former commissioner of the U.S. Securities and Exchange Commission; hence the name Treadway Commission.

The COSO ERM—Integrated Framework, discussed previously, was developed by the global accounting firm, PriceWaterhouseCoopers, and issued in September 2004. The ERM—Integrated Framework is an effective tool for senior management and the Board to set goals and strategies; identify, evaluate, and manage risk areas; select and implement controls to mitigate or address the risk areas; and ensure that the company ultimately achieves its objectives and goals. The ERM—Integrated Framework model is illustrated in Exhibit 6.1.

Insurance as Part of IT Risk Assessments

Risk assessments related to IT operations also include insurance. A clear understanding of insurance and risk management is necessary to review the adequacy of an organization's IT insurance. IT management and data security administrators must be aware of the relationship between risk and insurance to understand the reasons behind insurance choices and the types of insurance that are most applicable to the IT environment. This provides an overview of the reasons for and the methods of risk analysis, insurance alternatives, and what to look for in IT insurance coverage. The need for this review becomes apparent because of computer viruses, denial-of-service attacks, and so on, which can cost lost opportunities. Businesses must have a way to protect themselves and recover their losses.

Insurance distributes losses so that a devastating loss to an individual or business is spread equitably among a group of insured members. Insurance neither prevents loss nor reduces its cost; it merely reduces the risk. Risk is the possibility of an adverse deviation from a desired outcome (e.g., the possibility of dying before reaching age 72, an interruption in business operations, an e-commerce site overloaded with invalid transactions, IT business **spamming**, etc.). When not managed, risks may be assumed that should be insured and vice versa. Insurance policies often provide overlapping coverage in some areas and none in other critical ones.

IT Risks Typically Insured

In the IT environment, there are special risks that are commonly handled by insurance, including:

- Damage to computer equipment
- Cost of storage media
- Cost of acquiring the data stored on the media
- Damage to outsiders
- Business effects of the loss of computer functions

The types of insurance policies that cover these risks include property, liability, business interruption, and fidelity-bonding insurance. These policies, especially written for IT-related risks, should examine:

- Coverage of hardware and equipment (i.e., network, mass storage devices, terminals, printers, and central processing units).
- Coverage of the media and information stored thereon. For example, a disk drive that is destroyed can be replaced at the cost of a new drive. If the drive or mass storage device

contains important information, the value of the new replacement drive plus the value of the lost information must be recovered.

▪ Coverage of the replacement or reconstruction cost and the cost of doing business as usual (i.e., business interruption). This might involve renting time on equivalent equipment from a nearby company or outsourcing to a vendor, paying overtime wages for reconstruction, and detective work. In this area, logging of daily electronic business activity resulting in financial transactions is extremely important to identify business interruption or loss due to spamming or information theft.

▪ Coverage of items such as damage to media from magnets, damage from power failure (blackout) or power cut (brownout), and damage from software failure.

Cyber Insurance

Attempts to damage or destroy computer systems (also known as cyberattacks) are common today in organizations and can result in significant losses. For instance, in 2014, the Center for Strategic and International Studies estimated annual costs from cybercrime to range between $375 billion and $575 billion for mid-to-large organizations. Another study performed by Symantec in 2016 (and documented as part of its Internet Security Threat Report) indicated that 43% of all 2016 attacks targeted small businesses (i.e., organizations with less than 250 employees). Organizations must decide whether cyber insurance is now a viable option to mitigate such losses and their resulting excessive costs.

Typically, cyber insurance is either excluded from traditional commercial general liability policies, or not specifically defined in traditional insurance products. A cyber insurance policy (or cyber risk insurance) refers to an insurance product designed to protect organizations and individuals from risks relating to IT infrastructure and activities (e.g., cyber-related security breaches, Internet-based risks, etc.). Cyber insurance began catching on in 2005, with the total value of premiums forecasted to reach $7.5 billion by 2020. According to PriceWaterhouseCoopers, about one-third of U.S. companies currently purchase some type of cyber insurance.

This specific type of insurance covers expenses related to first-party losses or third-party claims. Coverage typically includes:

▪ losses from data destruction, extortion, theft, hacking, and denial of service attacks
▪ losses to others caused by errors and omissions, failure to safeguard data, or defamation

Before cyber insurance, most organizations did not necessarily report the full impact of their information security breaches in order to avoid negative publicity and damage the trust of their customers. Now they must strongly consider adding cyber insurance to their budgets, specially, if they store and maintain customer information, collect online payment information, or simply utilize the cloud to meet business goals and objectives. Because cyber risks change so frequently, adequate coverage of such IT risks must be in place. However, what if IT risks cannot be insured?

Reduction and Retention of Risks

Risks that are not insurable can be managed in other ways: reduced or retained. Just because a risk is insurable does not mean that insurance is the only way to handle it. Risk reduction can be accomplished through loss prevention and control. If the possibility of loss can be prevented, the risk is eliminated; even reducing the chance of the loss from occurring is a significant improvement.

If the chance cannot be reduced, at least the severity of the loss can often be controlled. The reduction method is frequently used with insurance to lessen the premiums. Examples of questions leading to determine whether IT risks can be reduced include:

■ Is there a comprehensive, up-to-date disaster recovery plan or business continuity plan?
■ What efforts have been made to check that both plans are workable?
■ Are there off-site backups of the appropriate file?
■ Are the procedures and practices for controlling accidents adequate?
■ Have practical measures been taken to control the impact of a disaster?
■ Is physical security effective to protect property and equipment?
■ Is software security adequate to protect confidential or sensitive information?
■ Are there appropriate balancing and control checks made at key points in the processing?
■ Are there appropriate control checks on the operations?
■ Are there appropriate control checks during the development and modification of systems?
■ Are network firewalls tested weekly?
■ Have firewalls been certified on a semiannual basis?
■ Do contracts for purchases or leases have terms and conditions and remedies that adequately protect the company if there is a problem?
■ Have contracts been prepared by legal counsel who has expertise in IT and legal issues?
■ Are facilities, equipment, and networks maintained properly?

Uninsurable risks can also be retained depending on the organization's awareness of the risks. If retained, risks should be consistent with management objectives and risk analyses. The retention method, which is sometimes referred to as self-insurance, should be voluntary and meet the following criteria:

■ The risk should be spread physically so that there is a reasonably even distribution of exposure to loss over several locations.
■ A study should be made to determine the maximum exposure to loss.
■ Consideration should be given to the possibility of unfavorable loss experience and a decision reached as to whether this contingency should be covered by provision for self-insurance reserves.
■ A premium charge should be made against operations that are adequate to cover losses and any increase in reserves that appear advisable.

Many companies, however, retain risks without estimating future losses or reserving funds to pay for these losses. Companies must carefully manage and assess their risks of significant losses in order to protect their business interests.

Conclusion

Organizations have recognized the benefit of protecting themselves from all types of potential risk exposures. Protection comes from effective management and evaluation of identified risks. Risk management refers to the process of identifying and assessing risk, followed by implementing the necessary procedures or controls to reduce such risk to acceptable levels. An example of a risk management tool is the ERM—Integrated Framework. The framework, developed by COSO, is an effective tool for senior management and the Board to set goals and strategies; identify, evaluate, and manage risk areas; select and implement controls to mitigate or address the risk areas; and ensure that the company ultimately achieves its objectives and goals.

Risk assessment forms the first step in the risk management methodology. They are used by organizations to determine the extent of potential threats and the risks associated with particular systems. Risk assessments should be completed by the line of business with assistance from the IT risk management coordinator or internal audit.

There are several professional standards from well-known organizations that provide guidance to auditors and managers involved in risk assessments. Standards provide a consistent quality measurement if adopted, maintained, and supported by the organization. Standards from organizations, such as COBIT, ISO/IEC, NIST, GAO, AICPA, ISACA, IIA, and COSO, are examples that relate to assessing risk in IT operations.

Organizations must develop a sound risk management program to be able to determine the adequacy of their IT insurance coverage. Insurance distributes losses so that a devastating loss to an individual or business is spread equitably among a group of insured members. Some of the IT risks covered by insurance policies include damage to computer equipment, cost of storage media, cost of acquiring the data stored on the media, and damage to outsiders, among others. One type of insurance against constant attempts made to organizations to damage or destroy their computer systems (cyberattacks) is called cyber insurance. A cyber insurance policy protects organizations and individuals from risks relating to IT infrastructure and activities (e.g., cyber-related security breaches, Internet-based risks, etc.).

Another major step in developing an effective risk management program is learning the methods of risk retention and reduction. Risks that are not insurable are either reduced or retained. Risk reduction can be accomplished through loss prevention and control, and typically lessens insurance premiums. Uninsurable risks can also be retained depending on the organization's awareness of the risks. If retained, risks should be consistent with management objectives and risk analyses. The development of a comprehensive risk management program requires significant effort from all parties; however, once established, the benefits of managing risk become invaluable.

Review Questions

1. Define Enterprise Risk Management (ERM) according to COSO. What is the ERM—Integrated Framework?
2. List the eight components of the ERM—Integrated Framework. List the management objectives typically related to the framework.
3. How does NIST define risk management? How does risk management protect the organization's information from IT threats?
4. Define risk assessment.
5. NIST is one of the several professional standards that provide guidance to auditors and managers involved in the risk assessment process. How does NIST guidelines have assisted federal agencies and organizations in significantly improving their overall IT security quality?
6. List and describe examples of four resources for tools and techniques used in the identification and evaluation of IT-related risks.
7. Explain what control activities refer to and describe the types of controls available.
8. What effect does insurance have on risk?

9. Describe what insurance policies for IT-related risks should typically include or cover.
10. Discuss what cyber insurance is. Why do you think cyber insurance is frequently excluded from traditional commercial general liability policies, or not specifically defined in traditional insurance products?

Exercises

1. Explain why the internal environment component of the ERM—Integrated Framework is critical for organizations.
2. One of the components of the ERM—Integrated Framework is *"Event (or Risk) Identification,"* where incidents (i.e., events or risks) could occur in the business organization and significantly affect its goals, objectives, and/or strategy. These incidents can be identified through responding to the following four management questions:
   a. What could go wrong?
   b. How can it go wrong?
   c. What is the potential harm?
   d. What can be done about it?
   The chapter described an example of a common business scenario: An office desk manufacturer that relies on sourcing the required wood to build the desks from specific regions in the Caribbean.

   Task: Identify two additional potential and common scenarios that may take place in a business environment. Then, provide responses to the above four questions to look for incidents that can significantly impact operations and company revenues.
3. Summarize the professional standards mentioned in the chapter that provide guidance to auditors and managers when conducting risk assessments.
4. Your organization has recently developed criteria for a risk management program. One goal of the program is to determine the adequacy and effectiveness of the company IT insurance coverage. Describe how an effective risk management program can enable a more cost-effective use of IT insurance.

GROUP PRESENTATION—RISK ASSESSMENT

INSTRUCTIONS: Read and study Appendix 5—IT Risk Assessment Example Using NIST SP 800-30. Appendix 5 includes nine steps to perform a risk assessment.

TASK: Class will be divided in groups. Groups will present and explain Steps 1–9 from the risk assessment example. Your presentation will be evaluated individually and as a group, when appropriate.

Further Reading

1. Bodin, L., Gordon, L., and Loeb, M. (2008). Information security and risk management. *Commun. ACM*, 51(1), 64–68.
2. Cavusoglu, H., Mishra, B., and Raghunathan, S. (2004). A model for evaluating IT security investments. *Commun. ACM*, 47(1), 87–92.

3. Deloitte, L. L. P. (2014). *ACL for Auditors*. Unpublished internal document.
4. Deloitte, L. L. P. (2014). *IT Audit Planning Work Papers*. Unpublished internal document.
5. Ernst & Young. (2010). Integrated risk management practices. Unpublished PowerPoint slides.
6. Fenz, S. and Ekelhart, A. (2010). Verification, validation, and evaluation in information security risk management. *IEEE Secur. Privacy*, 9, 1–14.
7. Institute of Internal Auditors. https://na.theiia.org/standards-guidance/Pages/Standards-and-Guidance-IPPF.aspx
8. Information Systems Audit and Control Association (ISACA). (2008). Is the IT risk worth a control? Defining a cost-value proposition paradigm for managing IT risks, www.isaca.org/Journal/archives/Pages/default.aspx
9. IS Audit Basics. The process of auditing information systems, http://www.isaca.org/knowledge-center/itaf-is-assurance-audit-/pages/is-audit-basics.aspx (accessed July 2017).
10. ISO/IEC 27005:2011 Information Technology—Security Techniques—Information Security Risk Management. ISO 27001 Security, www.iso27001security.com/html/27005.html
11. ISO/IEC 27005:2011 Information Technology—Security Techniques—Information Security Risk Management, www.iso.org/standard/56742.html
12. Keblawi, F. and Sullivan, D. (2007). The case for flexible NIST security standards. *IEEE Comput. Soc.*, 40(6), 19–26.
13. Lindros, K. and Tittel, E. (2016). What is cyber insurance and why you need it. CIO, www.cio.com/article/3065655/cyber-attacks-espionage/what-is-cyber-insurance-and-why-you-need-it.html
14. Mayo, J. W. (2009). Risk management for IT projects. ISACA, www.isaca.org/Groups/Professional-English/risk-management/GroupDocuments/Effective_Project_Risk_Management.pdf
15. McCafferty, J. (2016). *Five Steps to Planning an Effective IT Audit Program*. MIS Training Institute, http://misti.com/internal-audit-insights/five-steps-to-planning-an-effective-it-audit-program
16. National Association of Financial Services Auditors. (2002). *Enterprise Risk Management*, John Wiley & Sons, Inc., Hoboken, NJ, pp. 12–13.
17. Otero, A. R. (2015). An information security control assessment methodology for organizations' financial information. *Int. J. Acc. Inform. Syst.*, 18(1), 26–45.
18. Otero, A. R. (2015). Impact of IT auditors' involvement in financial audits. *Int. J. Res. Bus. Technol.*, 6(3), 841–849.
19. Otero, A. R., Tejay, G., Otero, L. D., and Ruiz, A. (2012). A fuzzy logic-based information security control assessment for organizations, IEEE Conference on Open Systems, Kuala Lumpur, Malaysia.
20. Otero, A. R., Otero, C. E., and Qureshi, A. (2010). A multi-criteria evaluation of information security controls using Boolean features. *Int. J. Netw. Secur. Appl.*, 2(4), 1–11.
21. Professional Risk Managers' International Association (PRMIA) (2008). *Enterprise Risk Management (ERM): A Status Check on Global Best Practices*, PRMIA, Northfield, MN.
22. Psica, A. (2007). Risk watch—Destination ahead. *Internal Auditor*, pp. 77–80.
23. Richardson, V. J., Chang, C. J., and Smith, R. (2014). *Accounting Information Systems*, McGraw Hill, New York.
24. Romney, M. B. and Steinbart, P. J. (2015). *Accounting Information Systems*, 13th Edition. Pearson Education, Upper Saddle River, NJ.
25. Ross, R. (2007). Managing enterprise security risk with NIST standards. *IEEE Comput. Soc.*, 40(8), 88–91. doi: 10.1109/MC.2007.284.
26. Senft, S., Gallegos, F., and Davis, A. (2012). *Information Technology Control and Audit*, CRC Press/Taylor & Francis, Boca Raton, FL.
27. Singleton, T. (2007). What every IT auditor should know about the new risk suite standards, Information Systems Audit and Control Association. *Inf. Syst. Control J.*, 5, 17–20.
28. Spacey, J. (2016). *Five Types of Risk Treatment*. Simplicable, http://simplicable.com/new/risk-treatment
29. United States General Accounting Office. (1999). *Information Security Risk Assessment Practices of Leading Organizations*, U.S. GAO, Washington, DC, www.gao.gov/special.pubs/ai00033.pdf (accessed January 2017).

30. Unknown. (2008). *GAIT for IT General Control Deficiency Assessment*, The Institute of Internal Auditors, Altamonte Springs, FL.
31. Unknown. (2008). *GAIT for Business and IT Risk*, The Institute of Internal Auditors, Altamonte Springs, FL.
32. U.S. General Accounting Office, *Assessing the Reliability of Computer Processed Data Reliability*, https://digital.library.unt.edu/ark:/67531/metadc302511/ (accessed November 2016).

Chapter 7

Project Management

LEARNING OBJECTIVES

1. Explain what project management is and list best practices for effective project management.
2. Discuss project management standards, leading authorities, and methodologies frequently used.
3. Describe key factors for effective project management.
4. Explain what program management is.
5. Discuss the role of the auditor in project management.
6. Explain the significance of big data in project management, and highlight the essential skills needed for project managers to effectively manage big data projects.

This chapter focuses on project management, particularly best practices, standards, and methodologies that are used by program managers in organizations to effectively and efficiently bring projects to an end. This chapter also discusses various success factors to put in place when conducting effective project management, and the auditor's role in project management. Topics such as program management and management of big data projects are also discussed. IT project management, for instance, refers to the processes and techniques used in the beginning-to-end development of software or other systems. Project management is one of the key controls for both auditors and organizations that ensures delivery of projects on time, on budget, and with full effective functionality.

Project Management

The purpose of project management is to identify, establish, coordinate, and monitor activities, tasks, and resources for a project that is consistent with the goals and objectives of the organization. Effectively controlling projects requires a disciplined approach to their various life cycles: project initiation, planning, execution, monitoring and controlling, and closing. These five cycles, domains, or process groups contain tasks, including knowledge and skills, that are actually assessed in the Project Management Professional (PMP) Certification. Overall, they relate to having the right people involved, following standard project management processes, and using a set of project management tools for effective execution.

COBIT recognizes project management as a process that impacts both the effectiveness and the efficiency of information systems. The process also involves IT resources that include people, applications, technology, operational facilities, and controls. Controls over managing projects that satisfy organizational business requirements typically consider:

- Business management sponsorship of project
- Project management capabilities
- User involvement
- Task breakdown, milestone definition, and phase approvals
- Allocation of responsibilities
- Rigorous tracking of milestones and deliverables
- Budgets and balancing internal and external resources
- Quality assurance plans and methods
- Program and project risk assessments
- Transition from development to operations

Project management has often been described as part art and part science. The art side involves the human element, the experience project managers bring to the project, the support they can muster from their management, and, a critical point, how project managers relate to the organization and their willingness to provide the right level of support to make the project succeed. Many times, the relationship between the project manager and the organization has not been built as a partnered approach. This can lead to loss of productivity by the project team and should be captured as a project risk as soon as recognized. The second part of the equation, the science side, is somewhat easier to deal with. The project manager should put in place the right project governance and project management life cycle (PMLC), and integrate these two elements with the appropriate project management methodology.

IT industry analysts have made general and specific recommendations on why projects are successful. Other IT industry organizations have built their own body of knowledge to document acceptable practices. The Gartner Group, for example, identifies Seven Best Practices for an Effective Project Management Office (PMO) that improve the effectiveness of project, portfolio and program management, as well as support the strategies and goals of the organization. The following recommendations are a good place to start.

1. *Acquire the Right People, Knowledge, Skills, and Collaborative Behaviors.* This is a key characteristic of a highly effective PMO that allows project managers to place emphasis on hiring only resources that will best fit the project.
2. *Identify and Execute High-Impact, High-Visibility Initiatives.* The focus here is on identifying and improving delivery of critical, highly visible projects to attract the attention of stakeholders and ensure their commitment and support for future PMO-driven initiatives.
3. *Report on What the Business Really Cares About.* PMOs should communicate and report on the status of relevant projects, portfolios, and programs in an effective and consistent manner. Such status should be systematically, forthrightly, and invariably reported to provide organizational leadership with appropriate information necessary to support effective decision-making.
4. *Build a Framework that Shows How the PMO Aligns with Strategic Enterprise Objectives.* A framework that articulates alignment between the PMOs and the continuously evolving organizational goals, milestones, and direction is essential to support the value of the PMO.

5. *Provide Senior Managers with Simple, Unambiguous Information.* Senior Managers are very busy people. PMOs should concentrate on providing them with relevant, accurate, and timely information to support effective decision-making. This type of informative reporting also avoids disconnection between expectations and perceived reality.

6. *Highlight the PMOs Achievements.* Success PMO stories, such as completion of projects on time and under budget and project contribution in solving a significant business problem, among others, should be shared, encouraged, and promoted throughout the organization.

7. *Evolve the PMO to Support Bimodal IT and Digital Business.* Effective PMOs must continuously reassess and adapt their service model, processes, and capabilities to ensure consistency with current goals, objectives, direction, and needs of the organization. PMOs that were focused on cost reduction and efficiency several years ago, may now need to shift gears to flexibility and delivery speed, for example.

Project Management Standards, Leading Authorities, and Methodologies

Strategic and tactical initiatives are dependent on effective and efficient project management. Project management applies skills, tools, techniques, and, most importantly, knowledge, in order to successfully initiate, plan, execute, manage, and complete projects. Knowledge is typically in the form of standards, guidance, and methodologies.

The primary standards organization for project management is the Project Management Institute (PMI). Founded in 1969, the PMI delivers value to professionals through global advocacy, collaboration, education, and research. The Institute's globally recognized standards as well as its certifications, tools, academic research, professional development courses, and networking opportunities have been key in the development and maturity of the project management profession.[*]

The PMI's developed project management standards or Project Management Body of Knowledge (PMBOK) comprises knowledge of innovative and advanced practices within the project management profession. PMI standards provide globally accepted guidelines, rules, and characteristics for project, program, and portfolio management. Included within the PMBOK are as follows:

- *Foundational standards*—reflect the continually evolving profession;
- *Practice standards*—describe the tools, techniques, processes, and/or procedures identified in the PMBOK or other foundational standards;
- *Practice guides*—provide supporting information and assistance when applying PMI standards; and
- *Lexicon of project management terms*—provide clear and concise definitions to project, program and portfolio management terms in order to improve understanding and consistent use of terminology.

Similar to the PMI is the Australian Institute of Project Management (AIPM). The AIPM is Australia's leading serving body for project management. It is considered a key promoter, developer,

---

[*] http://www.pmi.org/about/learn-about-pmi.

and leader in project management by Australian business, industry, and government. AIPM is the second-largest member of the International Project Management Association (IPMA). The IPMA is another relevant example of a leading authority on competent project, program, and portfolio management. IPMA, founded in 1965, is the world's first project management association established to advance the project management profession's achievements in project and business success.* Through IPMA's efforts, project management best practices are widely known and appropriately applied at all levels of public and private sector organizations. The Global Alliance for Project Performance Standards (GAPPS), also well known in the project management field, is a unique alliance of government, industry, professional associations, national qualification bodies, and training/academic institutions that have been working together since 2003. The Alliance helps practitioners and organizations make sense of the many standards and certifications available globally to guide the management of projects. GAPPS's primary aim is to facilitate mutual recognition and transferability of project management standards and qualifications by providing the global project management community with a reliable source of comparative information. Other common project management standards and guidance include:

- IEEE Standard 1490–2011: IEEE Guide. Adoption of the Project Management Institute Standard. A Guide to the PMBOK
- ISO 21500:2012. Guidance to Project Management
- ISO 10006:2003 Guidance. Quality Management Systems and Guidelines for Quality Management in Projects

The PMI defines methodology as a "system of practices, techniques, procedures and rules used by those who work in a discipline." Project managers employ methodologies for the design, planning, implementation, and achievement of project objectives. There are different project management methodologies to benefit different projects and organizations, including but not limited to: Traditional/Waterfall, Agile, Systems Development Life Cycle, PRojects IN Controlled Environments (PRINCE2), Portfolio, Program, and Project Management, Critical Chain/Path, Adaptive, Projects integrating Sustainable Methods (PRiSM), and the Crystal Method, among many others. Despite the methodology selected, overall project objectives, schedule, cost, and participants' roles and responsibilities must still be carefully considered. Exhibit 7.1 describes frequently used methodologies in the project management practice.

Other relevant project management methodologies include Scrum, Kanban, Extreme Programming (XP), etc., and they are discussed in a later chapter.

Simply put, no project management methodology can meet the objectives of all business organizations. A business can vary according to type, size, industry, business goals and objectives, and many other factors. As a result, organizations should learn about these project management methodologies, how they are used, and the potential benefits each method can offer. Common features to look for when selecting a project management methodology include:

- Organizational goals, objectives, and direction
- Core values
- Project constraints
- Project stakeholders
- Project size

---

* http://www.ipma.world/about/.

Exhibit 7.1   Project Management Methodologies

Project Management Methodology	*Description*
Traditional / Waterfall	Often referred to as the classic approach, the Traditional project management methodology works well as it simply evaluates the various project tasks, and provides a process to manage and monitor the completion of those tasks. In software development, this Traditional approach is referred to as the Waterfall model. The Waterfall methodology, built upon the framework of the Traditional method, includes fixed phases and linear timelines. In other words, it handles tasks sequentially, from the planning phase to development and quality assurance and finally to project completion and maintenance. Project requirements are usually defined at the beginning or planning phase, and there are typically little or no modifications to such requirements or to the plan. The Waterfall method is used often in large-scale software development projects where thorough planning and a predictable process are vital.
Agile Project Management (APM)	Agile means capable to move quickly and easily. The APM methodology is used on projects that need extreme agility in requirements (e.g., deliver products to the customer rapidly and continuously, etc.). APM focuses on adaptability to changing situations and constant feedback. In other words, with APM there is no clearly defined end product at the beginning stage. This is contrary to the Traditional/Waterfall methodology, which does require end-product detailed requirements to be set at the starting phase. APM's key features involve short-termed delivery cycles (or sprints), agile requirements, dynamic team culture, less restrictive project control, and emphasis on real-time communication. APM is most commonly used in software development projects, but the methodology can also assist other types of projects. The APM methodology is typically a good choice for relatively smaller software projects or projects with accelerated development schedules.
Systems Development Life Cycle (SDLC)	Project management methodology mostly used in software development projects that describes a process for planning, creating, testing, and deploying an information system. SDLC can be used individually or be combined with other project management methodologies in order to achieve the best results. That is, while there should only be one project management methodology highlighting the processes to be followed to ensure successful project implementation, other methodologies may also be in place to support specific needs of the application system or deliverable being considered. For instance, in an information systems project, a mainframe application may follow the Waterfall methodology while a Web-based application follows Agile. SDLC also heavily emphasizes on the use of documentation and has strict guidelines on it.

(Continued)

Exhibit 7.1 (*Continued*)   Project Management Methodologies

Project Management Methodology	Description
PRojects IN Controlled Environments (PRINCE2)	Internationally-known project management methodology endorsed, used, and supported by the United Kingdom government. The methodology, also used in the private sector, is a process-based approach that provides organizations a consistent and easily tailored and scalable method for the management of projects. PRINCE2 divides projects into multiple manageable and controllable phases, each with their own plans and processes to follow. PRINCE2 assures business justification; defines organization structure for the project management team (e.g., roles and responsibilities); and provides flexibility at all levels of the project. Because its adaptability and compatibility, PRINCE2 is useful for small and large organizations. There is a PRINCE2 certification which is process- and project-focused. The certification is administered in the United Kingdom by APMG-International Examination and Accreditation Institute.[a] To become certified, the project manager candidate must complete a PRINCE2 training from an accredited training organization, as well as pass the PRINCE2 Foundation and Practitioner Exams. The PRINCE2 certification is recognized in the United Kingdom, Europe, and Australia, whereas the PMP certification is preferred in the United States, Canada, Middle East, and Australia.
Portfolio, Program, and Project Management (PPPM)	PPPM comprises three management disciplines: Portfolio Management, Program Management, and Project Management. The PPPM methodology aligns resources and activities to (1) comply with organizational objectives and strategies; (2) increase portfolio potential; and (3) minimize risks. PPPM is used by project managers and PMO's to analyze and collectively manage current or proposed projects based on numerous key characteristics. Other potential benefits of the PPPM methodology include: • Guidance to establish consistency, transparency, and control over projects • Increased project value and likelihood of project success • Implementation of a comprehensive skill set to enable training and education based on standardized, scalable, and accessible guidance • Improved program and project delivery • Decreased cost and time overruns
Critical Chain/ Path (CC/P)	The CC/P methodology defines and focuses on the critical and non-critical activities (tasks) within a project to ensure its successful completion. Criticality is based on the timeline needed for completion of the tasks. Each project has a certain set of core elements/requirements, called a critical chain or critical path, that establish a project's minimum timeline. The CC/P methodology allows project managers to allocate resources to critical and/or non-critical tasks, and reassign them when necessary. This effective and efficient utilization of resources works well for projects that have tasks which are dependent on one another. IT also helps in measuring and prioritizing tasks, ultimately providing project managers with a well-defined description of the project's duration.

(*Continued*)

Exhibit 7.1 (*Continued*) **Project Management Methodologies**

Project Management Methodology	Description
Adaptive	When using the Adaptive methodology, the scope of a given project changes (adapts) though the time needed for completion and the cost of the project both remain constant. While managing and executing the specific project, its scope gets adjusted to achieve the maximum business value, such as when new ideas or opportunities are unlocked during the development of a project. Again, the project's costs and deadlines are not affected as a result of the adjusted scope.
PRojects integrating Sustainable Methods (PRiSM)	The PRiSM project methodology is mainly used in large-scale real estate development or construction/infrastructure projects. Similar to PRINCE2, PRiSM rewards project managers with accreditation.
Crystal Method (CM)	The CM project management methodology assigns a low priority to project processes, activities, and tasks. The methodology focuses instead on communication, interaction, and skills of team members. The idea of the CM is that by focusing on the skillsets and traits of team members, projects become more flexible and unique.
HERMES Method	HERMES is a project management method developed by the federal administration of Switzerland. HERMES supports the steering, management, and execution of projects in the context of IT, the development of services and products, and the changing of organizational structures. It is a clearly structured and simple method to understand with a modular, easily expandable design.
Total Cost Management Framework (TCM)	TCM, introduced by the AACE International (formerly the Association for the Advancement of Cost Engineering) in the 1990s, is a systematic approach/framework to managing costs throughout the life cycle of any organization, program, facility, project, product, or service. The *TCM Framework: An Integrated Approach to Portfolio, Program, and Project Management* is a structured, annotated process map that explains each practice area of the cost engineering field in the context of its relationship to the other practice areas including allied professions.
Program Evaluation and Review Technique (PERT)	The PERT project management methodology, often used with the Critical Chain/Path, is commonly found in developmental processes and manufacturing. It is especially useful for businesses like these who plan to expand in the near future, or would at least like to keep that possibility open. Project managers are expected to differentiate between events, and to measure the progress of activities and tasks being completed. By closely analyzing and estimating the amount of time it should take for each event to be completed, project managers can easily create realistic timelines and budgets for those aspects of the project.

(*Continued*)

Exhibit 7.1 (*Continued*) **Project Management Methodologies**

Project Management Methodology	Description
European Commission's Project Cycle Management	The European Commission's Project Cycle Management is the preferred project management methodology for designing, executing, and monitoring the progress of programs and projects funded by the European Commission and of many other international development institutions.

^a www.apmg-international.com/en/qualifications/prince2/prince2.aspx.

- Cost of the project
- Ability to take risks
- Need for flexibility

Adoption of a project management methodology is essential for today's businesses and organizations. Selecting the appropriate methodology can transform the way team members communicate, work on tasks, and accomplish project milestones.

Key Factors for Effective Project Management

Establishing and complying with a project management methodology will provide an adequate environment for the project, but will not guarantee its success. Project management has the ultimate accountability for the success or failure of any project through adequate planning, resource management, oversight and tracking (O&T), and the effective employment of management tools. Becoming certified as a PMP has also proven helpful in bringing projects to a successful end.

Planning

Effective project planning ensures that project tasks are adequately defined, resources are available and used efficiently, quality is maintained, and the project is completed on time and within budget. Auditors can assist by reviewing the project plan to ensure that tasks and deliverables are defined in sufficient detail, resource requirements are defined, time estimates are reasonable, resources are available at the right time, and project progress is regularly reported.

Depending on the organization, project planning may be formal or informal. In either case, basic project management techniques should be used to ensure that the project is well planned and effectively monitored. There are many tools available to assist the project manager in preparing a project plan. Project management tools allow the user to define tasks and dependencies, and track progress. A project plan should include interim milestones and regular review of project deliverables.

The objective of project planning is to be able to predict the project duration, resources required, and costs. The project manager should set up reasonable plans by establishing goals, estimating the tasks to be performed, assigning personnel responsible for those tasks, priorities, status, and task duration (i.e., start and end dates). Exhibit 7.2 illustrates an example of what a project plan for the development of a financial application may look like.

Exhibit 7.2 Project Planning

Project: Development of Financial Application				
Task	Personnel Responsible (Initials)	Priority/Status	Start Date	End Date
GOAL: With respect to the development of an application or the implementation of modifications to existing applications, management monitors that the project meets the objectives specified, is completed on budget, and meets the time requirements.				
1. Verify documentation submitted by IT department personnel to keep track of the established project's objectives, budget, and completion status.	ARO	High/ Completed	10/5/20XX	10/5/20XX
2. Corroborate information against the IT department administrative meeting minutes.	GMP	Medium/ Completed	10/5/20XX	10/7/20XX
3. Ensure that adequate monitoring and evaluation procedures are performed by IT department's management during development of the application, or implementation of modifications to existing applications.	EMO	High/Pending	10/10/20XX	10/31/20XX
GOAL: Users and other requests for the development of (or modifications to) the application are approved and implemented consistent with IT department's plans and management's intentions.				
1. Examine special access request forms to verify if these were adequately filled-out and signed by authorized personnel that requested, approved, and processed the forms.	LRO	High/ Completed	10/5/20XX	10/10/20XX

(Continued)

Exhibit 7.2 (*Continued*) Project Planning

Project: Development of Financial Application				
Task	Personnel Responsible (Initials)	Priority/Status	Start Date	End Date
2. Verify the IT department administrative meeting minutes and note that the impact of changes to hardware or software systems is adequately addressed by the department's technical and management staff.	GDO	Medium/ Completed	10/5/20XX	10/10/20XX
GOAL: Application is developed or modified, and tested in an environment separate from the production environment.				
1. Interview appropriate IT personnel regarding the new application developed in the test environment. Verify the Operations Manual to ensure consistency with development procedures.	ARO	High/ Completed	10/10/20XX	10/15/20XX
2. Ensure that demo environments are available to test the newly-developed or modified application previous to their installation in production environment.	GMP	High/ Completed	10/10/20XX	10/15/20XX
3. Obtain from the IT administrative personnel the special access request form to ensure proper authorization to work with the new or modified application system in the test environment.	EMO	High/Pending	10/10/20XX	10/15/20XX

(Continued)

Exhibit 7.2 (*Continued*)   Project Planning

Project: Development of Financial Application				
Task	Personnel Responsible (Initials)	Priority/Status	Start Date	End Date
GOAL: Access to the test and production environment is appropriately restricted.				
1. Verify that user-ids assigned to access test environments are not being utilized by users in live environments.	LRO	High/Pending	10/10/20XX	10/18/20XX
2. Verify that access to the test environment is restricted to authorized personnel from affected organization departments.	GDO	High/Pending	10/10/20XX	10/18/20XX
3. Examine user profile reports to verify that only users that perform special tasks or users that perform test procedures were included in the list.	ARO	High/Pending	10/12/20XX	10/20/20XX
GOAL: Management retains previous version of the application and/or data to allow for recovery of the IT environment in the event of processing problems.				
1. Corroborate with appropriate IT personnel that backup procedures are performed prior to any implementation or upgrade procedures performed to avoid disruption to operations.	GMP	High/Pending	10/20/20XX	10/31/20XX
2. Confirm that a test environment is created to perform test procedures previous to implementation to live environment of new applications and/or modifications.	ARO	High/Pending	10/10/20XX	10/5/20XX

Resource Management

There are many individual functions that are required to deliver a successful project. The business has to define the requirements, the application developers have to deliver the code, the quality assurance group and testers have to validate the code, and the infrastructure groups have to support the application. People with various skill sets may be assigned to a project team. Project assignments may be full or part time. Team members may be transferred or matrixed to the project team. The challenge for the project manager here is making sure that:

- Appropriate governance is in place.
- Right resources, such as money, people, and facilities are available at the right time.
- The project has a work breakdown structure that is sufficiently detailed to carry out.
- Project tasks are prioritized to prevent interference with other projects' due dates.
- Deliverables are produced successfully and in a timely fashion.
- Management is being communicated with and sufficiently involved.
- The end user is involved and takes delivery of the agreed-to-project results.

Oversight and Tracking

O&T helps ensure that a project lives up to its commitments. As with anything, the best laid plans can fail due to poor execution. Controls need to be put in place to identify projects that are running astray. O&T during all phases of the development process helps ensure that standard processes (or requirements) are followed and control is maintained. O&T continues after the project is implemented to ensure that all business benefits promised when the project was approved are realized as well as ongoing costs stay in line with the original estimates. The objective of O&T is to provide adequate visibility into actual progress so that management can take effective actions when the project's performance deviates significantly from the plans. These requirements must be documented and controlled. Exhibit 7.3 illustrates a sample checklist of project management requirements, per section, that are commonly tracked.

Exhibit 7.3 Project Oversight and Tracking

Project: [Project Name]
Section and Requirement
Goals
1. Actual results and performance are tracked against the plans.
2. Corrective actions are taken and managed to closure when actual results and performance deviate significantly from the plans.
3. All changes to commitments are agreed to by affected groups or parties.
Commitments
1. Project manager is designated to be responsible for the project's activities and results.

(Continued)

Exhibit 7.3 (*Continued*)   Project Oversight and Tracking

Project: [Project Name]
Section and Requirement
2. Project follows a documented organizational policy for managing software projects that includes a documented software development plan.
3. Project manager is informed of project status and issues.
4. Corrective action is taken as necessary.
5. Affected groups are involved and agree with all changes to commitments.
6. Senior management reviews all changes to commitments.
Abilities
1. Software development plan is documented and approved.
2. Project manager explicitly assigns responsibilities for work products and activities.
3. Adequate resources and funding are provided for tracking and oversight activities.
4. The management team is trained.
5. First-line managers understand the technical aspects of the project.
Activities
1. A documented development plan is used for tracking project activities and communicating status.
2. Revisions to the plan are documented.
3. Commitments and/or changes to commitments, either to individuals or to groups, are reviewed with senior management.
4. Changes to commitments are communicated to all affected individuals and groups according to a documented procedure.
5. The size estimates of work products or changes to existing work products are tracked and corrective action taken when necessary.
6. The effort and cost of the project are tracked and corrective action taken when necessary.
7. Project schedule and critical computer resources are tracked and corrective action taken when necessary.
8. Technical activities are tracked and corrective action taken when necessary.
9. Risks are tracked and corrective action taken when necessary.
10. Actual measurement and re-planning data are recorded.

(Continued)

Exhibit 7.3 (*Continued*)   Project Oversight and Tracking

Project: [Project Name]
Section and Requirement
11. Periodic internal reviews are conducted to track technical progress, plans, performance, and issues against the plan.
12. Formal reviews are conducted at selected project milestones according to a documented procedure.
Measurements
1. Measurements are devised and utilized to monitor management of all tracking and oversight activities.
Verification
1. Management activities are reviewed periodically with senior and project management.
2. The quality assurance team audits the management of planning activities and reports the results of such audit.

Source: Adapted from Senft, S., Gallegos, F., and Davis, A. 2012. *Information Technology Control and Audit*. Boca Raton: CRC Press/Taylor & Francis.

Project Management Tools

Effective project management requires the use of stand-alone and enterprise-wide project management tools. For enterprise-wide project development and tracking, for example, there are several functions that can be automated and integrated like:

- Project task planning and tracking
- Resource and time tracking
- Labor hour tracking
- Time capture and billing
- Time reporting
- Project budgeting
- Project communication
- Project documentation

Enterprise-wide project management tools allow for tracking people working on multiple projects and aid in identifying cross-project dependencies and issues. They also integrate tasks, resources, and costs into a single repository. If management has decided to use time and measurement tools, such as Critical Path Method (CPM), Program Evaluation Review Technique (PERT), or Gantt charts, then the auditor, for instance, must ensure that these tools are used according to the management's specifications. The use of one of these tools can help management or the auditor with time management for the entire project. The auditor can also use these tools to help get recommendations through and show management how much time is needed to implement recommended controls.

Additional project management tools are task sheets, which are used to allocate time (actual versus forecasted), assign personnel, and log the completion date and cost. In this way, the auditor and management can obtain a more detailed account of the time and money spent on a project and can track what is being worked on and what is finished. Future projects benefit most from these task sheets because management can base future project estimates on a history of times and costs.

The complexity of today's projects virtually requires the use of tools such as Microsoft Project and Deltek's Open Plan. For example, Microsoft Project provides a flexible tool designed to help project managers handle a full range of projects. The project manager can schedule and closely track all tasks, as well as use Microsoft Project Central, the Web-based companion to Microsoft Project, to exchange project information with the project team and senior management. Some of the main benefits and features of Microsoft Project include:

- *Personal Gantt chart.* Renders Gantt views such as those in Microsoft Project to outline each team member's own tasks across multiple projects.
- *Task delegation.* Once assigned by the project manager, tasks may be delegated from team leaders to team members or from peer to peer. The delegation feature can also be disabled if desired.
- *View nonworking time.* Team members can report nonworking time to the project manager, such as vacation or sick leave, and also report work time that cannot be devoted to the project.
- *Database performance.* Gets improved performance and access to data with changes to the Microsoft Project database.
- *Network diagram.* Customizes network diagrams with new filtering and layout options, increased formatting features, and enhanced box styles.

Another project management tool is Deltek's Open Plan, which allows the project manager to closely monitor and evaluate a project's priorities, risks and status from initiation to completion.* Open Plan is an enterprise project management system that substantially improves an organization's ability to complete multiple projects on time and on budget. With multi-project analysis, critical path planning, and resource management, Open Plan offers the power and flexibility to serve the differing needs of businesses, resources, and project managers.

Project Management Certification

The PMP Certification, a globally recognized certification offered by the PMI, validates the project manager's knowledge, experience, and skills in bringing projects to successful completion.† The certification recognizes the experience and capabilities of the project manager in leading and managing projects. As the demand for skilled project managers continues to increase, certified project manager professionals will not only become a necessity, but will be in a better position to command higher salaries over non-certified managers.

---

* https://www.deltek.com/en/products/project-and-portfolio-management/open-plan/benefits-by-role/
  project-management.
† http://www.pmi.org/-/media/pmi/documents/public/pdf/certifications/project-management-professional-
  handbook.pdf.

The PMP Certification evaluates candidates in five essential PMBOK process groups that outline the needed skills and competencies for managers to lead projects to successful solutions. These five project management areas or process groups include[*]:

1. *Initiating*—Involves the processes, activities, and skills needed to effectively define the start of a project. These would involve setting permits, authorizations, initial work orders, and clear phases for work to be completed, as well as initializing teams and having an approved budget before the project begins.
2. *Planning*—Process group that defines the scope of the project; sets strategic plans to maximize workflow; identifies project goals and expectations; assembles priority lists and plan team needs; and delineates the project infrastructure necessary to achieve desired goals based on timelines and budgetary constraints.
3. *Executing*—Process group with activities that involve managing teams effectively (i.e., addressing team concerns or other complex situations) while coordinating expectations and reaching benchmark goals on time and within budget.
4. *Monitoring and Control*—Involves processing change orders, addressing on-going budget considerations, and mitigating unforeseen circumstances that may impair a team's ability to meet initial project goals and expectations.
5. *Closing*—This process group relates to delivering a project to a successful close (i.e., on time and within budget). Good closure brings great reviews and can increase future word of mouth referrals.

Each process group above contains knowledge and skills required to perform effective and competent project management tasks, including providing the professional skills necessary to lead teams and achieve successful project results.

Program Management

Program management is the process required to coordinate multiple related projects with the purpose of delivering business benefits of strategic importance. Today's complex applications (i.e., enterprise resource planning systems) integrate requirements and functionality from multiple groups and multiple applications. Most new applications also require action from various functions within IT (e.g., software development, network engineering, security, production support, etc.). Program management brings all the pieces of a major program together. It does so by:

■ Defining a program management framework
■ Creating a program management office
■ Setting staffing requirements, processes, and metrics
■ Establishing consistent project management practices
■ Implementing technology for managing projects

---

[*] http://www.pmi.org/-/media/pmi/documents/public/pdf/certifications/project-managementprofessional-exam-outline.pdf.

According to research conducted by the Gartner Group, organizations that establish enterprise standards for program and project management, including a PMO with suitable governance, will experience half the major project cost overruns and delays and cancellations of those that fail to do so.

Project Management: Auditor's Role

The auditor's role in project management depends on the organization's culture, maturity of the information systems function, and philosophy of the auditing department. The objective of a project management audit is to provide an early identification of issues that may hinder successful implementation of an application that is controlled, documented, and able to be operated by a trained user community. Auditing project management requires specific knowledge about the project methodology and development process. Understanding these allows the auditor to identify key areas that would benefit from independent verification.

The scope of a project management audit can include an evaluation of the administrative controls over the project (e.g., feasibility results, staffing, budgeting, assignment of responsibilities, project plans, status reports, etc.) or an evaluation of specific deliverables to validate that the project is following established standards. By becoming involved at strategic points, the auditor can ensure a project that is well controlled. If the auditor feels that his/her knowledge, skills, and abilities are not current with the technologies being applied, he or she should request preliminary training on the technology prior to being assigned. Attending training sessions concurrent with the auditee can help gain understanding of the tools and techniques available to them. The following list highlights some of the key tasks the auditor may perform during a project's development:

- Gain the support and cooperation of the users and IT professionals
- Check project management tools for proper usage
- Perform project reviews at the end of each phase
- Assess readiness for implementation
- Present findings to management
- Maintain independence to remain objective

To determine the level of involvement, the auditor should first complete a risk assessment of the project development process and determine the amount of time to be allocated to a particular project. Next, the auditor should develop an audit plan that includes a schedule for the specific review points tied to the project schedule. The auditor then communicates the scope of his/her involvement as well as any findings to the project manager, users, and IT management.

Risk Assessment

Depending on the organization, auditors may not have enough time to be involved in all phases of every project. Project involvement will depend on the assessment of process and project risks. Process risks may include a negative organizational climate as well as the lack of strategic direction, project management standards, and formal project management process. Project risks, on the other hand, involve resource unavailability, budget overruns, project complexity and magnitude, inexperienced staff, and the lack of end-user involvement and management commitment; among others.

The level of risk may be a function of the size of the project, scope of organizational change, complexity of the application system being developed, the number of people involved, and the importance of the project to the organization. The scope of the audit involvement will depend on the maturity of project management in the organization. Audit involvement may be minimal if the IT group has a well-established project methodology and PMO that performs regular O&T activities. In this case, the auditor may focus more on project-specific risks rather than on project management risks. For less mature organizations, the auditors may also take the role of overseeing the project and tracking its tasks and activities.

Audit Plan

The audit plan will detail the objectives and the steps to fulfill the audit objectives. As in any audit, a project management audit will begin with a preliminary analysis of the control environment by reviewing existing standards and procedures. During the audit, these standards and procedures should be assessed for completeness and operational efficiency. The preliminary survey should identify the organization's strategy and the responsibilities for managing and controlling development.

The audit plan would include a project management process review to assess the adequacy of the control environment for managing projects. The review points listed represent checkpoints in the project management process. Auditors can use these checkpoints to determine both the status of the project's internal control system and the status of the development project itself. These reviews eliminate the necessity of devoting large amounts of audit resources to the development effort. As long as the development process is well controlled, the need for audit involvement is minimized.

The audit plan will further assist the project manager in identifying project risks and evaluating plans to mitigate and manage those risks, such as having trained and devoted resources, management support, and end-user commitment. Auditing can provide management with an independent review of project deliverables like project charter, tasks list, schedule, and budget. Auditing may also review the project tasks list and budget to verify that all project tasks are defined and all milestones have a deliverable.

During the planning phase, the auditor can facilitate communication between functions and raise issues that may impact the quality or timeliness of the project. In a software development project, for instance, resources from various departments need to come together to implement an automated process that may affect multiple-user functions. Because of various audit projects, auditors develop an overall knowledge of the organization and establish relationships with multiple groups and departments, including:

- Primary users
- Secondary users
- Vendors and consultants
- Programmers and analysts
- Database administrators
- Testing teams
- Computer operations
- Interfacing systems
- Implementation team
- Production support and maintenance programmers

These relationships are helpful in a software development project for making sure the information is flowing between the development team and other functionaries.

Communication of Scope of Involvement and Recommendations

The first area to communicate is the auditor's involvement role in the project management audit. It is very important to make sure that the expectations of the auditor's role are understood and communicated to all participants. The auditor must develop an open line of communication with both management and users. If a good relationship between these groups does not exist, information might be withheld from the auditor. This type of situation could prevent the auditor from doing the best job possible. In addition, the auditor must develop a good working relationship with the manager, the analysts, and the programmers. Although the auditor should cultivate good working relationships with all groups that have design responsibilities, he or she must remain independent.

Throughout the development of a project, the auditor will be making control recommendations. Depending on the organization's culture, these recommendations may need to be handled informally by reviewing designs with the project team or formally by presenting recommendations to the steering committee. In either case, the auditor must always consider the value of the control recommendation versus the cost of implementing such recommendation. Recommendations should be specific. They should identify the problem, not the symptom, and allow for the proper controls to be implemented and tested.

Recommendations are often rejected because of a time and cost factor. Project managers may sometimes feel that implementing an auditor's recommendations will put them behind schedule. The auditor must convince management of the value of the recommendations, and that if they are not put in place, more time and money will be spent in the long run. Informing management of the cost of implementing a control recommendation now rather than shutting down the system later to repair it or leaving possible exposures open will help convince management of the need to invest the required time and money.

Big Data Project Management

Big data offer significant benefits and challenges for professionals in the project management field. Big data, as defined by the TechAmerica Foundation's Federal Big Data Commission (2012), "describes large volumes of high velocity, complex and variable data that require advanced techniques and technologies to enable the capture, storage, distribution, management, and analysis of the information." Gartner, Inc. further defines it as "... high-volume, high-velocity and/or high-variety information assets that demand cost-effective, innovative forms of information processing that enable enhanced insight, decision making, and process automation."

Even though accurate big data may lead to more confident decision-making process, and better decisions often result in greater operational efficiency, cost reduction, and reduced risk, many challenges currently exist and must be addressed. Challenges of big data include, for instance, analysis, capture, data curation, search, sharing, storage, transfer, visualization, querying, as well as updating. Ernst & Young, on its EY Center for Board Matters' September 2015 publication, states that challenges for auditors include the limited access to audit relevant data; scarcity of available and qualified personnel to process and analyze such particular data; and the timely integration of analytics into the audit. These challenges of big data have also hit the project management field. The more data there are, the more data need to be analyzed, and the more projects will need to be managed.

Analysis of significant amounts of data is crucial to identify patterns and trends, as well as to create efficient and effective processes and solutions. Project managers must embark in this relatively new big data journey in order to make effective corporate decisions. Big data analyses allow project managers to pinpoint which processes, solutions, or technologies provide the best return or competitive advantage for the organization. As data continue to gain more intrinsic value, it will become a strategic and focal point for businesses. Exhibit 7.4 suggests some of the essentials skills required for project managers to successfully manage a big data project.

Exhibit 7.4   Big Data Essential Skills for Project Managers

Skill	Description
Cross-functional team management experience	Project managers must expand their team and skill sets to include a diverse group of professionals, including engineering disciplines, scientists, business analysts, business "super" users, operations, etc.
Ability to see the big picture	The ability to see the big picture when managing a big data project is critical for project managers to "make sense" of all the diverse and complex data.
Data-driven thinking	Project managers must employ data-driven thinking instead of relying on past experiences or instincts. The approach to managing a big data project should be discovery-driven where decisions are reached based on data and analysis and not on the intuition and experience of the project manager. This will be key to fully realize the potential value of the data.
Ability to deal with ambiguity	In big data projects, managers must embrace (and feel comfortable with) the unknown as they will not always have the right answers. Project managers must commit to uncovering problem solutions via experimentation and evidence-based findings.
Technical skills	Project managers must enhance existing technical skills to address big data tasks and challenges.
Process development skills	To adapt an organization's existing processes to big data, process development skills, such as utilizing resources efficiently (i.e., time, money, raw materials, and work); improving the quality of products, services, and data; and serving the needs of the markets all become necessary for project managers.
Soft skills	New soft skills must be acquired, or existing soft skills be polished, to include collaboration, curiosity, and creativity, which are vital qualities to deliver successful big data projects.
Good business judgment	Project managers will be required to switch from traditional project management (e.g., planning, identifying, mitigating risks, etc.) to focusing on delivering a quick, effective, and defined technology solution for business problems.

Conclusion

Project management is one of the key controls that ensures delivery of projects on time, on budget, and with full functionality. The purpose of project management is to identify, establish, coordinate, and monitor activities, tasks, and resources for a project that is consistent with the goals and objectives of the organization. Effectively controlling projects requires a disciplined approach to their various life cycles: project initiation; planning; execution; monitoring and controlling; and closing. These five cycles, domains, or process groups contain tasks, including knowledge and skills, relate to having the right people involved, following standard project management processes, and using a set of project management tools for effective execution.

The primary standards organization for project management is the PMI, which delivers value to professionals through global advocacy, collaboration, education, and research. The Institute's globally recognized standards have been key in the development and maturity of the project management profession. The PMI's PMBOK includes knowledge of innovative and advanced practices within the project management profession. Other leading authorities like the AIPM, the IPMA, and the GAPPS are well known and have been established to advance the project management profession. Project managers employ methodologies (best practices, techniques, or procedures) for the design, planning, implementation, and achievement of project objectives. There are different project management methodologies to benefit different projects and organizations.

Establishing and complying with a project management methodology will provide an adequate environment for the project, but will not guarantee its success. Project management has the ultimate accountability for the success or failure of any project through adequate planning, resource management, O&T, and the effective employment of management tools. Becoming certified as a PMP has also proven helpful in bringing projects to a successful end.

Program management is a helpful control used to coordinate multiple related projects with the purpose of delivering business benefits of strategic importance. It also offers many opportunities for auditors. Auditors need to develop the necessary skills and relationships to work with the project team to ensure that controls are considered and built into the system appropriately. Auditors can assist organizations by reviewing the project management environment, including tools, evaluating standards for project management, monitoring project progress, and evaluating phases in the overall project, among others.

Project managers must acquire new skills or polish existing ones when managing big data projects. Successful management of big data project allows project managers to make corporate decisions, as well as identify the processes, solutions, or technologies that provide the best return or competitive advantage for the organization. As data continue to gain more intrinsic value, it will become a strategic and focal point for businesses.

Review Questions

1. Explain what project management refers to.
2. List 10 controls normally considered when managing projects, according to COBIT.
3. Explain why/how project management has often been described as part art and part science.
4. What is the primary standards organization for project management and what is its purpose?
5. What is included within the Project Management Body of Knowledge (PMBOK)?
6. Differentiate between the traditional and agile project management methodologies.
7. List and briefly explain key success factors for effective project management.

8. What is the difference between program and project management? How does program management puts all program pieces together?
9. List key tasks the auditor may perform during a project's development.
10. Describe current challenges of big data to organizations. How do these challenges impact project managers and the project management field?

Exercises

1. List and describe seven best practices for an effective project management office (PMO), according to the Gartner Group.
2. Project Management Methodology ("Methodology")—Group Assignment and Presentation. Professor to divide the class in groups and assign Methodologies from Exhibit 7.1. Groups will go outside of the textbook (i.e., IT literature and/or any other valid external source) to summarize and present to the class the Methodology/ies assigned. The presentation should:
   a. Provide an overall explanation of the Methodology, including, but not limited to: definition; purpose and objectives; whether it is United States or International-based; industries where Methodology/ies has/have been used; etc.
   b. Highlight the benefits and challenges of the Methodology to project managers.
   c. Include examples of organizations that have used the particular Methodology and, if available, describe their overall experience.
   d. Be submitted in power-point-presentation format with a cover page and a reference section at the end. The submitted file should be between 8 and 10 pages long, including cover page and references.
3. Summarize the steps auditors should do in order to determine their level of involvement in a project management audit.
4. Exhibit 7.4 lists essential skills for big data project managers. Think of (and list) three to five additional skills that would assist project managers when dealing with big data projects. Explain your rationale for each skill added.

CASE—IT PROJECT MANAGEMENT FAILURES

INSTRUCTIONS: According to a 2012 study conducted by the firm McKinsey & Co. and the University of Oxford, "On average, large IT projects run 45% over budget and 7% over time, while delivering 56% less value than predicted." McKinsey's study focused on projects of $15 million or more.

TASK: Using an Internet web browser, search and examine three recent (within the last 5 years) IT projects that have failed. Summarize why they failed. Then, identify solutions or how these failures may had been avoided. Support your reasons with IT literature and/or any other valid external source. Include examples as appropriate to evidence your case point. Submit a word file with a cover page, responses to the tasks above, and a reference section at the end. The submitted file should be between 8 and 10 pages long (double line spacing), including cover page and references. Be ready to present your work to the class.

Further Reading

1. Australian Institute of Project Management (AIPM). www.aipm.com.au/about-us (accessed June 20, 2017).
2. Best, K., Zlockie, J., and Winston, R. (2011). International standards for project management. Paper presented at PMI® Global Congress 2011—North America, Dallas, TX. Newtown Square, PA: Project Management Institute.
3. Bloch, M., Blumberg, S., and Laartz, J. (2012). Delivering large-scale IT projects on time, on budget, and on value. McKinsey & Company - Digital McKinsey, www.mckinsey.com/business-functions/digital-mckinsey/our-insights/delivering-large-scale-it-projects-on-time-on-budget-and-on-value (accessed February 6, 2017).
4. Cavusoglu, H., Mishra, B., and Raghunathan, S. (2004). A model for evaluating IT security investments. *Commun. ACM*, 47(1), 87–92.
5. Crawford, T. (2013). *Big Data Analytics Project Management*. CreateSpace Independent Publishing Platform, Portsmouth, NH.
6. Doerscher, T. (2008). *PMO 2.0 Survey Report. The Continued Evolution of the Project, Program and Portfolio Management Office (PMO)*, Planview, Inc., 2009.
7. EY. Big data and analytics in the audit process. EY Center for Board Matters' September 2015. www.ey.com/Publication/vwLUAssets/ey-big-data-and-analytics-in-the-audit-process/$FILE/ey-big-data-and-analytics-in-the-audit-process.pdf
8. Flynn, T. A. (2007). Integration of the project management life cycle (PMLC) and the systems development life cycle (SDLC) in accelerated project efforts: adapting project management best practices to unreasonable requests. Paper presented at PMI® Global Congress 2007—North America, Atlanta, GA. *Newtown Square, PA: Project Management Institute.*
9. Fuster, J. E. (2006). Comparison of the European Commission's project cycle management/logical framework approach with international PM standards and methodologies: PMBOK, IPMA's ICB, ISO 10,006, PRINCE2 and TenStep. Paper presented at PMI® Global Congress 2006—EMEA, Madrid, Spain. Newtown Square, PA: Project Management Institute.
10. The Global Alliance for Project Performance Standards (GAPPS). http://globalpmstandards.org/about-us/ (accessed June 20, 2017).
11. Gartner identifies seven best practices for an effective project management office. April 2016. Press Release. Stamford, CT. www.gartner.com/newsroom/id/3294017
12. Gartner IT glossary. (n.d.) www.gartner.com/it-glossary/big-data/
13. Gilchrist, P. (2014). Project management skills for managing big data projects. *The Project Manager's Guide to Big Data*. www.freepmstudy.com/BigData/BigDataPMSkills.cshtml (accessed June 17, 2017).
14. Gomolski, B. and Smith, M. *Program and Portfolio Management: Getting to the Next Level*, Gartner Research, G00155601, Gartner Group, Stamford, CT, November 27, 2006.
15. HERMES Method Overview. www.hermes.admin.ch/onlinepublikation/index.xhtml (accessed June 15, 2017).
16. MC2 Group. (2016). Impact of big data in project management. www.mc2i.fr/Impact-of-Big-Data-in-Project-Management (accessed June 24, 2017).
17. International Project Management Association (IPMA). www.ipma.world/about/ (accessed June 20, 2017).
18. ISO 10006:2003 Guidance - Quality Management Systems and Guidelines for Quality Management in Projects. www.iso.org/standard/36643.html (accessed June 3, 2017).
19. Katcherovski, V. (2012). 5 effective project management methodologies and when to use them. Logic Software, Inc. https://explore.easyprojects.net/blog/project-management-methodologies
20. Project Management Institute (PMI). Learn about PMI. www.pmi.org/about/learn-about-pmi (accessed June 17, 2017).
21. Light, M. and Halpern, M. *Understanding Product vs. Project Portfolio Management*, Gartner Research, G00130796, Gartner Group, Stamford, CT, May 2, 2006.

22. Project Management Institute. Methodology. www.pmi.org/learning/featured-topics/methodology (accessed June 14, 2017).

23. Mullaly, M. (2013). Big data & project Management: Is there a point? Project Management.Com. www.projectmanagement.com/articles/281365/Big-Data---Project-Management--Is-There-a-Point- (accessed June 17, 2017).

24. Project Management Institute (PMI). *PMBOK® Guide and Standards.* www.pmi.org/pmbok-guide-standards (accessed June 17, 2017).

25. KPMG, LLP. Portfolio, program, and project management. https://advisory.kpmg.us/management-consulting/capabilities/portfolio-program-and-project-management.html (accessed June 2, 2017).

26. TutorialsPoint. Project management methodologies. https://www.tutorialspoint.com/management_concepts/project_management_methodologies.htm (accessed June 14, 2017).

27. Project Management Institute, Inc. (2017). Project management professional exam outline. www.pmi.org/-/media/pmi/documents/public/pdf/certifications/project-management-professional-exam-outline.pdf

28. Project Management Institute, Inc. (2017). Project management professional handbook. www.pmi.org/-/media/pmi/documents/public/pdf/certifications/project-management-professional-handbook.pdf

29. Scheid, J. (2015). *Project Management Methodologies: How Do They Compare?* Bright Hub Inc. www.brighthubpm.com/methods-strategies/67087-project-management-methodologies-how-do-theycompare/

30. Senft, S., Gallegos, F., and Davis, A. (2012). *Information Technology Control and Audit.* CRC Press/Taylor & Francis: Boca Raton.

31. Singleton, T. (2006). What every IT auditor should know about project risk management. *ISACA.*, 5, 17–20.

32. Smith, M. *Express IT Project Value in Business Terms Using Gartner's Total Value of Opportunity Methodology*, Gartner Research, G00131216, Gartner Group, Stamford, CT, January 11, 2006.

33. TechAmerica Foundation's Federal Big Data Commission. Demystifying big data: A practical guide to transforming the business of government. (2012). www.techamerica.org/Docs/fileManager.cfm?f=techamerica-bigdatareport-final.pdf

34. AACE International. Total cost management (TCM) framework. http://web.aacei.org/resources/publications/tcm (accessed June 14, 2017).

35. ILX Group 2017. What is PRINCE2? www.prince2.com/usa/what-is-prince2 (accessed June 15, 2017).

Chapter 8

System Development Life Cycle

LEARNING OBJECTIVES

1. Discuss the system development life cycle (SDLC) and its common phases.
2. Discuss additional risks and associated controls related to the SDLC phases.
3. Explain common approaches used for software development.
4. Discuss the IT auditor's involvement in the system development and implementation process.

Organizations are constantly building, replacing, and maintaining information systems. There are many different approaches to systems development, but the most successful systems follow a well-defined development methodology. The success of a systems development project is dependent on the success of key processes: project management, analysis, design, testing, and implementation. Because development efforts can be costly, organizations have recognized the need to build well-controlled quality systems. IT processes information that is integral to the financial stability and profitability of organizations. Therefore, these systems must be built with adequate internal controls to ensure the completeness and accuracy of transaction processing.

System Development Life Cycle

As discussed in the previous chapter, a project management life cycle provides guidelines to project managers on the processes that must be followed to ensure the overall success of a project. In a similar fashion, the system development life cycle (SDLC), also being referred to as the application development life cycle, provides a framework for effectively developing application systems. It specifically describes a standard process for planning, creating, testing, and deploying new information systems (i.e., new development or modified system). Either developing a new system or adding changes to an existing one, the SDLC provides the framework and steps necessary for an adequate implementation. Although there are many variations of the traditional SDLC, they all have the following common phases in one form or another (refer to Exhibit 8.1):

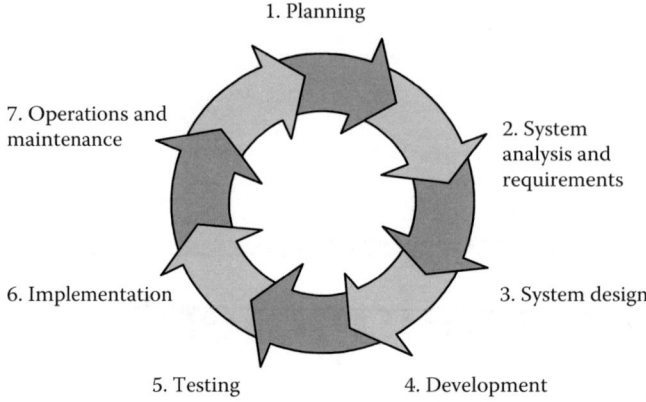

Exhibit 8.1   System development life cycle phases.

1. Planning
2. System Analysis and Requirements
3. System Design
4. Development
5. Testing
6. Implementation
7. Operations and Maintenance

Planning

The planning phase sets the stage for the success of the system development effort. It documents the reasons to develop the new system (as opposed to purchase it from an external source) in order to achieve the organization's strategic goals and objectives. During planning, organizations establish the scope of the work (considering costs, time, benefits, and other items), set initiatives to acquire the necessary resources, and determine solutions. If planning is not done properly, the budget and schedule may not be sufficient, the business problem or need to be addressed by the new system may not be adequately defined, the final product may not solve the problem or need, and the right people may not be involved. These are typical risks encountered by IT auditors and organization personnel during this phase. To be effective, planning should include and describe the following:

- *Needs for a new system analysis.* A study to determine whether a new system should either be developed internally or purchased from external sources.
- *Current system review.* A study of the current system to identify existing processes and procedures that will continue in the new system.
- *Conceptual design.* Preparation and assessment of the proposed design alternatives, system flows, and other information illustrating how the new system will operate.
- *Equipment requirements.* Identification of the hardware configuration needed to use the new system (e.g., processing speed, storage space, transmission media, etc.).
- *Cost/benefit analysis.* Detailed financial analysis of the cost to develop and operate the new system, the savings or additional expense, and the return on investment.

- *Project team formation.* Identification and selection of resources needed (e.g., programmers, end users, etc.) to develop and implement the new system.
- *Tasks and deliverables.* Establish defined tasks and deliverables to monitor actual results and ensure successful progress.

System Analysis and Requirements

In this phase, system analysts identify and assess the needs of end users with the ultimate purpose of ensuring that, once developed, the new system will meet their expectations. During this phase, end users and system analysts define **functional requirements** for the new software/system in terms that can be measured and functionally tested. The functionality of the existing system is matched with the new functionality and requirements are defined and validated with users so that they can become the basis for the system design phase. This phase also identifies and documents resources who will be responsible for individual pieces of the system, as well as the timeline expected. Common tools or practices used by organizations while on this phase include:

- Computer Aided Systems/Software Engineering (CASE)—software tool with methods to design, develop, and implement applications and information systems;
- Requirements Gathering—practice of collecting the requirements of a system from users, customers, and other stakeholders via meetings or interviews; and
- Structured Analysis—software engineering technique that uses graphical diagrams to analyze and interpret requirements, and to depict the necessary steps (and data) required to meet the design function of the particular system or software.

System Design

In the system design phase, the systems analyst defines and documents all system interfaces, reporting, screen layouts, and specific program logic necessary to build the new system consistent with the requirements. The system design phase describes, in detail, the specifications, features, and operations that will meet the requirements previously defined. At this phase, system analysts and end users, once again, review the specific business needs and determine (or confirm) what will be the final requirements for the new system. Technical details of the proposed system are also discussed with the various stakeholders, including the hardware/software needed, networking capabilities, processing and procedures for the system to accomplish its objectives, etc. Other more general and administrative topics within this phase include identifying existing risks, technologies to be used, capability of the team, project constraints, timing, and budget restrictions. Consideration of the aforementioned will aid in selecting the best design approach.

At the systems design stage, controls should be defined for input points and processing. Screen layouts, controls, and reports should be reviewed and approved by the end user before moving on to the next phase. Programmers will use the detailed specifications and output from the design phase to move on into the development or construction phase.

Development

In the development phase, programmers build or construct the new system based on analyses, requirements, and design previously agreed. The construction or coding phase is final once the programmer validates the new system code through individual unit testing (full testing of the system is performed in the next phase). The code is tested for both syntax and logic flow. All logic paths are exercised to ensure error routines work and the program terminates processing normally.

When new systems are developed, appropriate security access controls need to be developed as well to safeguard information against unapproved disclosure or modification, and damage or loss. Logical access controls, for instance, are used to ensure that access to systems, data, and programs are limited to appropriate users and IT support personnel.

Organizations must also keep in mind that development efforts generate code, and that this is where security and control of any system starts. In March 2011, the United States Computer Emergency Readiness Team (US-CERT) issued its top 10 Secure Coding Practices. These practices should be adhered to as one starts, designs, develops, tests, implements, and maintains a system:

1. *Validate input.* Validate input from all untrusted data sources. Proper input validation can eliminate the vast majority of software vulnerabilities. Be suspicious of most external data sources, including command line arguments, network interfaces, environmental variables, and user-controlled files.
2. *Heed compiler warnings.* Compile code using the highest warning level available for your compiler and eliminate warnings by modifying the code. Use static and dynamic analysis tools to detect and eliminate additional security flaws.
3. *Architect and design for security policies.* Create a software architecture and design your software to implement and enforce security policies. For example, if your system requires different privileges at different times, consider dividing the system into distinct intercommunicating subsystems, each with an appropriate privilege set.
4. *Keep the design as simple and small as possible.* Complex designs increase the likelihood that errors will be made in their implementation, configuration, and use. Additionally, the effort required to achieve an appropriate level of assurance increases dramatically as security mechanisms become more complex.
5. *Default deny.* Base access decisions on permission rather than exclusion. This means that, by default, access is denied and the protection scheme identifies conditions under which access is permitted.
6. *Adhere to the principle of least privilege.* Every process should execute with the least set of privileges necessary to complete the job. Any elevated permission should be held for a minimum time. This approach reduces the opportunities an attacker has to execute arbitrary code with elevated privileges.
7. *Sanitize data sent to other systems.* Sanitize all data passed to complex subsystems such as command shells, relational databases, and commercial off-the-shelf components. Attackers may be able to invoke unused functionality in these components through the use of SQL, command, or other injection attacks. This is not necessarily an input validation problem because the complex subsystem being invoked does not understand the context in which the call is made. Because the calling process understands the context, it is responsible for sanitizing the data before invoking the subsystem.
8. *Practice defense in depth.* Manage risk with multiple defensive strategies, so that if one layer of defense turns out to be inadequate, another layer of defense can prevent a security flaw from becoming an exploitable vulnerability and/or limit the consequences of a successful exploit. For example, combining secure programming techniques with secure runtime environments should reduce the likelihood that vulnerabilities remaining in the code at deployment time can be exploited in the operational environment.

9. *Use effective quality assurance techniques.* Good quality assurance techniques can be effective in identifying and eliminating vulnerabilities. **Fuzz testing**, **penetration testing**, and **source code audits** should all be incorporated as part of an effective quality assurance program. Independent security reviews can lead to more secure systems. External reviewers bring an independent perspective, for example, in identifying and correcting invalid assumptions.

10. *Adopt a secure coding standard.* Develop and/or apply a secure coding standard for your target development language and platform.

Other well-known practices referred to when developing and securing systems or applications include the secure coding principles described in the Open Web Application Security Project (OWASP) Secure Coding Guidelines. While OWASP's secure coding principles below specifically reference Web applications, such principles should be applied to non-Web applications as well.[*]

1. Input Validation
2. Output Encoding
3. Authentication and Password Management
4. Session Management
5. Access Control
6. Cryptographic Practices
7. Error Handling and Logging
8. Data Protection
9. Communication Security
10. System Configuration
11. Database Security
12. File Management
13. Memory Management
14. General Coding Practices

The Software Engineering Institute (SEI) has also developed US-CERT coding standards for common programming languages like C++, Java, Perl, and the Android platform. They include rules for developing safe, reliable, and secure systems. They identify sources for today's software vulnerabilities, and provide guidance on how to exploit them. Downloads of these standards are available to the community online.[†]

Testing

Testing is by far the most critical part of any system development and implementation. However, it is also the first to get short-changed when **go-live dates** get challenged. The primary purpose of system testing is to validate that the system works as expected and that it identifies errors, flaws, failures, or faults at an early stage because if discovered later, they will be costly to fix.

An overall testing strategy should be developed to define the individual test events, roles and responsibilities, test environment, problem reporting and tracking, and test deliverables. The testing process should be based on existing testing methodologies established by the organization. An effective testing process allows for documentation that will prevent duplicate testing efforts.

---

[*] https://security.berkeley.edu/secure-coding-practice-guidelines.
[†] www.securecoding.cert.org/confluence/display/seccode/SEI+CERT+Coding+Standards.

206 ■ *Information Technology Control and Audit*

A testing plan should be made in accordance with the organization's standards. The plan should include test scenarios, the role of the test participants, acceptance criteria, and testing logistics. It should also identify responsibility for documentation, review, and approval of tests and test results. End users and system owners should perform the required testing rather than programmers or developers. They should sign off that appropriate testing was performed with expected results for all requirements. Senior management sign-off is also required before programs are promoted to production environments.

Although each system may require different test events, in general, test events include unit testing, integration testing, technical testing, functional testing, performance load testing, and acceptance testing. Acceptance testing, for instance, verifies that acceptance criteria defined during the system definition stage are tested. Test cases should include system usability, management reports, performance measurements, documentation and procedures, training, and system readiness (operations/systems sign-off). Exhibit 8.2 summarizes the user acceptance testing event.

Exhibit 8.2   User Acceptance Testing

User Acceptance Testing
User acceptance testing (UAT) is key to a successful application system development and implementation. It ensures that the application fulfills the agreed-upon functional requirements (expectations) of the users, meets established usability criteria, and satisfies performance guidelines before being implemented into production. UAT minimizes the risks that the new application system will cause business interruptions or be disjointed with business processes. UAT should include inspections, functional tests, and workload trials. It should include all components of the application system (e.g., facilities, application software, procedures, etc.), and involve having the right team, agreeing on the testing requirements, and obtaining results approval from management.
Acceptance Team
The process owner should establish the acceptance team. The team is responsible for developing and implementing the acceptance process. The acceptance team should be composed of representatives from various functions including computer operators, technical support, capacity planning, help desk personnel, and database administrators.
Agreed-Upon Requirements
Requirements for UAT need to be identified, agreed upon, and prioritized. Acceptance requirements or criteria should be specific with detailed measures. Indirectly, the acceptance requirements become the criteria for making the "go/no-go decisions" or determining if the application system satisfies the critical requirements before being implemented into the live environment.
Management Approval
Acceptance plans and test results need to be approved by the affected functional department as well as the IT department. To avoid surprises, users should be involved in the application system testing throughout the development and implementation processes. This minimizes the risk of key functionality being excluded or not working properly.

Source: Adapted from Senft, S., Gallegos, F., and Davis, A. 2012. *Information Technology Control and Audit*. Boca Raton: CRC Press/Taylor & Francis.

Each test event should have a plan that defines the test-scope resources (i.e., people and environment) and test objectives with expected results. They should provide test case documentation and a test results report. It is often desirable to have the end user participate in the functional testing although all fundamental tests mentioned earlier should be applied and documented. At minimum, the thoroughness of the level of testing should be completed and reviewed by the development team and quality assurance staff. Quality of testing within each application and at the integration stage is extremely important.

Test scenarios, associated data, and expected results should be documented for every condition and option. Test data should include data that are representative of relevant business scenarios, which could be real or generated test data. Regardless of the type of test data chosen, it should represent the quality and volume of data that is expected. However, controls over the production data used for testing should be evaluated to ensure that the test data are not misused or compromised.

Testing should also include the development and generation of management reports. The management reports generated should be aligned with business requirements. The reports should be relevant to ensure effectiveness and efficiency of the report development effort. In general, report specifications should include recipients, usage, required details, and frequency, as well as the method of generation and delivery. The format of the report needs to be defined so that the report is clear, concise, and understandable. Each report should be validated to ensure that it is accurate and complete. The control measures for each report should be evaluated to ensure that the appropriate controls are implemented so that availability, integrity, and confidentiality are assured. Test events that may be relevant, depending on the type of system under development, are described in Exhibit 8.3.

Exhibit 8.3   System Test Events

System Test Event	Description
Unit testing	Verifies that stand-alone programs match specifications. Test cases should exercise every line of code.
Integration testing	Confirms that all software and hardware components work well together. Data are passed effectively from one program to the next. All programs and subroutines are tested during this phase.
Technical testing	Verifies that the application system works in the production environment. Test cases should include error processing and recovery, performance, storage requirements, hardware compatibility, and security (e.g., screens, data, programs, etc.).
Functional testing	Corroborates that the application system meets user requirements. Test cases should cover screens, navigation, function keys, online help, processing, and output (reports) files.
Performance load testing	Defines and tests the performance expectations of the application system in advance. It ensures that the application is scalable (functionally and technically), and that it can be implemented without disruption to the organization. The entire infrastructure should be tested for performance load to ensure adequate capacity and throughput at all levels: central processing, input and output media, networks, and so on. The test environment should also reflect the production/live environment as much as possible.

(Continued)

Exhibit 8.3 (*Continued*)   System Test Events

System Test Event	Description
Black-box testing	Software testing method that examines the overall operation and functionality of an application system without looking into its internal structure (e.g., design, implementation, internal paths, etc.). In other words, testers are not aware of the application's internal structure when employing black-box testing. Although black-box testing applies most to higher level testing, it can also cover virtually every level of software testing (i.e., unit, integration, system, and acceptance).
White-box testing	Software testing method that goes beyond the user interface and into the essentials of a system. It examines the internal structure of an application, as opposed to its operations and functionality. Contrary to black-box testing (which focuses on the application's operations and functionality), white-box testing allows testers to know about the application's internal structure (e.g., design, implementation, internal paths, etc.) when conducting tests.
Regression testing	Software testing method that follows the implementation of a change or modification to a given system. It examines implemented changes and modifications performed to ensure that the existing system (and its programming) is still functional and operating effectively. Once changes and modifications have been implemented, regression testing re-executes existing tests against the modified system's code to ensure the new changes or modifications do not break the previously working system.
Automated software testing	Software testing tools or techniques that simplify the testing process by automating the execution of pre-scripted tests on software applications before being implemented into the production environment. Automated software testing, such as automating the tests of units (e.g., individual program, class, method, function, etc.) that currently demands significant use of team's resources can result in a more effective and efficient testing process. Automated software testing can also compare current test results against previous outcomes.
Software performance testing	Software performance testing is key to determining the quality and effectiveness of a given application. The testing method determines how a system (i.e., computer, network, software program, or device) performs in terms of speed, responsiveness, and stability under a particular scenario.

Source: Adapted from Senft, S., Gallegos, F., and Davis, A. 2012. *Information Technology Control and Audit.* Boca Raton: CRC Press/Taylor & Francis.

Implementation

This phase involves the actual deployment and installation of the new system, and its delivery to end users. System implementation verifies that the new system meets its intended purpose and that the necessary process and procedures are in place for production. Implementing a system involves incorporating several controls (i.e., implementation plan, conversion procedures, IT disaster/continuity plans, system documentation, training, and support) to ensure a smooth installation and transition to the users. To ensure a smooth implementation, it is also important that users and technical support both be aware and on board with these controls.

Implementation Plan

An implementation plan should be documented to guide the implementation team and users in the implementation process. The documentation should cover the implementation schedule, the resources required, roles and responsibilities of the implementation team, means of communication between the implementation team and users, decision processes, issue management procedures, and a training plan for the implementation team and end users. In simple terms, the plan should cover the who, what, when, where, and how of the implementation process.

Data Conversion and Cleanup Processes

Unless a process is new, existing information will need to be converted to the new system. Conversion is the process where information is either entered manually or transferred programmatically from an old system into the new one. In either case, the existence of procedures should verify the conversion of all records, files, and data into the new system for completeness and accuracy purposes.

Data conversion procedures may fall into one of the following four generally recognized conversion methods:

- *Direct conversion.* Also referred to as "Direct cutover," it is a conversion method that involves shutting down the current system entirely and switching to a new system. The organization basically stops using the old system, say overnight, and begins using the new one the next day and thereafter. It is the riskiest of all methods because of the immediate learning curve required by users to effectively interact with the new system. A second risk would be the potential malfunction of the new system, which would significantly impact the organization as the old system is no longer available.
- *Pilot conversion.* Method where a small group of users and participants is established to interact with the new system whereas the rest continues to use the old/current one. This method assists organizations in identifying potential problems with the new system, so that they can be corrected before switching from the old one. Once corrected, the pilot/new system is installed for good and the old one is switched off. Retail chains typically benefit from this method. For example, installing a new point of sale system in one store for trial purposes and (upon operating properly) rolling out the new working system into the remaining stores.
- *Phased conversion.* Also referred to as the "Modular conversion," it is a method that gradually introduces the new system until the old system is completely replaced by the new system. This method helps organizations to identify problems early in the specific phase or module, and then schedule resources to correct them before switching over to the new system. Given that the current system is still partly operational when implementing this method, the risk tends to be relatively low compared to other methods. In the case of unexpected performance issues with the new system, the old system can still be used as it is still fully operational. In terms of disadvantages, the gradual replacement to the new system may be considered significant (i.e., implementation may take a longer period of time). Another disadvantage

would be training, which must be continuously provided to ensure that users understand the new system while it is being converted.

- *Parallel conversion.* Method that involves running both, the old and new system simultaneously for some pre-determined period of time. In this method, the two systems perform all necessary processing together and results are compared for accuracy and completeness. Once all issues have been addressed and corrected (if any) and the new system operates properly as expected, the old system is shut down, and users start interacting merely with the new system. The advantage of this conversion method is that it provides redundancy should the new system does not work as expected and/or system failures occur. Switching to the new system will only take place upon successfully passing all necessary tests, ensuring the new system will likely perform as originally designed and intended for. Common disadvantages of this method involve the financial burden of having two systems running simultaneously, the double-handling of data and associated operations, and the potential for data entry errors when users input data into the new system.

A conversion plan defines how the data are collected and verified for conversion. Before conversion, the data should be "cleaned" to remove any inconsistencies that introduce errors during the conversion or when the data are placed in the new application.

Tests to be performed while converting data include comparing the original and converted records and files, checking the compatibility of the converted data with the new system, and ensuring the accuracy and completeness of transactions affecting the converted data. A detailed verification of the processing with the converted data in the new system should be performed to confirm successful implementation. The system owners are responsible for ensuring that data are successfully converted.

The data conversion process often gets intermingled with data cleanup. Data cleanup is a process that organizations embark upon to ensure that only accurate and complete data get transferred into the new system. A common example is company names in a vendor file. A company can be entered into a vendor file multiple times in multiple ways. For example, "ABC Manufacturing" can be "ABC mfg," "abc Mfg.," and so on. Many of these data cleanup changes can be dealt with systematically because many errors happen consistently.

The data cleanup effort should happen before executing data conversion procedures. This allows the conversion programmers to focus on converting the data as opposed to coding for data differences. However, in reality, the exemptions from data conversion become issues for the data cleanup team to deal with. Data conversion and data cleanup teams should work closely with one another to ensure that only the most accurate and complete data are converted. Management should sign off on test results for converted data as well as approve changes identified by the data cleanup team.

IT Disaster Plan

This is another key review point for management and the IT auditor. As part of implementation, requirements for the system's recovery in the event of a disaster or other disruption should be accounted for. The IT disaster plan should be reviewed to ensure that the organization incorporates procedures and resources necessary to recover the new application system. Significant upgrades to existing applications may also require modification to disaster recovery requirements in areas such as processor requirements, disk storage, or operating system versions. Recovery procedures related to the new system should be tested shortly after it is put into production. Such recovery procedures must also be documented.

In the rush to implement a system, documentation can be the first to "slide." However, the price is paid when decisions to address problems become reactionary. Formalizing documentation and procedures is the difference between delivering a technology versus delivering a service. The disaster recovery plan should be in place at the point of implementation and carried through into operations.

System Documentation

System documentation ensures maintainability of the system and its components and minimizes the likelihood of errors. Documentation should be based on a defined standard and consist of descriptions of procedures, instructions to personnel, flowcharts, data flow diagrams, display or report layouts, and other materials that describe the system. System documentation should provide programmers with enough information to understand how the system works to decrease the learning cycle, as well as ensure effective and efficient analysis of program changes and trouble-shooting. Documentation should be updated as the system is modified.

The processing logic of the system should be documented in a manner that is understandable (e.g., using flowcharts, etc.), while containing sufficient detail to allow programmers to accurately support the application. The system's software must also include documentation within the code, with descriptive comments embedded in the body of the source code. These comments should include cross-references to design and requirements documentation. The documentation should describe the sequence of programs and the steps to be taken in case of a processing failure.

User documentation should include automated and manual workflows for initial training and ongoing reference. User reference materials (processes and procedures) should be included as part of the development, implementation, and maintenance of associated application systems. They should be reviewed and approved as part of the acceptance testing. User reference materials should be designed for all levels of user expertise and should instruct them on the use of the application system. Such documentation should be kept current as changes are made to the dependent systems.

Training

Training is an important aspect of any project implementation. Training provides users with the necessary understanding, skills, and tools to effectively and efficiently utilize a system in their daily tasks. Training is critical to deliver a successful implementation because it introduces users to the new system and shows them how to interact with it. Delivering an effective training engages users, motivates them to embrace change, and ultimately assists the organization in achieving its desired business results. On the other hand, the cost of not training users may exceed the investment organizations would make for training purposes in the new system. One reason for this paradox is that it may take users longer times to learn the system on their own and become productive with it.

Effective training and education also enable organizations to realize financial gains in the long term, reducing support costs significantly. This results from users making fewer mistakes and having fewer questions. Training and education along with effective project management are critical factors for a successful implementation of any system.

Support

Continuing user support is another important component needed to ensure a successful implementation. Support includes having a help desk to provide assistance to users, as well as problem reporting solutions allowing the submission, search, and management of problem reports.

Effective support involves strategies to work closely with the users in order to ensure issues are resolved promptly, ultimately enhancing productivity and user experience.

Help desk support ensures that problems experienced by the user are appropriately addressed. A help desk function should provide first-line support to users. Help requests should be monitored to ensure that all problems are resolved in a timely manner. **Trend analysis** should be conducted to identify patterns in problems or solutions. Problems should be analyzed to identify root causes. Procedures need to be in place for escalating problems based on inadequate response or level of impact. Questions that cannot be resolved immediately should be escalated to higher levels of management or expertise.

Organizations with established help desks will need to staff and train help desk personnel to handle the new application system. Good training will minimize the volume of calls to the help desk and thereby keep support costs down. Help desks can be managed efficiently with the use of problem management software, automated telephone systems, expert systems, e-mail, voicemail, etc.

Ongoing user support allows organizations to handle and address incoming user requests in a timely and accurate fashion. For instance, support can be provided by establishing a centralized call center (similar to having a help desk) that not only reports issues with the new system, but also finds the right solution. By assisting users in the appropriate use of the new system, organizations can ensure a successful system implementation.

Operations and Maintenance

No matter how well a system is designed, developed, and/or tested, there will always be problems discovered or enhancements needed after implementation. In this phase, programmers maintain systems by either correcting problems and/or installing necessary enhancements in order to fine-tune the new system, improve its performance, add new capabilities, or meet additional user requirements. Maintenance of systems can be separated into three categories:

■ *Corrective maintenance*—involves resolving errors, flaws, failures or faults in a computer program or system, causing it to produce incorrect or unexpected results. These are commonly known as "bugs." The purpose of corrective maintenance is to fix existing functionality to make it work as opposed to providing new functionality. This type of maintenance can occur at any time during system use and usually is a result of inadequate system testing. Corrective maintenance can be required to accommodate a new type of data that were inadvertently excluded, or to modify code related to an assumption of a specific type of data element or relationship. As an example of the latter, it was assumed in a report that each employee's employment application had an employee requisition (or request to hire) associated with it in the system. However, when users did not see a complete listing of their entire employee applications listed, they discovered that not every employee application had an associated hiring request. In this case, the requirement for each application to be associated with a hiring request was a new system feature provided in the latest software release. As a result, employee applications entered into the system previous to the installation of the new release did not have hiring requests associated with them.
■ *Adaptive maintenance*—results from regulatory and other environmental changes. The purpose of adaptive maintenance is to adapt or adjust to some change in business conditions as opposed to fix existing or provide new functionality. An example of adaptive maintenance is modifications to accommodate changes in tax laws. Annually, federal and state

laws change, which require changes to financial systems and their associated reports. A past example of this type of issue was the Year 2000 (Y2K) problem. Many software programs were written to handle dates up to 1999 and were rewritten at significant costs to handle dates beginning January 1, 2000. Although these changes cost organizations many millions of dollars in maintenance effort, the goal of these changes was not to provide users with new capabilities, but simply to allow users to continue using programs the way they are using them today. Some people argue that fixing code to accommodate Y2K was actually corrective maintenance, as software should have been designed to accommodate years beyond 1999. However, due to the expense and limitations of storage, older systems used two digits to represent the year as a means to minimize the cost and limits of storage.

▪ *Perfective maintenance*—includes incorporation of new user needs and enhancements not met by the current system. The goal of perfective maintenance is to modify software to support new requirements. Perfective maintenance can be relatively simple, such as changing the layout of an input screen or adding new columns to a report. Complex changes can involve sophisticated new functionality. In one example, a university wanted to provide its students with the ability to pay for their fees online. A requirement for such a system involves a number of complexities including the ability to receive, process, and confirm payment. These requirements include additional requirements such as the ability to secure the information and protect the student and institution by maintaining the integrity of the data and information. Along with this, additional requirements are necessary to protect the process in its ability to recover and continue processing, as well as the ability to validate, verify, and audit each transaction.

A reporting system should be established for the users to report system problems and/or enhancements to the programmers, and in turn for the programmers to communicate to the users when they have been fixed or addressed. Such a reporting system should consist of **audit trail**s for problems, their solutions, and enhancements made. The system should document resolution, prioritization, escalation procedures, incident reports, accessibility to configuration, information coordination with change management, and a definition of any dependencies on outside services, among others.

Reporting systems should ensure that all unexpected events, such as errors, problems, etc. are recorded, analyzed, and resolved in a timely manner. Incident reports should be established in the case of significant problems. Escalation procedures should also be in place to ensure that problems are resolved in the most timely and efficient way possible. Escalation procedures include prioritizing problems based on the impact severity as well as the activation of a **business continuity plan** when necessary. A reporting system that is also closely associated with the organization's change management process is essential to ensure that problems are resolved or enhancements being made and, most importantly, to prevent their reoccurrence.

Maintaining a system also requires keeping up-to-date documentation related to the new system. Documentation builds at each phase in the SDLC. System documentation can be created as flowcharts, graphs, tables, or text for organization and ease of reading. System documentation includes:

▪ Source of the data
▪ Data attributes
▪ Input screens
▪ Data validations

- Data selection criteria
- Security procedures
- Description of calculations
- Program design
- Interfaces to other applications
- Control procedures
- Error handling
- Operating instructions
- Archive, purge, and retrieval
- Backup, storage, and recovery

There is a definite correlation between a well-managed system development process and a successful system. A system development process provides an environment that is conducive to successful systems development. Such process increases the probability that a new system will be successful and its internal controls will be effective and reliable.

Additional Risks and Associated Controls Related to the SDLC Phases

Additional risks attributable to the SDLC phases just discussed, and that are significant to the organization and the IT auditor are listed below. These may result in invalid or misleading data, bypassed automated controls, and/or fraud.

- Developers or programmers with unauthorized access to promote incorrect or inappropriate changes to data, application programs, or settings into the production processing environment.
- Changes to applications, databases, networks, and operating systems are not properly authorized and/or their testing is not appropriately performed before implementation into the production environment.
- Change management procedures related to applications, databases, networks, and operating systems are inadequate, ineffective, or inconsistent, thus affecting the stability or manner in which data are processed within the production environment.
- Existing controls and procedures related to data conversion are non-existent, inadequate, or ineffective, thus affecting the quality, stability, or manner in which data are processed within the production environment.

Relevant IT controls and procedures to assess the SDLC process just discussed include ensuring that:

- Business risks and the impact of proposed system changes are evaluated by management before implementation into production environments. Assessment results are used when designing, staffing, and scheduling implementation of changes in order to minimize disruptions to operations.
- Requests for system changes (e.g., upgrades, fixes, emergency changes, etc.) are properly documented and approved by management before any change-related work is done.
- Documentation related to the change implementation is accurate and complete.

- Change documentation includes the date and time at which changes were (or will be) installed.
- Documentation related to the change implementation has been released and communicated to system users.
- System changes are successfully tested before implementation into the production environment.
- Test plans and cases involving complete and representative test data (instead of production data) are approved by application owners and development management.

Additional controls over the change control process are shown in Appendix 3 from Chapter 3. The appendix lists controls applicable to most organizations that are considered guiding procedures for both, IT management and IT auditors.

Approaches to System Development

There are various approaches applicable to system development. Although each approach is unique, they all have similar steps that must be completed. For example, each approach will have to define user requirements, design programs to fulfill those requirements, verify that programs work as intended, and implement the system. IT auditors need to understand the different approaches, the risks associated with the particular approach, and help ensure that all the necessary components (controls) are included in the development process. Following are descriptions of common system development approaches starting from the traditional waterfall system development method. Other modern and non-sequential methods, such as agile and lightweight methodologies (e.g., Scrum, Kanban, Extreme Programming (XP), etc.) are also discussed.

Waterfall System Development

The waterfall (also referred to as the traditional method) approach to system development is a sequential process with defined phases beginning with the identification of a need and ending with implementation of the system. The traditional approach uses a structured SDLC that provides a framework for planning and developing application systems. Although there are many variations of this traditional method, they all have the seven common phases just discussed: Planning, System Analysis and Requirements, System Design, Development, Testing, Implementation, and operations and Maintenance. Refer to Exhibit 8.4 for an illustration of the Waterfall development approach.

Although the waterfall development process provides structure and organization to systems development, it is not without risks. The waterfall approach can be a long development process that is costly due to the amount of resources and length of time required. The business environment may change between the time the requirements are defined and when the system is implemented. The users may have a long delay before they see how the system will look and feel. To compensate for these challenges, a project can be broken down into smaller subprojects where modules are designed, coded, and tested. The challenge in this approach is to bring all the modules together at the end of the project to test and implement the fully functional system.

Agile System Development

Agile system development practices are transforming the business of creating and/or maintaining information systems. Agile means capable to move quickly and easily. The Agile System

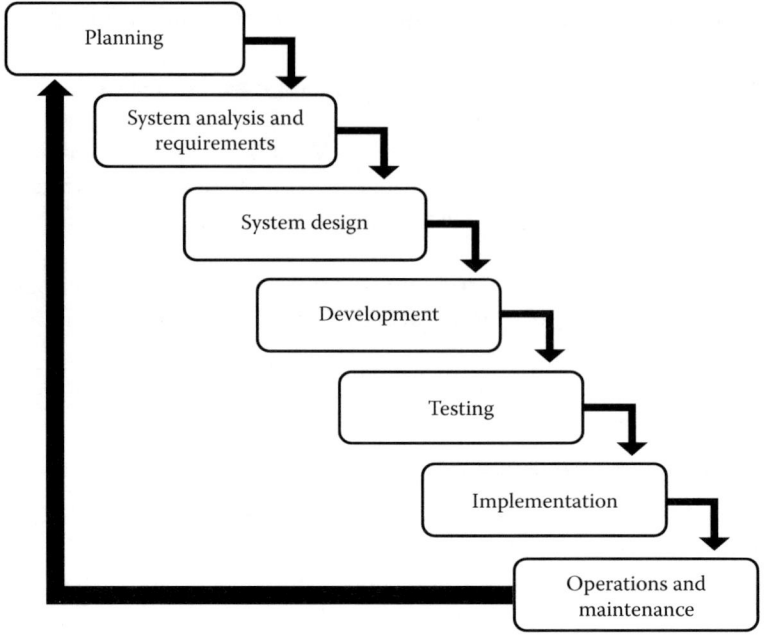

Exhibit 8.4   Waterfall system development.

Development methodology (ASD) is used on projects that need extreme agility in requirements (e.g., to deliver products to the customer rapidly and continuously, etc.). ASD focuses on the adaptability to changing situations and constant feedback. With ASD, there is no clearly defined end product at the beginning stage. This is contrary to the traditional waterfall approach, which requires end-product detailed requirements to be set at the starting phase. ASD's key features involve short-termed delivery cycles (or sprints), agile requirements, a dynamic team culture, less restrictive project control, and emphasis on real-time communication. Even though ASD is most commonly used in software development projects, the approach can also assists other types of projects. The ASD approach is typically a good choice for relatively smaller software projects or projects with accelerated development schedules. Refer to Exhibit 8.5 for an illustration of the Agile development approach.

Agile practices are growing in use by industry. In a study performed by Protiviti in 2016, 44% of companies overall, including 58% of technology companies and 53% of consumer products and retail, were investing in and adopting these practices. Thus, it is safe to conclude that ASD will continue to be a standard practice for a significant percentage of IT functions. Common ASD methodologies include Scrum, Kanban, and Extreme Programming, and they are discussed next.

1. *Scrum*. A derivative of the ASD approach, Scrum is an iterative and incremental software development framework for managing product development. Its main goal is to engage a flexible, holistic product development strategy that improves productivity by enabling small, cross-functional, and self-managing teams work as a unit to reach a common goal. As an iterative/agile approach, Scrum promotes various "sessions" (also referred to as "sprints"), which typically last for 30 days. These sessions promote prioritization of tasks and ensure they are completed on a timely manner. Because of this, teams switching to Scrum tend to

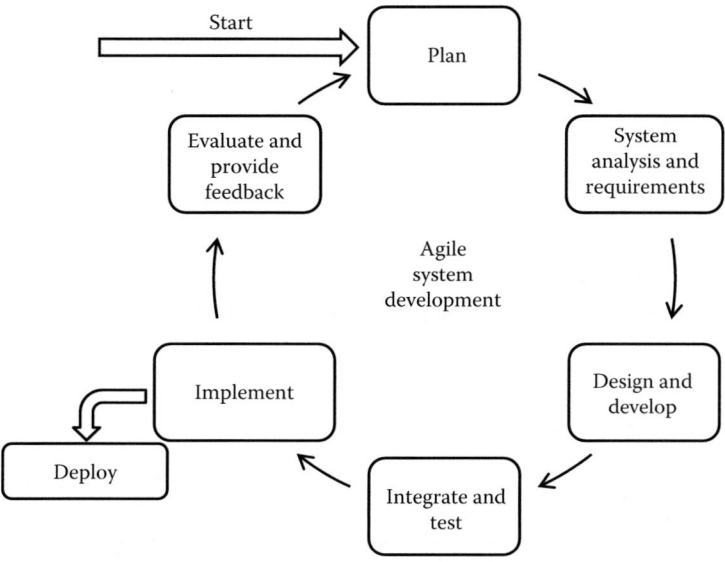

Exhibit 8.5 Agile system development.

see great gains in productivity. Scrum's project manager is referred to as a Scrum Master. The Scrum Master's main responsibility is to enable daily project communications and deal with distractions between team members preventing the successful completion of the job at hand. The Scrum Master conducts regular meetings with the teams to discuss status, progress, results, and timelines, among others. These meetings are also very useful to identify either new tasks or existing ones that need to be reprioritized. Scrum is applicable in certain types of environments, particularly those with members located at the same physical location where face-to-face collaboration among the team members is possible and practiced (i.e., co-located teams).

2. *Kanban.* Kanban is also a type of agile methodology that is used to increase visibility of the actual development work, allowing for better understanding of the work flow, as well as rapid identification of its status and progress. Visualizing the flow of work is also useful in order to balance demand with available capacity. With Kanban, development work items and tasks are pictured to provide team members a better idea of "what's going on" and "what's left to finish." Kanban diagrams or graphs are also typically used to depict general categories of activities or tasks, such as "activities-in-progress," "activities-in-queue," or "activities that have been just completed." This visualization allows team members, including management personnel, to view current work and what is left to complete; reprioritize if necessary; and assess the effect of additional, last-minute tasks should their incorporation become required. Kanban focuses on the actual work from small, co-located project teams rather than on individuals' activities (though many individuals also promote the use of personal Kanban boards). It is argued that Kanban exposes (visualizes) operational problems early, and stimulates collaboration to correct them and improve the system. There are six general practices used in Kanban: visualization, limiting work in progress, flow management, making policies explicit, using feedback loops, and collaborative or experimental evolution.

3. *Extreme Programming.* Extreme Programming (XP) is another type of agile software development methodology intended to improve productivity and quality by taking traditional

software engineering basic elements and practices to "extreme" levels. For instance, incorporating continuous code review checkpoints (rather than having just the traditional, one-time-only code review) on which new customer requirements can be evaluated, added, and processed. Another example of reaching "extreme" levels would be the implementation of automated tests (perhaps inside of software modules) to validate the operation and functionality of small sections of the code, rather than testing only the larger features. XP's goal is to increase a software organization's responsiveness while decreasing development overhead. Similar to Scrum and other agile methods, XP focuses on delivering executable code and effectively and efficiently utilizing personnel throughout the software development process. XP emphasizes on fine scale feedback, continuous process, shared understanding, and programmer welfare.

Adaptive Software Development

Adaptive Software Development (ASWD) is a development approach designed for building complex software and systems. It is focused on rapid creation and evolution of software systems (i.e., consistent with the principle of continuous adaptation). ASWD follows a dynamic lifecycle instead of the traditional, static lifecycle *Plan-Design-Build*. It is characterized by constant change, re-evaluation, as well as peering into an uncertain future and intense collaboration among developers, testers, and customers.

ASWD is similar to the rapid application development approach. It replaces the traditional waterfall cycle with a repeating series of speculate, collaborate, and learn cycles. These dynamic cycles provide for continuous learning and adaptation to the emergent state of the project. During these cycles or iterations, knowledge results from making small mistakes based on false assumptions (speculate), re-organizing teams to work together in finding a solution (collaborate), and finally correcting (and becoming proficient with) those mistakes (learn), thus leading to greater experience and eventually mastery in the problem domain. Refer to Exhibit 8.6 for an illustration of the ASWD approach.

Joint Application Development

Joint Application Development (JAD) is an approach or methodology developed in the late 1970s that involves participation of either the client or end user in the stages of design and development of an information system, through a succession of collaborative workshops called JAD sessions. Through these JAD sessions, end users, clients, business staff, IT auditors, IT specialists, and other technical staff, among others are able to resolve their difficulties or differences concerning

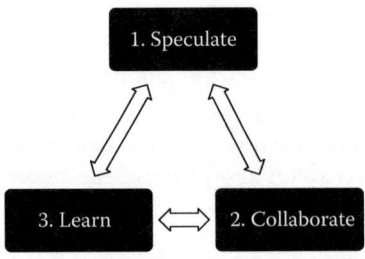

Exhibit 8.6   Adaptive software development.

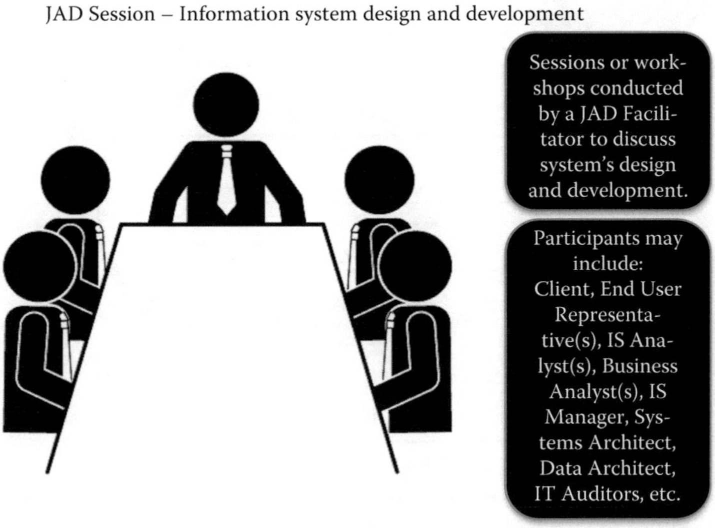

JAD Session – Information system design and development

Sessions or workshops conducted by a JAD Facilitator to discuss system's design and development.

Participants may include: Client, End User Representative(s), IS Analyst(s), Business Analyst(s), IS Manager, Systems Architect, Data Architect, IT Auditors, etc.

Exhibit 8.7   Joint application design/development.

the new information system. The sessions follow a detailed agenda in order to prevent any miscommunications as well as to guarantee that all uncertainties between the parties are covered. Miscommunications can carry far more serious repercussions if not addressed until later on in the process.

The JAD approach leads to faster development times and greater client satisfaction than the traditional approach because the client is involved throughout the whole design and development processes. In the traditional approach, on the other hand, the developer investigates the system requirements and develops an application with client input typically consisting of an initial interview.

A variation on JAD is prototyping and rapid application development, which creates applications in faster times through strategies, such as using fewer formal methodologies and reusing software components. In the end, JAD results in a new information system that is feasible and appealing to both the client and end users. Refer to Exhibit 8.7 for an illustration of the JAD approach.

Prototyping and Rapid Application Development

In general, Prototyping and Rapid Application Development (RAD) includes:

- the transformation and quick design of the user's basic requirements into a working model (i.e., prototype);
- the building of the prototype;
- the revision and enhancement of the prototype; and
- the decision to whether accept the prototype as the final simulation of the actual system (hence, no further changes needed), or go back to redesign and work with the user requirements.

Exhibit 8.8 illustrates this Prototyping and RAD process.

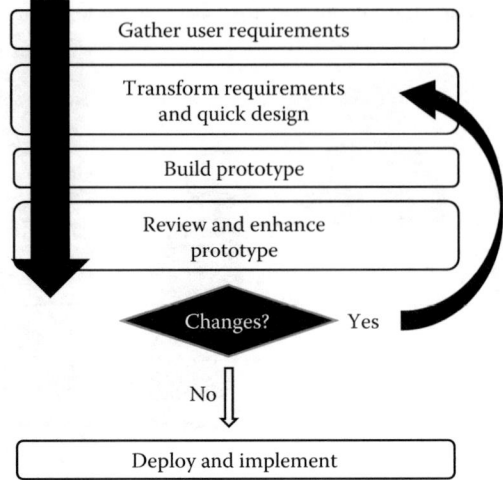

Exhibit 8.8 Prototyping and RAD process.

Prototyping and RAD can facilitate interaction between the users, system analysts, and the IT auditor. These techniques can be applied to production report development, a specific application module, or the entire support system. Some advantages of prototyping and RAD include:

- Prototypes can be viewed and analyzed before commitment of large funding for systems.
- User approval and final satisfaction is enhanced because of increased participation in the design of the project.
- The cost of modifying systems is reduced because users and designers can foresee problems earlier and are able to respond to the users' rapidly changing business environment.
- A rudimentary prototype can be redesigned and enhanced many times before the final form is accepted.
- Many systems are designed "from scratch" and no current system exists to serve as a guide.

On the other hand, because prototypes appear to be final when presented to the users, programmers may not be given adequate time to complete the system and implement the prototype as the final product. Often the user will attempt to use the prototype instead of the full delivery system. The user must understand that the prototype is not a completed system. Risks associated with prototyping and RAD include:

- Incomplete system design
- Inefficient processing performance
- Inadequate application controls
- Inadequate documentation
- Ineffective implementations

Lean Software Development

Lean Software Development (LSD) is a translation of lean manufacturing and lean IT principles and practices to the software development domain. LSD is a type of agile approach that can be summarized by seven key principles, which are very close in concept to lean manufacturing principles. They are:

1. Eliminate waste—identify what creates value to the customer
2. Build quality in—integrate quality in the process; prevent defects
3. Create knowledge—investigate and correct errors as they occur; challenge and improve standards; learn from mistakes
4. Defer commitment—learn constantly; perform only when needed, and perform fast
5. Deliver as fast as possible—deliver value to customer quickly; high quality and low cost
6. Empower team and respect people—engage everyone; build integrity; provide stable environment
7. Optimize the whole—deliver complete product; monitor for quality; continuous improvement

Refer to Exhibit 8.9 for an illustration of the LSD approach.

End-User Development

End-user development (EUD) (also known as end-user computing) refers to applications that are created, operated, and maintained by people who are not professional software developers (end users). There are many factors that have led the end user to build their own systems. First, and probably foremost, is the shift in technology toward personal computers (PCs) and generation programming languages (e.g., **fourth-generation languages** [4GL], 5GL, etc.). This shift has been due, in part, to the declining hardware and software costs that have enabled individuals to own computers. Because of this, individuals have become more computer literate. At the same time, users are frustrated with the length of time that it takes for traditional systems development efforts to be completed. Fourth-generation programming languages, for example, have provided users with the tools to create their own applications. Examples of such tools include:

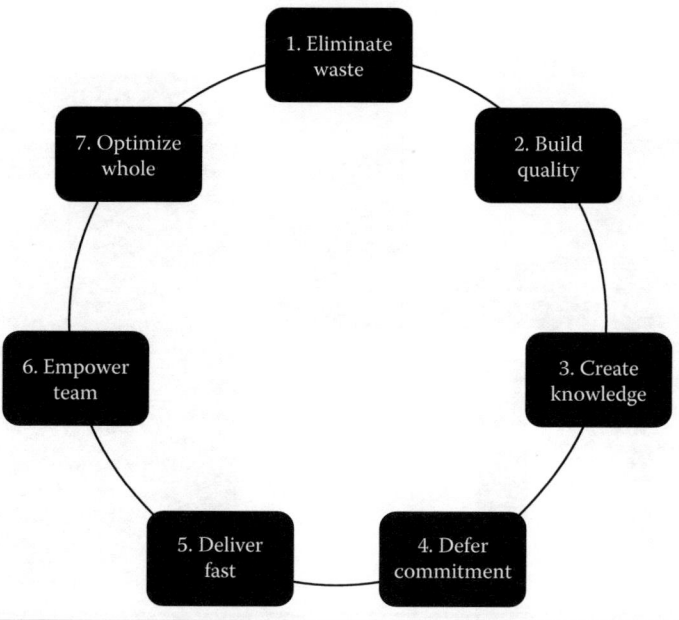

Exhibit 8.9 Lean software development.

- Mainframe-based query tools that enable end users to develop and maintain reports. This includes fourth-generation languages such as EZ-TRIEVE and SAS or programmer-developed report generation applications using query languages.
- Vendor packages that automate a generic business process. This includes accounting packages for generating financial statements and legal packages for case management.
- EUD applications using PC-based tools, databases, or spreadsheets to fulfill a department or individual information processing need.

Because PCs seem relatively simple and are perceived as personal productivity tools, their effect on an organization has largely been ignored. In many organizations, EUD applications have limited or no formal procedures. End users may not have the background knowledge to develop applications with adequate controls or maintainability. This becomes an issue when organizations rely on user-developed systems for day-to-day operations and important decision making. Simultaneously, end-user systems are becoming more complex and are distributed across platforms and organizational boundaries. Some of the risks associated with EUD applications include the following.

- Higher organizational costs
- Incompatible systems
- Redundant systems
- Ineffective implementations
- Absence of segregation of duties
- Incomplete system analysis
- Unauthorized access to data or programs
- Copyright violations
- Destruction of information by computer viruses
- Lack of back-up and recovery options

Exhibit 8.10 summarizes the EUD approach to system development.

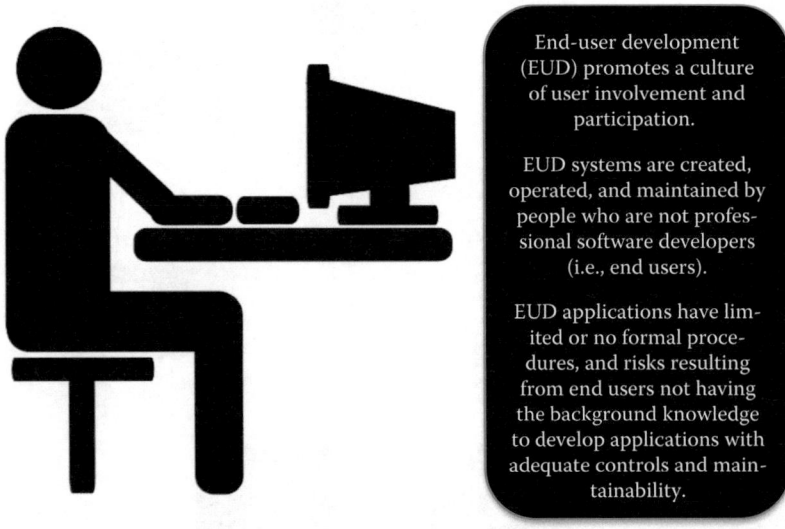

End-user development (EUD) promotes a culture of user involvement and participation.

EUD systems are created, operated, and maintained by people who are not professional software developers (i.e., end users).

EUD applications have limited or no formal procedures, and risks resulting from end users not having the background knowledge to develop applications with adequate controls and maintainability.

Exhibit 8.10   End-user development.

IT Auditor's Involvement in System Development and Implementation

IT auditors can assist organizations by reviewing their systems development and implementation (SD&I) projects to ensure that new systems comply with the organization's strategy and standards. Each SD&I project will need to be risk assessed to determine the level of audit's involvement. The type of review will also vary depending on the risks of a particular project. IT auditors may only be involved in key areas or the entire SD&I project. In any case, IT auditors need to understand the process and application controls to add value and ensure adequate controls are built into the system.

SD&I audits are performed to evaluate the administrative controls over the authorization, development, and implementation of new systems (i.e., applications), and to review the design of the controls/audit trails of the proposed system. The scope of a SD&I audit includes an evaluation of the overall SDLC approach or methodology. The audit also focuses on the evaluation of the quality of the deliverables from each system development phase (e.g., evaluation of the controls design and audit trails, system test plan and results, user training, system documentation, etc.). Recommendations from SD&I audits might include improvements in user requirements, application controls, or the need to document test plans and expected test results.

Developing and implementing new systems can be a costly and time-consuming endeavor. A well-controlled environment with an overall strategy, standards, policies, and procedures helps ensure the success of development efforts. There are many processes that need to be well controlled to ensure the overall success of a system. Because of the significant cost to implement controls after a system has already gone into production, controls should be defined before a system is built.

There are many opportunities for auditor involvement in the SD&I process. IT auditors need to develop the skills and relationships to work with the SD&I team to ensure that controls are built into the system. IT auditors can assist organizations by:

- reviewing the SD&I environment
- evaluating standards for SD&I
- evaluating phases in the SD&I process
- reviewing critical systems for input, processing, and output
- verifying that the new system provides an adequate audit trail

The IT auditor's role in a SD&I project depends on the organization's culture, maturity of the IS function, and philosophy of the auditing department. Auditing SD&I requires specific knowledge about the process (i.e., development and implementation) and application controls. Understanding the process allows the auditor to identify key areas that would benefit from independent verification. Understanding application controls allows the auditor to evaluate and recommend controls to ensure complete and accurate transaction processing.

IT auditors can take on two different roles in a SD&I project: control consultant or independent reviewer.

- As a control consultant, the auditor becomes a member of the SD&I team and works with analysts and programmers to design application controls. In this role, the auditor is no longer independent of the SD&I team.
- As an independent reviewer, the auditor has no design responsibilities and does not report to the team, but can provide recommendations to be acted on or not by the project/system manager.

By becoming involved at strategic points, the auditor can ensure that a system is well controlled and auditable. The following highlights some of the key responsibilities for the auditor when involved in a SD&I project:

1. Review user requirements
2. Review manual and application controls
3. Check all technical specifications for compliance with company standards
4. Perform design walk-throughs at the end of each development phase
5. Submit written recommendations for approval after each walk-through
6. Ensure implementation of recommendations before beginning the next phase
7. Review test plans
8. Present findings to management
9. Maintain independence to remain objective

These can help minimize control weaknesses and problems before the system is implemented in production and becomes operational rather than after it is in use.

IT auditors determine their level of involvement in a SD&I audit by completing a risk assessment of the SD&I process. Results from the risk assessment also prompt the amount of time necessary to allocate to the particular project, required resources, etc. Preparation of an audit plan follows. The plan describes the audit objectives and procedures to be performed in each phase of the SD&I process. IT auditors communicate not only the scope of his/her involvement, but also findings and recommendations to development personnel, users, and management resulting from the audit.

Risk Assessment

IT auditors may not have enough time to be involved in all phases of every SD&I project. Involvement will depend on the assessment of process and application risks. Process risks may include negative organizational climate, as well as the lack of strategic direction, development standards, and of a formal systems development process. Application risks, on the other hand, relate to application complexity and magnitude; inexperienced staff; lack of end-user involvement; and lack of management commitment.

The level of risk may be a function of the need for timely information, complexity of the application, degree of reliance for important decisions, length of time the application will be used, and the number of people who will use it.

The risk assessment defines which aspects of a particular system or application are covered by the audit. Depending on the risk, the scope of the assessment may include evaluating system requirements, as well as reviewing design and testing deliverables, application controls, operational controls, security, problem management, change controls, or the post-implementation phase.

Audit Plan

IT auditors may be involved in the planning process of a SD&I project in order to: develop an understanding of the proposed system; ensure time is built into the schedule to define controls; and verify that all the right people are involved. The audit plan will also detail the steps and procedures to fulfill the audit objectives. As in any audit, a SD&I audit begins with a preliminary analysis of the control environment by reviewing existing standards, policies, and procedures.

During the audit, these standards, policies, and procedures should be assessed for completeness and operational efficiency. The preliminary analysis should identify the organization's strategy and the responsibilities for managing and controlling applications.

The audit plan will further document the necessary procedures to review the SD&I process to ensure that the system is designed consistent with user requirements, that management approves such design, and that the system or application is adequately tested before implementation. An additional focus of the audit plan is making certain that the end user is able to use the system based on a combination of skills and supporting documentation.

A SD&I audit assesses the adequacy of the control environment for developing effective systems to provide reasonable assurance that the following tasks are performed:

- Comply with standards, policies, and procedures
- Achieve efficient and economical operations
- Conform systems to legal requirements
- Include the necessary controls to protect against loss or serious errors
- Provide controls and audit trails needed for management, auditor, and for operational review purposes
- Document an understanding of the system (also required for appropriate maintenance and auditing)

For any kind of a partnership involving IT auditors, users, and IS management, it is important that the organization plans for and establishes a formal procedure for the development and implementation of a system. Auditor influence is significantly increased when there are formal procedures and required guidelines identifying each phase and project deliverable in the SDLC and the extent of auditor involvement. Without formal SDLC procedures, the auditor's job is much more difficult and recommendations may not be as readily accepted. Formal procedures in place allow IT auditors to:

- Review all relevant areas and phases of the SDLC
- Identify any missing areas for the development team
- Report independently to management on the adherence to planned objectives and procedures
- Identify selected parts of the system and become involved in the technical aspects based on their skills and abilities
- Provide an evaluation of the methods and techniques applied in the SD&I process, as defined earlier

The audit plan must also document the auditor's activities and responsibilities (tasks) to be performed within each remaining SD&I phase (i.e., System Analysis and Requirements, System Design, Development, Testing, Implementation, and Operations and Maintenance). These are described below.

Auditor Task: System Analysis and Requirements

The project team typically expends considerable effort toward the analysis of the business problem and what the system is to produce without initially attempting to develop the design of the system. The IT auditor should observe that the primary responsibility is not to develop a product but to satisfy the user. Often, the user does not understand what is truly needed. Only by understanding the user's business, its problems, goals, constraints, weaknesses, and strengths can the project team deliver the product the user needs. IT auditors can participate by reviewing requirements, and verifying user understanding and sign-off. A good checkpoint for IT auditors is to ensure that in the System Analysis and Requirements phase, defining security requirements is included. The IT auditor should identify and document security requirements early in the development life cycle, and make sure that subsequent development artifacts are evaluated for compliance with those requirements. When security requirements are not defined, the security of the resulting system cannot be effectively evaluated and the cost to retrofit into the system later in its life cycle can be extremely costly.

Auditor Task: System Design

The IT auditor may review the design work for any possible exposures or forgotten controls, as well as for adherence with company standards, policies, and procedures. Standards, policies, and procedures should be documented as part of the SDLC methodology and defined before the beginning of the project. If exposures, missing controls, and/or lack of compliance are identified, the IT auditor should recommend the appropriate controls or procedures.

As seen earlier, a methodology or technique that brings users and project team members together for an intensive workshop in which they create a system proposal into a detail design is JAD. Usually a trained JAD facilitator, having some claim to neutrality, takes the group through formatted discussions or sessions of the system. The IT auditor may be an active participant in this process. The result of the JAD session is a user view of the system for further development. This is an excellent setting for the discussion of the advantages and cost effectiveness of controls. In addition, analysis time is compressed, discrepancies resolved, specification errors reduced, and communications greatly enhanced. IT auditors can review deliverables and recommend application controls. Application controls are discussed in more detail in a later chapter.

Auditor Task: Development

The IT auditor may review the new system's programs to verify compliance with programming and coding standards. These standards help ensure that the code is well-structured, tracks dependencies, and makes maintenance easier. The IT auditor may review a sample of programs to verify that the standards are being followed and that the programs conform to systems design. In addition, programs may be checked for possible control exposures and for the placement of proper controls per design. If it is determined that controls are needed, the IT auditor should make recommendations, following the same criteria that were used during the System Design phase. During this Development phase, however, cost and time factors must be carefully considered because the cost of changing programs to include controls increases as the project progresses.

Auditor Task: Testing

The IT auditor may be called on to assure management that both, developers and users, have thoroughly tested the system to ensure that it:

- possesses the built-in controls necessary to provide reasonable assurance of proper operation;
- provides the capability to track events through the systems and, thus, supports audit review of the system in operation; and
- meets the needs of the user and management.

If the level of testing does not meet standards, the IT auditor must notify the development team or management who should then take corrective action.

Auditor Task: Implementation

System implementation is a key IT audit review point because implementation is often where critical controls may be overwritten or deactivated to bring the system up and operational to meet organizational needs and requirements. The IT auditor should review implementation materials related to strategy, communication, training, documentation, and conversion procedures, among others. Production readiness should also be reviewed, which may include evaluating the readiness of the system in relation to the results of testing, the readiness of production support programmers, computer operations, and users in terms of training, and the readiness of the help desk with trained staff and a problem-tracking process.

Once the system is implemented in production, the IT auditor may survey users to: evaluate its effectiveness from a workflow perspective; review error detection and correction procedures to confirm they are working as intended; and perform tests of data to confirm completeness of transaction processing and audit trail.

Auditor Task: Operations and Maintenance

In this phase, the IT auditor evaluates post-implementation processes and procedures. For instance, code modifications and testing procedures should be assessed in order to determine whether the organization's standards, policies, and/or procedures are being followed. The IT auditor also conducts procedures to ensure systems are well maintained; that is, programmers correct problems or make necessary enhancements in a timely and adequate fashion. When maintaining application systems, the IT auditor must ensure that correction of problems and/or installation of enhancements are both worked in a separate test environment, and upon successful results, promoted to the production environment. There are other common metrics that should be reviewed by the IT auditor in order to evaluate the effectiveness and efficiency of the maintenance process:

■ The ratio of actual maintenance cost per application versus the average of all applications.
■ Requested average time to deliver change requests.
■ The number of change requests for the application that were related to bugs, critical errors, and new functional specifications.
■ The number of production problems per application and per respective maintenance changes.
■ The number of divergence from standard procedures, such as undocumented applications, unapproved design, and testing reductions.
■ The number of modules returned to development due to errors discovered in acceptance testing.
■ Time elapsed to analyze and fix problems.
■ Percent of application software effectively documented for maintenance.

Another relevant procedure performed during this phase is the review and assessment of all related system, user, or operating documentation. Such documentation should be evaluated for completeness and accuracy. Documentation should be practical and easily understandable by all user types (e.g., end users, programmers, senior management, etc.). Diagrams of information flow, samples of possible input documents/screens, and output reports are some of examples of information that enhance user understanding of the system and, therefore, should be documented.

Exhibit 8.11 illustrates a template of a standard audit checklist that can be used as a starting point when assessing the SDLC phases for a project. The checklist is based on standard *ISO/IEC 12207:2013- Systems and Software engineering Software Life Cycle Processes*, which provides guidance for defining, developing, controlling, improving, and maintaining system/software life cycle processes. The standard can also be adapted according to the particular system/software project.

Exhibit 8.11   Sample SDLC Audit Checklist

Sample SDLC Audit Checklist: Development and Implementation of a Financial Application System		
Task	Yes, No, N/A	Comments
Phase 1: Planning		
1. Establish and prepare an overall project plan with defined tasks and deliverables.		
2. Plan includes scope of the work (e.g., period, name of the new system, schedule, restrictions, etc.), necessary resources, and required deadlines.		
3. Plan describes the extent of the responsibilities of all involved personnel (e.g., management, internal audit, end users, quality assurance (QA), etc.)		
4. Plan identifies equipment requirements, such as the hardware configuration needed to use the new system (e.g., processing speed, storage space, transmission media, etc.).		
5. Plan includes detailed financial analyses, including costs to develop and operate the new system and the return on investment.		
6. Plan is reviewed and approved at appropriate levels.		
Phase 2: System Analysis and Requirements		
1. Analysis includes a study to determine whether the new system should be developed or purchased.		
2. Analysis includes a study of the current system to identify existing processes and procedures that may continue in the new system.		
3. Procedures for performing a needs analysis are appropriately assessed, and conform to the organization's standards, policies, and/or procedures.		
4. Expectations from end users and system analysts for the new or modified system are clearly translated into requirements.		

(Continued)

Exhibit 8.11 (*Continued*) Sample SDLC Audit Checklist

Sample SDLC Audit Checklist: Development and Implementation of a Financial Application System		
Task	*Yes, No, N/A*	*Comments*
5. Requirements of the new software/system are defined in terms that can be measured.		
6. Requirements are quantifiable, measurable, relevant, and detailed.		
Phase 3: System Design		
1. Design describes the proposed system flow and other information on how the new system will operate (i.e., conceptual design).		
2. System design specifications, features, and operations meet the requirements previously defined.		
3. System design specifications are approved and comply with the organization's standards, policies, and/or procedures.		
4. The systems analyst defines and documents all system interfaces, reporting, screen layouts, and specific program logic necessary to build the system consistent with the requirements.		
5. The systems analyst and end users review, and ensure that, the specific business needs have been translated into the final requirements for the new system.		
6. Technical details of the proposed system (e.g., hardware and/or software needed, networking capabilities, procedures for the system to accomplish its objectives, etc.) have been discussed with appropriate stakeholders.		
7. The design of the system describes controls for input points, processing, and screen layout/output.		
8. The design of the system incorporates audit trails and programmed controls.		
Phase 4: Development		
1. Obtain and review the source/program code for the new or modified system or application.		
2. Determine whether the source/program code meets the organization's programming standards, policies, and/or procedures.		
3. The source/program code is tested for both syntax and logic flow.		

(Continued)

Exhibit 8.11 (*Continued*)   Sample SDLC Audit Checklist

Sample SDLC Audit Checklist: Development and Implementation of a Financial Application System		
Task	*Yes, No, N/A*	*Comments*
4. All logic paths within the source/program code are exercised to ensure error routines work and the program terminates processing normally.		
5. Logical security access controls are configured and incorporated within the source/program code.		
6. Security controls configured are designed to address requirements related to the confidentiality, integrity, and availability of information.		
7. Security controls configured are designed to address authorization and authentication processes, as well as business access requirements and monitoring.		
8. The source/program code is validated through individual unit testing (thorough system testing is assessed in the *Testing* phase). The Development phase is final once the source/program code is validated.		
Phase 5: Testing		
1. A plan for system testing is prepared that conforms with organization's standards, policies, and/or procedures.		
2. The testing plan defines: individual test events and scenarios; roles and responsibilities of the test participants; test environments; acceptance criteria; testing logistics; problem reporting and tracking; test deliverables; and personnel responsible for review and approval of tests and test results.		
3. Testing is based on existing testing methodologies established by the organization.		
4. Tests include (real or generated) test data that are representative of relevant business scenarios.		
5. Test scenarios, associated data, and expected results are documented for every test condition.		
6. Testing performed is documented to prevent duplicate testing efforts.		
7. End users perform the testing; not developers or programmers.		
8. Testing is performed in separate/development environments; not in production environments.		

(*Continued*)

Exhibit 8.11 (*Continued*) Sample SDLC Audit Checklist

Sample SDLC Audit Checklist: Development and Implementation of a Financial Application System		
Task	*Yes, No, N/A*	*Comments*
9. Test results are signed off and approved, as applicable, to support that appropriate testing was performed and ensure such testing results are consistent with the requirements.		
10. Systems with unsuccessful test results are not implemented in the production environment.		
11. Upon successful test results, management personnel sign-off and approve promotion of new or modified systems into production environments.		
12. The system is assessed by a QA professional to verify that it works as intended and that it meets all design specifications.		
13. Documented testing procedures, test data, and resulting outputs are reviewed to determine if they are comprehensive and follow the organization's standards, policies, and/or procedures.		
14. System testing results validate that the system works as expected and that all errors, flaws, failures or faults identified have been corrected and do not prevent the system from operating effectively.		
Phase 6: Implementation		
1. An implementation plan is documented and put in place to guide the implementation team and users throughout the implementation process.		
2. The implementation plan covers the implementation schedule, conversion procedures (if any), resources required, roles and responsibilities of team members, means of communication, issue management procedures, and training plans for the implementation process.		
3. The implementation plan includes the date and time at which the new system or changes to an existing one will be installed.		
4. Documentation related to system implementation has been released and communicated to the users.		
5. System was successfully tested before its implementation into the production environment consistent with organization standards, policies, and procedures.		
6. Documentation related to implementation provides programmers with enough information to understand how the system works, and to ensure effective and efficient analysis of changes and troubleshooting.		

(*Continued*)

Exhibit 8.11 (*Continued*)   Sample SDLC Audit Checklist

Sample SDLC Audit Checklist: Development and Implementation of a Financial Application System		
Task	*Yes, No, N/A*	*Comments*
7. Documentation related to implementation is updated as the system is modified.		
8. Training procedures have been incorporated in the system implementation process.		
9. Support (e.g., via a help desk, etc.) is provided to users following implementation.		
10. Determine if the organization's standards, policies, and/or procedures are followed and if documentation supporting compliance with the standards is available.		
Phase 7: Operations and Maintenance		
1. Review and evaluate the procedures for performing post-implementation reviews.		
2. Review system modifications, testing procedures, and supporting documentation to determine if the organization's standards, policies, and/or procedures have been followed.		
3. A reporting system is established for users to communicate problems or enhancements needed.		
4. Procedures are in place for programmers to correct problems with the new system or make necessary enhancements.		
5. Correction of problems or enhancements to the new system are worked in a separate/test environment, and upon successful results, promoted to the production environment.		

Communication

The first area to communicate is the IT auditor's scope of involvement in the SD&I project. It is very important to make sure that the management and development teams' expectations of the IT auditor's role are understood and communicated to all participants. To influence the SD&I effort, the IT auditor must develop an open line of communication with both management and users. If a good relationship between these groups does not exist, information might be withheld from the IT auditor. This type of situation could prevent the IT auditor from doing the best job possible. In addition, the IT auditor must develop a good working relationship with analysts and programmers. Although the IT auditor should cultivate good working relationships with all groups with design responsibilities, the IT auditor must remain independent.

Throughout the SD&I project, the IT auditor will be making control recommendations resulting from identified findings. Depending on the organization's culture, these recommendations may need to be handled informally by reviewing designs with the project team or formally by presenting them to the steering committee. In either case, the IT auditor must always consider the value of the control recommendation versus the cost of implementing the control. Recommendations should be specific. They should identify the problem and not the symptom, and allow for the proper control(s) to be implemented and tested. Findings, risks as a result of those findings, and audit recommendations are usually documented in a formal letter (i.e., Management Letter). Refer to Exhibit 3.9 on Chapter 3 for an example of the format of a Management Letter from an IT audit.

On receipt of the Management Letter, IT management and affected staff should review the document. Issues and matters not already completed should be handled and followed-up. Within a relatively short time, the fact that all discrepancies have been corrected should be transmitted to the audit staff in a formal manner. These actions are noted in the audit files, and such cooperation reflects favorably in future audits.

Recommendations may often be rejected though because of a time and cost factor. Managers may sometimes feel that implementing an auditor's recommendations will delay their schedule. The IT auditor must convince management of the value of the recommendations, and that if they are not implemented, more time and money will be spent in the long run. Informing management of the cost of implementing a control now, rather than shutting down the system later (leaving potential exposures open), will help convince management of the need to take appropriate and immediate action.

Conclusion

Developing new systems can be a costly and time-consuming endeavor. A well-controlled environment with an overall strategy, standards, policies, and procedures in place helps ensure the success of system development and implementation efforts. There are many processes that need to be followed to ensure the overall success of a system. These processes or phases are provided by a SDLC. The SDLC provides a framework for effectively developing application systems. It specifically describes a standard process for planning, analyzing, designing, creating, testing, deploying, and maintaining information systems (i.e., new development or modified system).

Risks related to the SDLC phases should constantly be assessed by the organization. These risks are significant to both the organization and the IT auditor and should prompt for the identification (and implementation) of controls that can mitigate them. Because of the cost to implement controls after a system has already been implemented into production, controls should be defined before a system is built.

There are various approaches applicable to system development. Although each approach is unique, they all have similar steps that must be completed. For example, each approach will have to define user requirements, design programs to fulfill those requirements, verify that programs work as intended, and implement the system. IT auditors need to understand the different approaches, the risks associated with the particular approach, and help ensure that all the necessary procedures and controls are included in the development process.

There are many opportunities for auditor involvement in the SD&I process. IT auditors can assist organizations by reviewing the SD&I environment; evaluating standards for SD&I; monitoring project progress; evaluating phases in the SD&I process; reviewing critical systems for input, processing, and output; verifying that the new system provides an adequate audit trail; and by ensuring that risks are identified and proper controls are considered during the implementation process.

SD&I audits, for example, are performed to evaluate the administrative controls over the authorization, development, and implementation of new systems (i.e., applications), and to review the design of the controls/audit trails of the proposed system. The scope of a SD&I audit includes an evaluation of the overall SDLC approach or methodology. The audit also focuses on the evaluation of the quality of the deliverables from each system development phase (e.g., evaluation of the controls design and audit trails, system test plan and results, user training, system documentation, etc.). Recommendations from SD&I audits might include improvements in user requirements, application controls, or the need to document test plans and expected test results.

Review Questions

1. How does a system development life cycle (SDLC) provide an environment that is conducive to successful systems development?
2. Describe the purpose of test data.
3. Explain what conversion procedures referred to as part of implementing a new system.
4. Why should disaster recovery plans be addressed during an implementation as opposed to after?
5. Why is a help desk function critical to system development? Discuss its interrelationship with the problem management and reporting system.
6. Why is it necessary for programmers to have good documentation as part of the operations and maintenance phase of the SDLC?
7. Discuss how the IT auditor can benefit an organization's system development and implementation process.
8. Differentiate between the two roles IT auditors can take on in a SD&I project.
9. What methodology or technique is used to bring users and project team members together to create a detail design?
10. Throughout the system development and implementation project, the IT auditor will make control recommendations to management resulting from identified findings. Explain why recommendations from IT auditors may often be rejected.

Exercises

1. Summarize the common phases in the traditional system development life cycle (SDLC) approach.
2. A company is developing a new system. As the internal IT auditor, you recommend that planning for the new system development should be consistent with the SDLC framework. IT personnel have identified the following as major activities to be completed within the upcoming system development.
   - Ensure Help desk is in place to provide support
   - Integration of security access controls within the code
   - Correct problems and implement enhancements

   – Data conversion procedures
   – Discussion of technical details and specification
   – Functional requirements definition
   – Conduct tests of various scenarios
   – Implementation planning
   – Document user procedures, deliver training
   – Current system review

Arrange these 10 activities in the sequence in which you believe they should take place. In addition, IT personnel is not quite sure about what process to follow when converting existing data files from the old system to the new one. Describe the four data conversion methods available that may be applicable in this case

3. Prepare a one-page, two-column audit program table listing all risks you can think of that are significant to any organization when implementing the SDLC phases. Next to the risks, list relevant IT controls and procedures that should be in place to mitigate the risks listed. Make sure you document at least one IT control for every risk listed.

4. List advantages and disadvantages for each of the System Development approaches discussed in the chapter.

5. Differentiate between the various system test events. Describe what aspects of the system are covered during each event.

6. The chapter highlights nine key responsibilities for auditors when involved in a SD&I project. By becoming involved at strategic points during such process, auditors can ensure that the system being developed and implemented is well controlled and auditable. List and explain in your own words the significance of each of these nine responsibilities.

CASE—COMPLIANCE WITH SDLC STANDARDS AND CONTROLS

INSTRUCTIONS: In order to gain competitive advantage and keep up with business growth, the company you work for just decided to develop and implement a new system. The new state-of-the-art system is also expected to increase productivity and automate current operations, among others. It is a very exciting time for your company. You are the IT Lead in charge of this project, and just finished a planning meeting with senior management. At the meeting, both the CIO and CEO requested that in order to speed up the implementation process, you and your team delay following common SDLC standards as well as adding necessary controls to the new system until after it is implemented in the production environment. This request caught you by surprise and made you feel uncomfortable. You do not feel you must comply with the request.

TASK: Document, in memo format, reasons as to why you should not comply with the CIO and CEO request. You must back up your response (reasons) with IT literature and/ or any other valid external source. Include examples, as appropriate, to evidence your case point. Submit a word file with a cover page, responses to the tasks above, and a reference section at the end. The submitted file should be between 8 and 10 pages long (double line spacing), including cover page and references. Be ready to present your work to the class.

CASE—IT AUDIT INVOLVEMENT IN SYSTEM DEVELOPMENT AND IMPLEMENTATION

INSTRUCTIONS: The president of the company just asked you, Internal IT Audit Manager, to be involved in the upcoming development and implementation of a new information system. The new system is expected to increase productivity by automating current operations. The new system is to be implemented within the next 6 months. The president also asked that the implemented system be ready for (and successfully pass) the next IT external audit.

TASK: Outline the audit procedures you would perform to complete the assigned task. Include, as a minimum, a description of the following:

- Audit plan, including purpose and scope of the involvement; role and responsibilities, etc.
- Audit procedures to be performed on each SDLC phase, including documenting the source of information needed; methods of documenting such information; methods of verifying the information collected, etc.
- Communication of audit involvement, results, and recommendations to management resulting from the audit procedures performed.

Submit a word file with a cover page, responses to the tasks above, and a reference section at the end. The submitted file should be at least eight-page long (double line spacing), including cover page and references. Be ready to present your work to the class.

Further Reading

1. Adaptive software development. *The Ultimate Guide to the SDLC*. http://ultimatesdlc.com/adaptive-software-development/ (accessed June 30, 2017).
2. TechTarget. Automated software testing. http://searchsoftwarequality.techtarget.com/definition/automated-software-testing (accessed July 1, 2017).
3. Software Testing Fundamentals. Black box testing. http://softwaretestingfundamentals.com/black-box-testing/ (accessed July 1, 2017).
4. Deloitte LLP. (2014). *IT Audit Planning Work Papers*. Unpublished internal document.
5. Hettigei, N. (2005). The auditor's role in IT development projects, Information Systems Audit and Control Association. *Inf. Syst. Con. J.*, 4, 44.
6. ISO/IEC 12207:2013- Systems and Software Engineering Software Life Cycle Processes. International Organization for Standardization. www.iso.org/standard/43447.html (accessed July 6, 2017).
7. Information Systems Audit and Control Foundation, *IS Audit and Assurance Guidelines*, ISACA, September 2014.
8. JAD (Joint Application Development). http://searchsoftwarequality.techtarget.com/definition/JAD (accessed June 30, 2017).
9. Jones, D.C., Kalmi, P., and Kauhanen, A. (2011). Firm and employee effects of an enterprise information system: Micro-econometric evidence. *International Journal of Production Economics*, 130(2), 159–168.
10. Kanban. *PM Methodologies*. www.successfulprojects.com/PM-Topics/Introduction-to-Project-Management/PM-Methodologies (accessed June 29, 2017).

11. Mallach, Efrem G. (2011). Information system conversion strategies: A unified view. In *Managing Adaptability, Intervention, and People in Enterprise Information Systems*, ed. Madjid Tavana, 91–105 (accessed July 05, 2017). doi:10.4018/978-1-60960-529-2.ch005.

12. Merhout, J. and Kovach, M. (2017). Governance practices over agile systems development projects: A research agenda. Proceedings of the Twelfth Annual Midwest Association for Information Systems Conference (MWAIS 2017), Springfield, Illinois, May 18–19, 2017.

13. OWASP. Secure coding practices—Quick reference guide. www.owasp.org/index.php/OWASP_Secure_Coding_Practices_-_Quick_Reference_Guide (accessed June 28, 2017).

14. Protiviti. (2016). *From Cloud, Mobile, Social, IoT and Analytics to Digitization and Cybersecurity: Benchmarking Priorities for Today's Technology Leaders*. www.knowledgeleader.com/KnowledgeLeader/Content.nsf/Web+Content/SRFromCloudMobileSocialIoTandAnalytics!OpenDocument (accessed June 28, 2017).

15. Rama, J., Corkindaleb, D., and Wu, M. (2013). Implementation critical success factors (CSFs) for ERP: Do they contribute to implementation success and post-implementation performance? *International Journal of Production Economics*, 144(1), 157–174.

16. Microsoft Developer Network. Regression testing. https://msdn.microsoft.com/en-us/library/aa292167(v=vs.71).aspx (accessed July 1, 2017).

17. Romney, M. B. and Steinbart, P. J. (2015). *Accounting Information Systems*, 13th Edition, Pearson Education, Upper Saddle River, NJ.

18. Schiesser, R. *Guaranteeing Production Readiness prior to Deployment*. Prentice Hall PTR, New York, www.informit.com/isapi/product_id·%7B0CF23CBC-CDCC-4B50-A00E-17CBE595AA31%7D/content/index.asp (accessed August 1, 2003).

19. Scrum. *PM Methodologies*. www.successfulprojects.com/PM-Topics/Introduction-to-Project-Management/PM-Methodologies (accessed June 29, 2017).

20. Berkeley Information Security and Policy. Secure coding practice guidelines. https://security.berkeley.edu/secure-coding-practice-guidelines (accessed July 1, 2017).

21. SEI CERT Coding Standards. (2017). Software Engineering Institute. www.securecoding.cert.org/confluence/display/seccode/SEI+CERT+Coding+Standards

22. Senft, S., Gallegos, F., and Davis, A. (2012). *Information Technology Control and Audit*, CRC Press/Taylor & Francis: Boca Raton.

23. TechTarget. Software Performance Testing. http://searchsoftwarequality.techtarget.com/definition/performance-testing (accessed July 1, 2017).

24. Innovative Architects. The seven phases of the system-development life cycle. www.innovativearchitects.com/KnowledgeCenter/basic-IT-systems/system-development-life-cycle.aspx (accessed June 27, 2017).

25. Waters, K. (2010). *7 Key Principles of Lean Software Development*. Lean Development. www.101ways.com/7-key-principles-of-lean-software-development-2/

26. Agile Alliance. What is agile software development? www.agilealliance.org/agile101/ (accessed June 28, 2017).

27. Software Testing Fundamentals. White box testing. http://softwaretestingfundamentals.com/white-box-testing/ (accessed July 1, 2017).

28. US-CERT, *Top 10 Coding Practices*, Software Engineering Institute, Carnegie Mellon University, www.securecoding.cert.org/confluence/display/seccode/Top+10+Secure+Coding+Practices March 2011.

AUDITING IT ENVIRONMENT

Chapter 9

Application Systems: Risks and Controls

LEARNING OBJECTIVES

1. Discuss common risks associated with application systems.
2. Discuss common risks associated with end-user development application systems.
3. Discuss risks to systems exchanging business information and describe common standards for their audit assessments.
4. Describe Web applications, including best secure coding practices and common risks.
5. Explain application controls and how they are used to safeguard the input, processing, and output of information.
6. Discuss the IT auditor's involvement in an examination of application systems.

Application systems provide automated functions to effectively support the business process. Applications also introduce risks to organizations in the form of increased costs, loss of data integrity, weaknesses in confidentiality, lack of availability, and poor performance, among others. Further, once implemented, applications may be periodically modified to either correct errors or just implement upgrades and enhancements (maintenance). Such maintenance will need to be consistent with business or IT strategies; otherwise, it may cause performance issues and inefficient use of resources.

This chapter discusses common risks to various types of application systems and provide examples of such potential risks. It also touches upon relevant application controls that can be implemented by organizations in order to mitigate the risks discussed. Lastly, involvement of the IT auditor when examining applications is discussed.

Application System Risks

Application systems include concentrated data in a format that can be easily accessed. Such concentration of data increases the risks by placing greater reliance on a single piece of data or on a single computer file or on a database table. If the data entered are erroneous, the impact of error would be significant as applications rely on such piece of data. Similarly, the higher the number of applications that use the concentrated data, the greater the impact when that data become unavailable due to hardware or software problems. A good example to further the discussion of application systems is an Enterprise Resource Planning (ERP) system.

ERP systems provide standard business functionality in an integrated IT environment system (e.g., procurement, inventory, accounting, and human resources). ERP systems allow multiple functions to access a common database—reducing storage costs and increasing consistency and accuracy of data from a single source. In fact, having a single database improves the quality and timeliness of financial information. However, processing errors can quickly impact multiple functions as the information is shared but sourced from the same database. According to the June 2016 edition of Apps Run the World, a technology market-research company devoted to the applications space, the worldwide market of ERP applications will reach $84.1 billion by 2020 versus $82.1 billion in 2015. Some of the primary ERP suppliers today include SAP, FIS Global, Oracle, Fiserv, Intuit, Inc., Cerner Corporation, Microsoft, Ericsson, Infor, and McKesson.

Despite the many advantages of ERP systems, they are not without risks. ERP systems are not much different than purchased or packaged application systems, and may therefore require extensive modifications to new or existing business processes. ERP modifications (i.e., software releases) require considerable programming to retrofit all of the organization-specific code. Because packaged systems are generic by nature, organizations may need to modify their business operations to match the vendor's method of processing, for instance. Changes in business operations may not fit well into the organization's culture or other processes, and may also be costly due to training. In addition, some integration may be required for functionality that is not part of the ERP, but provides integral information to the ERP functions. Moreover, as ERP systems are offered by a single vendor, risks associated with having a single supplier apply (e.g., depending on a single supplier for maintenance and support, specific hardware or software requirements, etc.).

Another risk with ERP systems is the specialized nature of the resources required to customize and implement. In most organizations, these specialized resources must be procured from high-priced consulting firms. To decrease the dependency on high-priced consultants, organizations need to invest in educating their own staff to take over responsibility for maintaining the ERP system. As these resources are in high-demand, the challenge is to keep these resources once they are fully trained.

ERP systems can be quite complex with the underlying database, application modules, and interfaces with third-party and legacy applications. The complexity of ERP systems may actually cost more than the multiple application environments it was intended to replace.

Application systems like ERP systems are frequently exposed to many types of risks. Additional common risks associated with application systems include:

- Weak information security
- Unauthorized access to programs or data
- Unauthorized remote access
- Inaccurate information
- Erroneous or falsified data input
- Incomplete, duplicate, and untimely processing
- Communications system failure
- Inaccurate or incomplete output
- Insufficient documentation

Weak Information Security

Information security should be a concern of IT, users, and management. However, it has not been a consistent top priority for many organizations. Past surveys and reports have shown that organizations are more concerned with budgets and staff shortages than information security. Respondents to such surveys still continue to identify obstacles to reducing information security risks, such as lack of human resources, funds, management awareness, and tools and solutions. Meanwhile, advanced technology and increased end-user access to critical and sensitive information continue to proliferate information security risks.

Unauthorized Access to Programs or Data

Application systems should be built with various levels of authorization for transaction submission and approval. Once an application goes into production, programmers should no longer have access to programs and data. If programmers are provided access, such access should be a "read-only" access for the purpose of understanding issues reported by the user.

Similarly, users' access should be limited to a "need-to-know" basis. This means that the information made available to a user, whether it is "read-only" or with open access for modification, should be in accordance with the user's job functions and responsibilities. For example, a payroll clerk needs access to the payroll system, but not to the billing system.

Unauthorized Remote Access

More and more users are demanding remote access to organizations' computer resources. Remote access allows users within an organization to access its network and computer resources from locations outside the organization's premises. Remote access, if unauthorized, does represent a risk because **client devices** (used for the remote access) tend to have weaker protection than standard or organization-based client devices. These devices may not necessarily be managed by the organization and, hence, not be defined under the firewalls and antivirus rules, for example.

Remote access communications may be carried over untrusted networks, subjecting the communication to unauthorized monitoring, loss, or manipulation. In other words, if users within (or outside) the organization's network were to gain unauthorized access:

- sensitive and confidential information may be at risk and negatively impacted; and
- **computer viruses** can be introduced affecting company files, performance of computer systems, or just slowing down the network and its resources.

To combat unauthorized remote access, at a minimum, user IDs and passwords should use **encryption** when transmitted over public lines. Furthermore, confidential data that are transmitted over public lines should also be encrypted. The encryption security solution depends on the sensitivity of the data being transmitted. Lastly, user access reviews should be periodically performed by IS security personnel, and approved by management, to ensure the remote access granted is accurate and consistent with job tasks and responsibilities.

Inaccurate Information

Accurate information must be ensured whether the end user is accessing data from an application, a departmental database, or information on the cloud. End users may be asked to generate reports for analysis and reporting without fully understanding the downloaded information. Departmental **data repositories** (e.g., databases, data clouds, etc.) may have redundant information with different timeframes. The result is waste of time in reconciling these repositories to determine which data are accurate. Another major area of concern is that management may fail to use information properly due to failures in identifying significant information; interpreting meaning and value of the acquired information; and/or communicating critical information to the responsible manager or chief decision maker on a timely basis.

Erroneous or Falsified Data Input

Erroneous data input is when inaccurate data are inputted in the application system unintentionally due to human error. Preventative measures include built-in application controls, such as check digits and double entry. Falsified data input, on the other hand, is when inaccurate data are inputted to the application system intentionally to defraud the organization or its stakeholders. In this case, preventative measures may include safeguarding access to programs and data through user authentication and authorization mechanisms.

Incomplete, Duplicate, and Untimely Processing

Incomplete processing occurs when transactions or files are not processed due to errors. It may occur in **batch processing** when a file is not present, or during **online processing** when a request or trigger fails to kick off a transaction. Duplicate transaction processing includes executing transactions more than once. It can occur during batch processing if files are executed multiple times, or during online processing when a transaction trigger kicks off a transaction more than once. Duplicate processing may also occur when a job abnormally terminates (abends), but some of the records that have been processed are not reset. Untimely processing includes delayed processing due to production problems or missing a time cutoff. For example, financial processes must occur at month-end closing to ensure that the detailed transactions processed in one application system match the transaction posting to the **general ledger**. In addition, when an online system post transactions to a batch system, there is usually a cutoff time where processing ends on day one and begins for day two.

Communications System Failure

Today, application systems within IT environments are responsible for many critical services, including communication services (e.g., e-mail, intranets, Internet, instant messaging, etc.). Because of this increasing reliance on IT communication services, the potential failure of these services presents an increasing source of risk to organizations. Information that is routed from one location to another over communication lines is vulnerable to accidental failures, intentional interception, and/or modification by unauthorized parties.

Inaccurate or Incomplete Output

If not adequately safeguarded, output reports may contain errors after processing (compromising their integrity) and also be distributed improperly. Output controls should be in place, for example, to verify that the data are accurate and complete (i.e., properly recorded), and that report distribution and retention procedures are effective. Examples of output controls involve performing reviews, reconciliations, and verifying for data transmission. Additionally, access to reports should be based on a "need-to-know" basis in order to maintain confidentiality.

Insufficient Documentation

End users typically focus on solving a business need and may not recognize the importance of documentation. Any application system that is used by multiple users or has long-term benefits must be documented, particularly if the original developer or programmer is no longer available. Documentation provides programmers with enough information to understand how the application works, and assists in solving problems to ensure effective and efficient analysis of program changes and troubleshooting. Documentation should be updated as the application system is modified.

Documentation further ensures maintainability of the system and its components and minimizes the likelihood of errors. Documentation should be based on a defined standard and consist of descriptions of procedures, instructions to personnel, flowcharts, data flow diagrams, display or report layouts, and other materials that describe the application system.

End-User Development Application Risks

End-user development (EUD) (also known as end-user computing) generally involves the use of department-developed applications, such as spreadsheets and databases, which are frequently used as tools in performing daily work. These spreadsheets and databases are essentially an extension of the IT environment and output generated from them may be used in making business decisions impacting the company. As a result, the use of EUD has extended the scope of audits outside the central IS environment. The level of risk and the required controls to be implemented depend on the criticality of the EUD application. For example, an EUD application that consolidates data from several departments that will later be an input into the financial reporting system is a prime target for an audit.

EUD application risks are not easily identified because of lack of awareness and the absence of adequate resources. For instance, personal computers or PCs, notebooks, laptops, and mobile devices hosting relevant department-developed spreadsheets and/or databases may be perceived as personal productivity tools, and thus be largely ignored by the organization. Similarly, many organizations have limited or no formal procedures related to EUD. The control or review of reports produced by EUD applications may be limited or nonexistent. The associated risk is that management may be relying on end-user-developed reports and information to the same degree as those developed under traditional centralized IS environment. Management should consider the levels of risk associated with EUD applications and establish appropriate levels of controls. Common risks associated with EUD application systems include:

- Higher organizational costs
- Incompatible systems
- Redundant systems
- Ineffective implementations
- Absence of segregation of duties
- Incomplete system analysis
- Unauthorized access to data or programs

- Copyright violations
- Lack of back-up and recovery options
- Destruction of information by computer viruses

Higher Organizational Costs

EUD may at first appear to be relatively inexpensive compared to traditional IT development. However, a number of hidden costs are associated with EUD that organizations should consider. In addition to operation costs, costs may increase due to lack of training and technical support. Lack of end-user training and their inexperience may also result in the purchase of inappropriate hardware and the implementation of software solutions that are incompatible with the organization's systems architecture. End users may also increase organizational costs by creating inefficient or redundant applications.

Incompatible Systems

End-user-designed application systems that are developed in isolation may not be compatible with existing or future organizational IT architectures. Traditional IT systems development verifies compatibility with existing hardware and related software applications. The absence of hardware and software standards can result in the inability to share data with other applications in the organization.

Redundant Systems

In addition to developing incompatible systems, end users may be developing redundant applications or databases because of the lack of communication between departments. Because of this lack of communication, end-user departments may create a new database or application that another department may have already created. A more efficient implementation process has end-user departments coordinating their systems development projects with IT and meeting with other end-user departments to discuss their proposed projects.

Ineffective Implementations

End users typically use fourth-generation programming languages, such as database or Internet Web development tools to develop applications. In these cases, the end user is usually self-taught. However, they lack formal training in structured applications development, do not realize the importance of documentation, and tend to omit necessary control measures that are required for effective implementations. In addition, there is no segregation of duties. Because of insufficient analysis, documentation, and testing, end-user developed systems may not meet management's expectations.

Absence of Segregation of Duties

Traditional application systems development is separated by function, and tested and completed by trained experts in each area. In many EUD projects, one individual is responsible for all phases, such as analyzing, designing, constructing, testing, and implementing the development life cycle. There are inherent risks, such as overlooking errors, when having the same person creating and testing a program. It is more likely that an independent review will catch errors made by the end-user developer, and such a review helps to ensure the integrity of the newly designed application system.

Incomplete System Analysis

End-user departments eliminate many of the steps established by central IT departments. For example, the analysis phase of development may be incomplete and not all facets of a problem appropriately identified. In addition, with incomplete specifications, the completed system may not meet objectives nor solve the business problem. End users must define their objectives for a particular application before they decide to purchase existing software, have IT develop the application, or use their limited expertise to develop the application. Incomplete specifications will likely prompt system deficiencies.

Unauthorized Access to Data or Programs

Access controls provide the first line of defense against unauthorized users who gain entrance to an application system's programs and data. The use of access controls, such as user IDs and passwords, are typically weak in user-developed systems. In some cases, user IDs and passwords may not even be required, or they would be very simple and easily guessed. This oversight can subject applications to accidental or deliberate changes or deletions that threaten the reliability of any information generated. Systems require additional protection to prevent any unexpected changes. To prevent any accidental changes, the user should be limited to execute only.

Copyright Violations

Organizations are responsible for controlling the computing environment to prevent **software piracy** and copyright violations. However, some organizations may not specifically address software piracy in training, in policy and procedures, or in the application of general internal controls. Since software programs can easily be copied or installed on multiple computers, many organizations are in violation of copyright laws and are not even aware of the potential risks.

Organizations face a number of additional risks when they tolerate software piracy. Copied software may be unreliable and carry viruses. Litigation involving copyright violations is highly publicized, and the organization is at risk of losing potential goodwill. Furthermore, tolerating software piracy encourages deterioration in business ethics that can seep into other areas of the organization.

Organizations should inform end users of the copyright laws and the potential consequences that result from violations of those laws. To prevent installation of unauthorized software, organizations may restrict user's ability to install software by disabling administrative access to their PCs. Additionally, when users are given access to a personal or desktop computer, they should sign an acknowledgment that lists the installed software, the individual's responsibilities, and any disciplinary action for violations. Written procedures should clearly define user's responsibility for maintaining a software inventory, auditing compliance, and removing unlicensed software.

Lack of Back-Up and Recovery Options

Organizations that fail to maintain a copy of their data are really asking for trouble. Nowadays, it is extremely easy to lose data and all but impossible to rebuild that data if backups had not been performed. EUD applications are often stored in one's PC and not properly backed-up. In case of a disaster or virus attack, these applications (and their data) may not be recoverable because of the lack of backups. It may therefore not be possible for the end user to recreate the application and its data within a reasonable period of time.

The absence of a back-up and recovery strategy results in computer data loss. Unbacked up data is constantly subject to risks, such as accidental deletion of files, viruses and damaging malware, hard drive failures, power failures or crashes, theft of computer, water damage, fire, and many others.

Destruction of Information by Computer Viruses

Most end users are knowledgeable about computer virus attacks, but the effect of a virus remains only a threat until they actually experience a loss. Based on the McAfee Labs Threats Report for December 2016, the number of attacks approximates 650 million. For mobile devices, the number for 2016 is also significant, almost approaching the 13.5 million mark. Further, in its 2017 Threats Predictions report, McAfee Labs predicts the following, among others:

■ Attackers will continue to look for opportunities to break traditional (non-mobile) computer systems, and exploit vulnerabilities. Attackers are well able to exploit systems whose firmware (permanent software programmed into a read-only memory) controls input and output operations, as well as other firmware, such as **solid-state drives**, network cards, and Wi-Fi devices. These types of exploits are probable to show in common **malware** attacks.
■ **Ransomware** on mobile devices will continue its growth though attackers will likely combine these mobile device locks with other forms of attack, such as credential theft, allowing them to access such things as banks accounts and credit cards.

A virus is the common term used to describe self-reproducing programs, **worms**, **moles**, **holes**, **Trojan horses**, and **time bombs**. In today's environment, the threat is high because of the unlimited number of sources from which a virus can be introduced. For example, viruses can be copied from a disk or downloaded from an infected Web page. They spread to other files, those files in turn spread to other files, and so on. The **boot sector** of a disk is one of the most susceptible to virus infection because it is accessed every time that the computer is turned on, providing easy replication of the virus. When a virus is activated, it copies code to the hard drive, and it can spread to additional media by executing a common application such as a word processor or mail program. The media that contain the virus will continue to infect other computers and spread the virus throughout an organization.

Viruses can also spread among computers connected within a network (local, Internet, etc.). They can spread when infected files or programs are downloaded from a public computer through attachments to e-mails, hidden code within hyperlinks, and so on. Viruses can cause a variety of problems such as:

■ Destroying or altering data
■ Destroying hardware
■ Displaying unwanted messages
■ Causing keyboards to lock and become inactive
■ Slowing down a network by performing many tasks that are really just a continuous loop with no end or resolution

- Producing spamming
- Launching denial-of-service attacks

The risk to organizations is the time involved in removing the virus, rebuilding the affected systems, and reconstructing damaged data. Organizations should also be concerned about sending virus-infected programs to other organizations. Viruses cause significant financial damage, and recipients may file lawsuits against the instituting organization.

Risks to Systems Exchanging Electronic Business Information

Electronic Data Interchange (EDI) refers to the electronic exchange of business documents between business (or trading) partners using a standardized format. EDI allows organizations to electronically send and receive information in a standard format so that computers are able to read and understand the documents being interchanged. A standard format describes the type, as well as the design, style, or presentation (e.g., integer, decimal, mmddyy, etc.) of the information being traded. Common examples of business information exchanged through EDI includes invoices and purchase orders. Exhibit 9.1 describes risks associated with EDI or systems exchanging electronic business information.

Exhibit 9.1 Risks to EDI or Systems Exchanging Electronic Business Information

Risk	Description
Loss of business continuity/going-concern problem	Inadvertent or deliberate corruption of EDI-related applications could affect every EDI transaction entered into by an organization, impacting customer satisfaction, supplier relations, and possibly business continuity eventually.
Interdependence	There is increased dependence on the systems of trading partners, which is beyond the control of the organization.
Loss of confidentiality of sensitive information	Sensitive information may be accidentally or deliberately divulged on the network or in the mailbox storage system to unauthorized parties, including competitors.
Increased exposure to fraud	Access to computer systems may provide an increased opportunity to change the computer records of both a single organization and that of its trading partners by staff of the trading parties or by third-party network staff. This could include the introduction of unauthorized transactions by user organization or third-party personnel.
Manipulation of payment	A situation where amounts charged by or paid to suppliers are not reviewed before transmission. Therefore, there is a risk that payments could be made for goods not received, payment amounts could be excessive, or duplicate payment could occur.
Loss of transactions	Transactions could be lost as a result of processing disruptions at third-party network sites or en route to the recipient organization, which could cause losses and inaccurate financial reporting.

(Continued)

Exhibit 9.1 (*Continued*) Risks to EDI or Systems Exchanging Electronic Business Information

Risk	Description
Errors in information and communication systems	Errors in the processing and communications systems, such as incorrect message repair, can result in the transmission of incorrect trading information or inaccurate reporting to management.
Loss of audit trail	EDI eliminates the need for hard copy. There will be less paper for the auditors to check. The EDI user may not provide adequate or appropriate audit evidence, either on hard copy or on electronic media. The third-party vendor may not hold audit trails for a significant length of time, or audit trails could be lost when messages are passed across multiple networks.
Concentration of control	There will be increased reliance on computer controls where they replace manual controls, and they may not be sufficiently timely. The use of EDI with its greater reliance on computer systems concentrates control in the hands of fewer staff, increases reliance on key people, and increases risk.
Application failure	Application or EDI component failures could have a significant negative impact on partner organizations within the respective business cycles, especially for **Just-In-Time inventory** management, production, and payment systems. In addition, there is a possibility of error propagation across other systems due to integration with other business applications.
Potential legal liability	A situation where liability is not clearly defined in trading partner agreements, legal liability may arise due to errors outside the control of an organization or by its own employees. There is still considerable uncertainty about the legal status of EDI documents or the inability to enforce contracts in unforeseen circumstances.
Overcharging by third-party service providers	Third-party suppliers may accidentally or deliberately overcharge an organization using their services.
Manipulation of organization	The information available to the proprietors of third-party networks may enable them or competitors to take unfair advantage of an organization.
Not achieving anticipated cost savings	Happens where the anticipated cost savings from the investment in EDI are not realized for some reason by an organization.

Source: Adapted from Senft, S., Gallegos, F., and Davis, A. 2012. *Information Technology Control and Audit.* Boca Raton: CRC Press/Taylor & Francis.

Implications arising from these risks include:

■ Potential LOSS of transaction audit trail, thereby making it difficult or impossible to reconcile, reconstruct, and review records. This could possibly be a breach of legislation and result in prosecution and fines.

- Increased exposure to ransom, blackmail, or fraud through potential disruption of services or increased opportunities to alter computer records in an organization and its trading partners' IS.
- Disruption of cash flows when payment transactions are generated in error or diverted or manipulated.
- Loss of profitability occurring through increased interest charges or orders going to a competitor due to lack of receipt of EDI messages.
- Damage to reputation through loss of major customers, especially if EDI problems are widely publicized.

Standards for EDI Audit Assessments

Auditors, management, developers, and security consultants must need to be aware of the business risks associated with EDI systems. Some well-known standards that provide a basis for EDI audit assessments include: the *Accredited Standards Committee (ASC) X12* group of standards supported by North America's American National Standards Institute (ANSI) and the *Electronic Data Interchange for Administration, Commerce and Transport (EDIFACT)* international standards supported by the United Nations' Economic Commission for Europe.

- ASC X12 standards facilitate the electronic interchange of business transactions, such as placing and processing orders, shipping, receiving, invoicing, and payment. Specifically, ASC X12 standards identify the data being used in the transaction, the order in which such data must appear, whether data are mandatory or optional, when data can be repeated, and how loops, if applicable, are structured and used.
- EDIFACT standards provide a set of common international standards for the electronic transmission of commercial data. EDIFACT international standards deal with the electronic interchange of structured data, such as the trade in of goods and services between independent computerized information systems.

Other common major standards for EDI assessments include:

- Tradacoms standard, which is predominant in the United Kingdom retail sector. The standard is currently referred to as GS1 UK.
- The Organization for Data Exchange by Tele Transmission (ODETTE) standards, which represent the interests of the automotive industry in Europe. ODETTE creates standards, develops best practices, and provides services which support logistics management, e-business communications, and engineering data exchange throughout the European automotive industry.
- The Verband der Automobilindustrie (VDA) standards and best practices are also applicable for the European automotive industry. VDA standards particularly focus on serving the needs of German automotive companies.
- Health Level-7 (HL7) international standards relate to the electronic exchange of clinical and administrative data among healthcare providers. HL7 standards have been adopted by other standard issuing bodies like ANSI and International Organization for Standardization.
- GS1 EDI global standards guide the electronic communication and automation of business transactions that typically takes place across the entire supply chain. These standards apply to retailers, manufacturers, material suppliers, and logistic service providers, for example.

Web Application Risks

PC Magazine defines a Web application as "an application in which all or some parts of the software are downloaded from the Web each time it is run."[*] Use of Web applications has become a key strategy for direction for many companies. Some companies simply use Web applications for marketing purposes, but most use them to replace traditional client-server applications. Other characteristics of Web applications include the use of a Web browser on the client side that is usually platform independent, and requires less computing power. Some other benefits of Web applications include reduced time to market, increased user satisfaction, and reduced expenses related to maintenance and supports.

From a development point of view, a Web application should be designed to perform the specific tasks agreed upon and documented as part of the functional requirements. When developing Web applications, teams must understand that client-side controls like input validation, hidden fields, and interface controls, for example, are not fully dependable for security purposes. Attackers may easily bypass these client-side controls and gain access to analyze or manipulate application traffic, submit requests, etc. Well-known practices referred to when developing Web application systems or applications include the Top 10 Secure Coding Practices, issued in March 2011 by the United States Computer Emergency Readiness Team (US-CERT). Other common practices include the secure coding principles described in the Open Web Application Security Project (OWASP) Secure Coding Guidelines. While the following OWASP secure coding principles specifically reference Web applications, these principles may also apply to non-Web applications.

- Input validation
- Output encoding
- Authentication and password management
- Session management
- Access control
- Cryptographic practices
- Error handling and logging
- Data protection
- Communication security
- System configuration
- Database security
- File management
- Memory management
- General coding practices

OWASP offers a practical checklist[†] that focuses on implementing secure coding practices and principles. The checklist is designed to serve as a secure coding kick-start tool to help development teams understand (and comply with) secure coding practices.

Risks attributable to Web applications, as stated on the 2017 OWASP Top 10 Most Critical Web Application Security Risks,[‡] include:

---

[*] www.pcmag.com/encyclopedia/term/54272/web-application.

[†] www.owasp.org/images/0/08/OWASP_SCP_Quick_Reference_Guide_v2.pdf.

[‡] www.owasp.org/index.php/Category:OWASP_Top_Ten_Project#tab=OWASP_Top_10_for_2017_Release_Candidate.

- Injection
- Broken authentication and session management
- Cross-site scripting
- Broken access control
- Security misconfiguration
- Sensitive data exposure
- Insufficient attack protection
- Cross-site request forgery
- Using components with known vulnerabilities
- Under protected application program interfaces

The OWASP secure coding principles and practices checklist is one effective way to minimize risks and ensure that the organization develops successful Web applications. However, auditors, management, developers, and security consultants must consider the levels of risks associated with all types of applications in order to design and implement appropriate application controls.

Application Controls

There are two broad groupings of computer controls that assist in mitigating the application risks discussed above, and are essential to ensure the continued proper operation of application systems. They are: General Computer Controls and Application Controls. General computer controls ("general controls" or "ITGC") include examining policies and procedures that relate to many applications and supports the effective functioning of application controls. General controls cover the IT infrastructure and support services, including all systems and applications. General controls commonly include controls over (1) information systems operations; (2) information security; and (3) change control management (i.e., system software acquisition, change and maintenance, program change, and application system acquisition, development, and maintenance).

Application controls examine procedures specific and unique to the application. Application controls are also referred as "automated controls." They are concerned with the accuracy, completeness, validity, and authorization of the data captured, entered, processed, stored, transmitted, and reported. Examples of application controls include validating data input, checking the mathematical accuracy of records, and performing numerical sequence checks, among others. Application controls are likely to be effective when general controls are effective. Exhibit 1.3 from Chapter 1 illustrates general and application controls, and how they should be in place in order to mitigate risks and safeguard applications. Notice in the exhibit that the application system is constantly surrounded by risks. Risks are represented in the exhibit by explosion symbols. These risks could be in the form of unauthorized access, loss or theft or equipment and information, system shutdown, etc. The general controls, shown in the hexagon symbols, also surround the application and provide a "protective shield" against the risks. Lastly, there are the application or automated controls which reside inside the application and provide first-hand protection over the input, processing, and output of the information.

Application controls implemented at organizations may include, among others, system and/or application configuration controls; security-related controls enforcing user access, roles, and segregation of duties; and automated notification controls to alert users that a transaction or process is awaiting their action. Application controls also check for mathematical calculations, balancing totals between jobs, reasonableness against expected volumes or values, reconciliations between systems, and the controlled distribution of output to ensure accuracy and completeness of transactions. Application controls can be described as techniques used to control the input, processing, and output of information in an application. They are broken down into three main categories: input, processing, and output controls.

Input Controls

Input controls are meant to minimize risks associated with data input into application systems. The "user interface" is the means by which the user interacts with the system. In most cases, this is the computer screen, mouse, and keyboard. An effective interface for the users will help reduce desk costs and improve accuracy and efficiency. Also, a user interface should provide a means for the user to obtain context-sensitive help.

Defining input requirements ensures that the method of capturing the data is appropriate for the type of data being input and how it is subsequently used. Performance problems and accuracy issues can be introduced with inappropriate methods for capturing data. Input requirements should specify all valid sources for data as well as the method for validating the data. Input controls prevent invalid transactions from being entered and prevent invalid data within valid transactions. They specifically ensure the authenticity, accuracy, and completeness of data entered into an application.

Authenticity

NIST defines authenticity as "the property of being genuine and being able to be verified and trusted."* Authenticity ensures that only authorized users have access to entering transactions. During an application system development process, authorized users should be defined along with their security levels for data access. This information can be used when designing input screens to limit screens or fields to particular user groups. Controls can also be designed to enforce separation of duties. For example, a user may be able to enter a transaction, but a supervisor may need to approve the transaction before it is submitted for processing.

Authentication must also be considered when automated applications interface with other applications. Authentication, according to NIST, verifies "the identity of a user, process, or device often as a prerequisite to allowing access to resources in an information system."* Often, scheduled batch jobs operate under the authority with specified access privileges to the database. Risks associated with these access accounts as well as the access privileges need to be reviewed. **Generic accounts** should not be used. The batch jobs should be given minimal privileges and system-level accounts should not be used.

Accuracy

Accuracy is ensured through edit checks that validate data entered before accepting the transaction for processing. Accuracy ensures that the information entered into an application is consistent and complies with policies and procedures. This is accomplished by designing input screens with edits and validations that check the data being entered against predefined rules or values.

---

* http://nvlpubs.nist.gov/nistpubs/SpecialPublications/NIST.SP.800-53Ar4.pdf.

The accuracy of transactions processed can be ensured by having all inputted transactions go through data validation checks, whether coming from an online screen, an interface from another application, or generated by the system. Programs that automatically generate transactions (i.e., time triggered programs) should have built-in edits that validate transaction accuracy similar to transactions entered by a user. It is also important to track transaction volume and frequency against expected trends to ensure that transactions are triggered properly. Missing and duplicate checks should also be programmed in case an error occurs in the triggering logic.

Edit and validation routines are generally unique to the application system being used, although some general-purpose routines may be incorporated. Exhibit 9.2 lists common edit and validation routine checks or controls when inputting data. Edit and validation routines are placed in a system to aid in ensuring the completeness and accuracy of data. Therefore, overriding edit routines should not be taken lightly. In most systems, the user is not provided this capability. Overriding edit routines is allowed only to privileged user department managers or supervisors, and from a master terminal. Overrides should be automatically logged by the application so that these actions can be analyzed for appropriateness and correctness.

Completeness

Completeness confirms that all data necessary to meet current and future business needs are actually ready and available. Having complete data to work with assists management when making business decisions impacting the organization. Complete data, in the form of financial statements, vendor lists, customer receivable reports, loan reports, etc., reflect an accurate status of the organization and how it is coping with competitors and industry trends and patterns. Completeness is ensured, for Instance, through error-handling procedures that provide logging, reporting, and correction of errors.

Exhibit 9.2   Edit and Validation Controls When Inputting Data

Control	Description
Field check	Confirms that characters in a field are of a proper type.
Sign check	Validates that the data in a field has the appropriate positive or negative sign.
Limit or range check	Verifies that the numerical amount entered is within acceptable minimum and maximum values.
Size check	Checks that the size of the data entered fits into the specific field.
Completeness check	Corroborates that all required and necessary data is entered.
Validity check	Compares data from transaction file to that of master file to verify for existence.
Reasonableness check	Checks for correctness of logical relationship between two data items.
Check digit verification	Recalculates a check digit to verify data entry error has not been made.

Processing Controls

Processing controls prevent, detect, and/or correct errors while data processing (batch or online) takes place. These controls help ensure that the data are accurately and completely processed through the application (e.g., no data are added, lost, or altered during processing, etc.).

Jobs scheduled within an application should be reviewed to ensure that changes being made are appropriate and do not introduce risks. As an example, in an ERP application system, a **Structure Query Language** program can be written to modify data directly against the database, avoiding the controls within the application and operating against the database with system administrator privileges. However, from the screen, this program can look like a report if the underlying code is not evaluated.

Accuracy and Completeness

To ensure data accuracy and completeness (A&C), programs should be built with logic to prevent, detect, and/or correct errors. Error handling procedures should include:

- Logging error activity
- Approval of error correction and resubmission
- Defined responsibility for suspense files
- Reports of unresolved errors
- Aging and prioritization of unresolved errors

A&C can also be achieved by balancing batch transactions going in with transactions going out of a predecessor. Balancing steps should occur in major job processing points. The following control points are examples of major job processing points:

- *Input points.* Programs that accept transactions from input processing (e.g., user interface, etc.)
- *Major processing modules.* Programs that modify the data (e.g., perform calculations, etc.)
- *Branching points.* Programs that split or merge data (e.g., a programs merging data from two or more different input sources into one file; file is then used as data feed for the financial reporting system, etc.)
- *Output points.* Result of data processing (e.g., financial or operational reports, printed checks, output data file, etc.)

Designed properly, balancing totals for transaction count and amount can detect missing or duplicate transactions (see Exhibit 9.3, part A). Additionally, balancing and reconciliation should occur between applications that share common data. This can be achieved by creating a reconciliation report that lists data from all application systems involved, and reports on any differences for a user group to review and follow upon any exceptions. Exhibit 9.3, part B is used to illustrate a sample reconciliation between the Billing, Payment, and Accounts Receivable systems. Notice how the Accounts Receivable system confirms (or reconciles) all 17 records involved and, most importantly, the balance of $400 pending after billings were sent out and collections (or payments) were received.

Balancing totals should also include a transaction (quantity) count, totals for all amount fields for each type of transaction, and cross-foot totals for detail fields to total fields. Notice in Exhibit 9.4 how the total number of quantity is verified, and how the total amount per part

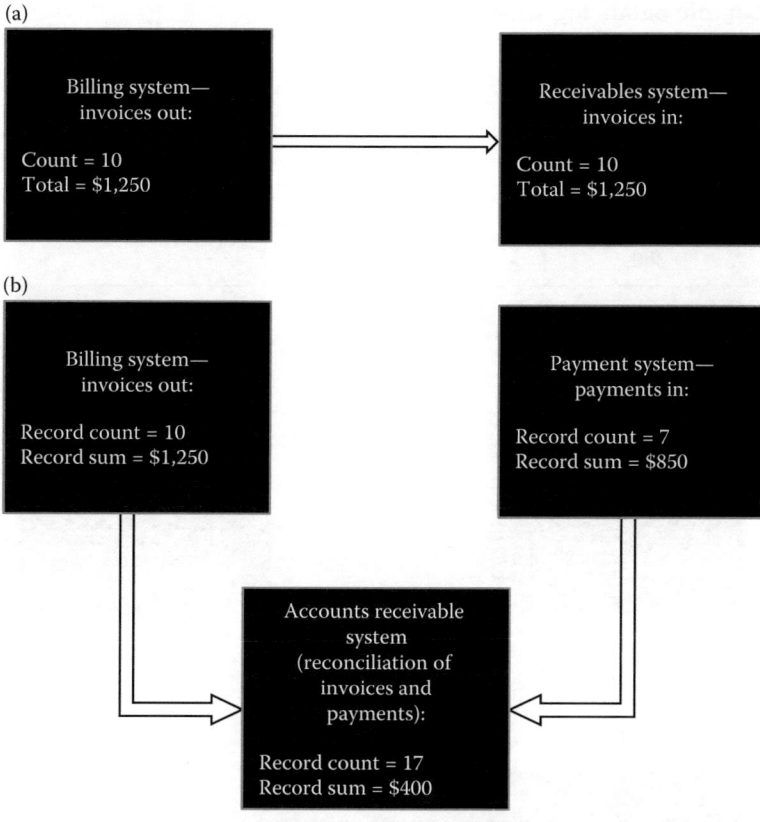

Exhibit 9.3   Batch balancing totals (A) and cross-application reconciliation (B).

results from cross-footing Quantity and Unit Price. In files where there are no meaningful totals, **hash totals** can be created that add all of the figures in a column to verify that the same total is accepted by the next process. For example, totaling part numbers does not provide any meaning. However, this total can be used to verify that all the correct part numbers were received. Transaction flows should be balanced on a daily basis and cumulatively to monthly jobs before the register closes. Balancing totals should also consider both error transactions leaving and entering the processing flow. In Exhibit 9.5, for example, 10 total transactions (with amount of $1,250) minus 2 transactions written to an error file (with amount of $250) were processed in the Accounts Receivable System. The reconciled/balanced total count and amount in the Accounts Receivable System is now eight and $1,000, respectively.

Other common examples of processing controls include:

- *Data matching.* Matches two or more items before executing a particular command or action (e.g., matching invoice to purchase order and receiving report before making the payment, etc.).
- *File labels.* Ensure that the correct and most updated file is being used.
- *Cross-footing.* Compares two alternative ways of calculating the same total in order to verify for accuracy (e.g., adding by rows and by columns in a spreadsheet, etc.).

Exhibit 9.4   Sample Balancing Totals

Order	Quantity	Part Number	Unit Price	Total
Part A	100	1288543	$1.20	$120.00
Part B	80	0982374	$0.60	$48.00
Part C	<u>200</u>	5436682	$0.45	<u>$90.00</u>
Total	380			$258.00

Source: Adapted from Senft, S., Gallegos, F., and Davis, A. 2012. *Information Technology Control and Audit.* Boca Raton: CRC Press/Taylor & Francis.

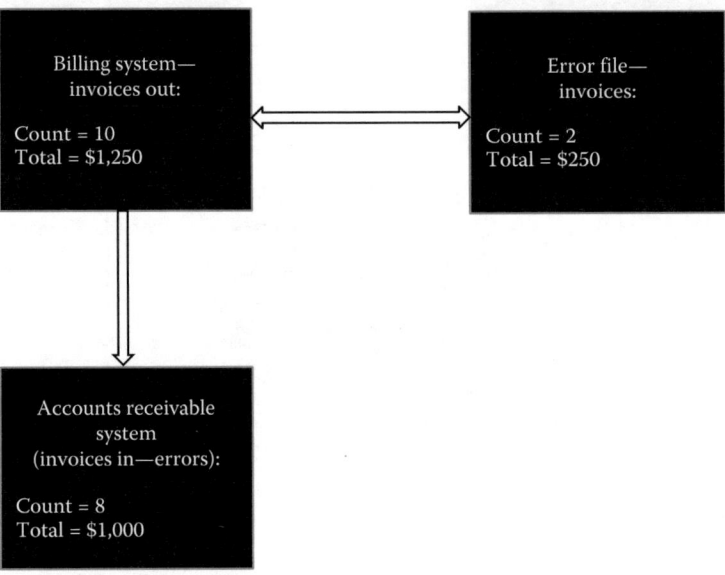

Exhibit 9.5   Balancing totals with error transactions.

- *Zero-balance tests.* Check that a particular account (e.g., payroll clearing account, etc.) maintains a balance of zero. This test assists organizations in eliminating excess balances in separate accounts and maintaining greater control over disbursements.
- *Write-protection mechanisms.* Safeguard against overwriting or erasing data.
- *Concurrent update controls.* Prevent errors of two or more users updating the same record at the same time.

Output Controls

Output controls are designed to detect and correct errors after processing is completed, ensuring the integrity of the output produced. In particular, output controls include: (1) procedures to verify if the data are accurate and complete (i.e., properly recorded) and (2) procedures for adequate report distribution and retention. If outputs are produced centrally, then conventional controls, such as having a security officer and distributing logs may be appropriate. If output is distributed over a data communication network, control emphasis shifts to access controls for individual workstations. To maintain confidentiality, access to reports should be based on a "need-to-know" basis.

Accuracy and Completeness

Output should be verified against an independent source to verify its accuracy and completeness. Three common types of output controls related to accuracy and completeness are user reviews, reconciliations, and data transmission controls. User reviews ensure outputs (reports) generated are secure, confidential, and private through performing balancing and completeness checks; comparisons of key data fields; checks for missing information; and document recreation. Reconciliations include procedures to reconcile control reports. For example, transaction totals posted to the general ledger should be reconciled against the detailed balance due in the accounts receivable **subsidiary ledger**. Another example includes external data reconciliations, such as reconciliation of bank/cash accounts. As mentioned earlier, data that are common to two or more applications should be reconciled to verify consistency. Data transmission controls are implemented to protect the physical transfer of data over a point-to-point or point-to-multipoint communication channel. An example here would be the implementation of encryption techniques over transmitted data.

Distribution and Retention

Distribution of output should be clearly defined, and physical and logical access should be limited to authorized personnel. The need for output should be regularly reviewed as reports may be requested at the time an application is developed, but may no longer be useful. Also, the same information may be used for more than one system with different views, organization, and use. For example, the marketing department may use sales information to pay commission and monitor sales quotas, whereas the accounting department uses the same information to prepare financial statements and reports. These two systems should be reconciled to make sure that the amount reported for paying sales staff is the same as the amount reported on the financial statements and reports.

Because storage space (online or physical) is expensive, retention periods and storage requirements should be defined for programs, data, and reports. Critical information should be stored securely (i.e., encrypted) and its destruction should be permanent and conducted in such a way as to prevent unauthorized viewing. Organizations should consider any law and regulations that may govern the duration of retention periods.

IT Auditor's Involvement

IT auditors can assist organizations by reviewing their application systems to ensure they comply with the organization's strategy and standards, as well as provide automated functions to effectively support the business process. Applications will need to be risk assessed to determine the level of audit involvement. The type of assessment will also vary depending on the risks of the particular application. Applications introduce risks to organizations in the form of increased costs, loss of data integrity, weaknesses in confidentiality, lack of availability, poor performance, and others. These risks should be addressed with adequate selection and implementation of controls.

Auditing application systems requires specific knowledge about application risks and controls. Understanding those allows the IT auditor to identify key areas that would benefit from independent verification. Moreover, understanding application controls allows the IT auditor to evaluate and recommend the ones that will ensure complete and accurate transaction processing.

As stated before, IT auditors can be involved as control consultants or independent reviewers. The level of involvement is determined by completing a risk assessment. Results from the risk assessment also prompt the amount of time necessary to allocate to the particular application, required resources, etc. Preparation of an audit plan then follows. The plan describes the audit objectives and procedures to be performed to ensure applications are adequately implemented, and safeguard the information. IT auditors finally communicate findings identified throughout the audit plus recommendations to management.

Risk Assessment

IT auditors may not have enough time to assess every particular application system in the organization. Involvement within a particular application will depend on the assessment of the application risks. Application risks relate to application complexity and magnitude, inexperienced staff, lack of end-user involvement, and lack of management commitment. The level of risk may be a function of the need for timely information, complexity of the application, degree of reliance for important decisions, length of time the application will be used, and the number of people who will use it. The risk assessment defines which aspects of a particular application will be covered by the audit. The scope of the audit may vary depending on the risks identified.

Audit Plan

The audit plan details the steps and procedures to fulfill the audit objectives. As in any audit, an audit of application systems begins with a preliminary analysis of the control environment by reviewing existing standards, policies, and procedures. During the audit, these standards, policies, and procedures should be assessed for completeness and operational efficiency. The preliminary analysis should identify the organization's strategy and the responsibilities for managing and controlling applications. Documenting an understanding of the application system is also a must at this stage.

The audit plan will further document the necessary procedures to carry on the examination to ensure that the application system is designed and implemented effectively, as well as operates consistent with the organization policies and procedures. Procedures performed by IT auditors should provide reasonable assurance that applications have been adequately designed and implemented, and:

- Comply with standards, policies, and procedures
- Achieve efficient and economical operations
- Conform to legal requirements
- Include the necessary controls to protect against loss or serious errors
- Provide controls and audit trails needed for management, auditor, and for operational review purposes

NIST's Special Publication 800-53A, Revision 4, Assessing Security and Privacy Controls in Federal Information Systems and Organizations (2014), provides all-inclusive assessment procedures for examining security and privacy controls in federal information systems and organizations.* It is important to note that these procedures also apply to non-federal application systems.

Communication

The first area to communicate is the IT auditor's scope of involvement. It is very important to make sure that management's expectations of the IT auditor's role are understood and communicated to all participants. IT auditors must develop an open line of communication with both management and users. If a good relationship between these groups does not exist, information might be withheld from the IT auditor. Although the IT auditor should cultivate good working relationships with all groups with design responsibilities, the IT auditor must remain independent.

Throughout the audit, the IT auditor will be making control recommendations resulting from identified findings. Depending on the organization's culture, these recommendations may need to be handled informally with each application owner in charge of the deficient area or process, or formally by presenting them to the steering committee. In either case, the IT auditor must always consider the value of the control recommendation versus the cost of implementing the control. Recommendations should be specific. They should identify the problem and not the symptom, and allow for the proper controls to be implemented and tested. Findings, risks as a result of those findings, and audit recommendations are usually documented in a formal letter (i.e., Management Letter). Refer to Exhibit 3.9 from Chapter 3 for an example of the format of a Management Letter from an IT audit.

Conclusion

Applications are critical for organizations in conducting their business. They represent a significant investment for many organizations as they provide automated functions to effectively support the business process. Applications also introduce risks to organizations. These risks should be addressed with adequate selection and implementation of application controls.

EUD involves the use of department-developed applications, such as spreadsheets and databases, which are frequently used as tools in performing daily work. These spreadsheets and databases are essentially an extension of the IT environment and output generated from them may be used in making business decisions impacting the company. The level of risk and the required controls to be implemented depend on the criticality of the EUD application.

Auditors, management, developers, and security consultants need to be aware of the business risks associated with systems exchanging electronic business information. Such electronic exchange of business documents between business (or trading) partners using a standardized format is referred to as EDI. Common examples of business information exchanged through EDI includes invoices and purchase orders, and risks like loss of business continuity, increased dependence on systems, loss of confidentiality of sensitive information, and increased exposures to frauds are some of many. Some standards that provide a basis for EDI audit assessments include ANSI's ASC X12 (North America) and EDIFACT (International).

Use of Web applications has become key for direction for many companies. Companies may use Web applications for marketing purposes, and others for replacing their traditional client–server applications. Web applications include the use of a Web browser on the client side that is

---

* http://nvlpubs.nist.gov/nistpubs/SpecialPublications/NIST.SP.800-53Ar4.pdf.

usually platform independent, and requires less computing power. Some benefits of Web applications include reduced time to market, increased user satisfaction, and reduced expenses related to maintenance and supports. Web applications are also subject to risks similar to those traditional applications systems are exposed to.

Owing to these risks, controls need to be implemented to ensure that applications continue to meet the business needs in an effective and efficient manner. Application controls are specific and unique to applications. They are concerned with the accuracy, completeness, validity, and authorization of the data captured, entered, processed, stored, transmitted, and reported. Application controls are broken down into input, processing, and output controls.

IT auditors can assist organizations by reviewing their application systems to ensure that they comply with the organization's strategy and standards, as well as provide automated functions to effectively support the business process. Applications will need to be risk assessed to determine the level of audit's involvement. Results from the risk assessment also prompt the amount of time necessary to allocate to the particular application, required resources, etc. Preparation of an audit plan then follows describing the audit objectives and procedures to be performed. Lastly, IT auditors communicate findings identified throughout the audit plus recommendations to management.

Review Questions

1. Explain why unauthorized remote access represents a risk to applications.
2. Explain how incomplete, duplicate, and untimely processing can negatively impact applications.
3. List seven common risks associated with EUD application systems.
4. How can EUD applications become incompatible systems?
5. In today's environment, the threat of computer viruses is high because of the unlimited number of sources from which they can be introduced. Computer viruses can be copied from a disk, downloaded from an infected Web page, spread among computers connected within a network, etc. Describe the risks or problems that may result from computer viruses.
6. Explain what EDI means. Describe potential implications resulting from risks related to application systems exchanging electronic business information.
7. List and explain five secure coding principles and practices according to OWASP for Web applications.
8. Application controls can be described as techniques used to control the input, processing, and output of information in an application. What do *input* controls refer to? Briefly describe what input controls ensure.
9. Application controls can be described as techniques used to control the input, processing, and output of information in an application. What do *processing* controls refer to? Briefly describe what processing controls ensure.
10. Application controls can be described as techniques used to control the input, processing, and output of information in an application. What do *output* controls refer to? Briefly describe what output controls ensure.

Exercises

1. A company allows orders to be placed directly through its Web site. Describe the three most prominent application system risks that could contribute to unauthorized access to a customer's order information. Identify controls to put in place to mitigate those risks.

2. A payroll department has a time sheet application where employees enter their hours worked. Describe the two most prominent application system risks and the controls that would help mitigate those risks.

3. Departments within a company have their own technical support person who creates and maintains the applications. Describe three risks associated with this practice. What controls would you recommend to help minimize those risks?

4. Explain the significance of application controls and provide examples on how they are used to safeguard the input, processing, and output of information.

CASE—EUD APPLICATIONS

INSTRUCTIONS: A company uses EUD applications, particularly, a spreadsheet to maintain its budget. The spreadsheet is used to solicit the budget from each of the company's departments. The budget department subsequently compiles the individual spreadsheets into a master sheet, reviews and revises the budget based on its constraints, and then uses it to load the budget values into the company's finance system where the department can then view its finalized budget.

TASK: List and describe at least seven prominent application risks associated with the use of a spreadsheet system. You are also required to explain how each of the risks you list may impact the spreadsheet system. Search beyond the chapter for specific examples, IT literature, and/or any other valid external source to support your response. Submit a word file with a cover page, responses to the task above, and a reference section at the end. The submitted file should be between 5 and 7 pages long (double line spacing), including cover page and references. Be ready to present your work to the class.

CASE—INPUT CONTROLS

INSTRUCTIONS: A company has a centralized accounting system. Each individual department currently compiles its accounting paper transactions from its local accounting system. To eliminate the paper and increase efficiency, the Audit Manager of the company just asked you, IT auditor, to help him come up with a plan to implement an interface from each individual department's accounting system to the centralized accounting system.

TASK: Prepare a memo to the Audit Manager naming and describing the most critical controls that you would recommend in this particular case. You are required to search beyond the chapter (i.e., IT literature and/or any other valid external source) to support your response. Include examples, as appropriate, to evidence your case point. Submit a word file with a cover page, responses to the task above, and a reference section at the end. The submitted file should be 5 pages long (double line spacing), including cover page and references. Be ready to present your work to the class.

Further Reading

1. A survey of key concepts and issues for electronic recordkeeping. (2003). *Electronic Data Interchange.* www.ctg.albany.edu/publications/reports/key_concepts?chapter=3&PrintVersion=2

2. Baker, S., Waterman, S., and Ivanov, G. (2010). *In the Crossfire: Critical Infrastructure in the Age of Cyber War*, McAfee.
3. Berkeley Information Security and Policy. Secure coding practice guidelines. https://security.berkeley.edu/secure-coding-practice-guidelines (accessed July 2017).
4. Federal Bureau of Investigation (FBI), *Financial Crimes Report to the Public Fiscal Years 2007 through 2011*, Department of Justice, United States. www.fbi.gov/stats-services/publications/financial-crimes-report-2010-2011
5. *Global Technology Audit Guide (GTAG) 8: Auditing Application Controls*. The Institute of Internal Auditors. https://na.theiia.org/standards-guidance/recommended-guidance/practice-guides/Pages/GTAG8.aspx (accessed July 2017).
6. GS1 EDI. GS1. www.gs1.org/edi (accessed July 2017).
7. ISACA. (2017). *COBIT and Application Controls: A Management Guide*, www.isaca.org/knowledge-center/research/researchdeliverables/pages/cobit-and-application-controls-a-management-guide.aspx
8. ISACA. (2017). *Web Application Security: Business and Risk Considerations*, www.isaca.org
9. Jones, D.C., Kalmi, P., and Kauhanen, A. (2011). Firm and employee effects of an enterprise information system: Micro-econometric evidence. *Int. J. Prod. Econ.*, 130(2), 159–168.
10. McAfee Labs 2017 threats predictions, report issued on November 2016. www.mcafee.com/au/resources/reports/rp-threats-predictions-2017.pdf
11. McAfee Labs threats report––December 2016 www.mcafee.com/ca/resources/reports/rp-quarterly-threats-dec-2016.pdf
12. Morella, R. (August 2015). Auditing web applications. IT audit strategies for web applications. ISACA Geek Week. www.isaca.org/chapters3/Atlanta/AboutOurChapter/Documents/GW2015/081115-10AM-WebAppSecurity.pdf
13. National Institute of Standards and Technology. Special Publication 800-53A, Revision 4, Assessing security and privacy controls in federal information systems and organizations, December 2014. http://nvlpubs.nist.gov/nistpubs/SpecialPublications/NIST.SP.800-53Ar4.pdf
14. Odette. EDI basics. www.edibasics.com/edi-resources/document-standards/odette/ (accessed July 2017).
15. Otero, A. R. (2015). An information security control assessment methodology for organizations' financial information. *Int. J. Acc. Inform. Syst.*, 18(1), 26–45.
16. Otero, A. R. (2015). Impact of IT auditors' involvement in financial audits. *Int. J. Res. Bus. Technol.*, 6(3), 841–849.
17. Otero, A. R., Tejay, G., Otero, L. D., and Ruiz, A. (2012). A fuzzy logic-based information security control assessment for organizations, IEEE Conference on Open Systems. 21–24 Oct. 2012, Kuala Lumpur, Malaysia.
18. Otero, A. R., Otero, C. E., and Qureshi, A. (2010). A multi-criteria evaluation of information security controls using Boolean features. *Int. J. Network Secur. Appl.*, 2(4), 1–11.
19. OWASP. (2013). OWASP top 10–2013: Top 10 most critical web application security risks. www.owasp.org/index.php/Top_10_2013-Risk
20. OWASP. Secure coding practices—Quick reference guide. www.owasp.org/index.php/OWASP_Secure_Coding_Practices_-_Quick_Reference_Guide (accessed June 2017).
21. Romney, M. B. and Steinbart, P. J. (2015). *Accounting Information Systems*, 13th Edition. Pearson Education, Reading, UK.
22. Berkeley Information Security and Policy. Secure coding practice guidelines. https://security.berkeley.edu/secure-coding-practice-guidelines (accessed June 2017).
23. Senft, S., Gallegos, F., and Davis, A. (2012). *Information Technology Control and Audit*, CRC Press/Taylor & Francis: Boca Raton.
24. Tradacoms. EDI basics. www.edibasics.co.uk/edi-resources/document-standards/tradacoms/ (accessed July 2017).

Chapter 10

Change Control Management

LEARNING OBJECTIVES

1. Describe the importance of a change control system.
2. Explain the change control management process.
3. Discuss change control management procedures.
4. Define configuration management, and describe sample activities conducted as part of a configuration management plan.
5. Describe organizational change management.
6. Describe the audit involvement in a change control examination.

Change control management is the process that ensures the effective implementation of changes in an IT environment. The purpose of change control management is to minimize the likelihood of disruption and unapproved changes as well as errors. A change control management process consists of analysis, review, approval, and implementation of changes. From an IT perspective, change control management is thought of in terms of changes made to the existing IT systems. However, changes affecting the organization are also a factor. In many cases, organizational changes are the ones that introduce changes to the IT systems.

This chapter describes change control management, in terms of both IT and organizational change. IT change control management is one of the single most important control areas to ensure the integrity, availability, reliability, security, confidentiality, and accuracy of an organization or IT system supporting the organization. Change control management is also one of the three major general computer controls that assess organization's policies and procedures, related to application systems, in order to support the effective functioning of application controls. Examples of general controls within change control management may include change request approvals; application and database upgrades; and network infrastructure monitoring, security, and change management.

Organizational change also deserves consideration, due to its potential impact to the organization and the increased relationships with changes in the IT environment. Organizational change is impacted by limitations introduced by the technology and the organization's culture. Research debates whether the technology is a product of the culture or whether the organization's practices are dictated by technology. Regardless, it is safe to say that they are interdependent. Consequently, the discussion of change management has been expanded to also include those changes related to the organization.

Importance of a Change Control System

Application software is designed to support a specific function, such as payroll or loan processing. Typically, several applications may operate under one instance of operating system software. Establishing controls over the modification of application software programs helps to ensure that only authorized programs and authorized modifications are implemented. Instituting policies, procedures, and techniques help make sure all programs and program modifications are properly authorized, tested, and approved. Such policies, procedures, and techniques also ensure that access to distribution of the programs is carefully controlled. Without proper controls, there is a risk that security features could be inadvertently or deliberately omitted or "turned off." Such lack of controls could also increase the risk of introducing processing irregularities or malicious code. For example:

- A knowledgeable programmer could secretly modify program code to provide a means of bypassing controls to gain access to sensitive data.
- The wrong version of a program could be implemented, thereby perpetuating outdated or erroneous processing that is assumed to have been updated.
- A virus could be introduced, inadvertently or on purpose, that disrupts processing.

The primary focus of a change control system is on controlling the changes that are made to software systems in operation, as operational systems produce the financial statements and a majority of program changes are made to maintain operational systems. However, the same risks and mitigating controls apply to changes associated with systems under development, once both user management and the project development team have formally approved their baseline requirements.

An IT change control system ensures that there is proper segregation of duties between who initiates the change, who approves the change, and who implements the change into a **production environment**.

Change Control Management Process

The most important area of control in any information-processing environment is the management of changes to existing systems. Given the complexity of hardware, software, and application relationships in the operating environment, each change must be properly defined, planned, coordinated, tested, and implemented.

An effective change control management process reduces the risk of disruption of IT services. Once a change has been proposed, it must be evaluated for risk and impact. If a proposed change introduces significant risk to the operating environment, all parties affected must be notified, the appropriate level of management must approve the implementation schedule, and back out plans must be developed to remove the change from the system if necessary. The proposed change must first be reviewed by change management personnel to identify potential conflicts with other systems. The change control management process should be reviewed periodically to evaluate its effectiveness. A well-defined, structured, and well-implemented change control management process benefits organizations by:

- Reducing system disruptions that can lead to business losses
- Minimizing the number of back outs caused by ineffective change implementation

■ Providing consistent change implementation that permits management to allocate staff and system time efficiently and meet scheduled implementation dates
■ Providing accurate and timely documentation to minimize the impact to change-related problems

Changes can be introduced to fix a bug or add new functionality. Changes can also be introduced from new software releases and the distribution of new software. Additionally, changes can result from configuration management and business process redesign. Large enterprises tend to employ automated tools to help ensure effective change management.

There are three types of changes: routine, nonroutine, and emergency. Routine changes typically have minimal impact on daily operations. They can be implemented or backed out quickly and easily. Nonroutine changes potentially have a greater impact on operations. They affect many users and frequently have lengthy, complex implementation and back out procedures. An emergency change is any change, major or minor, that must be made quickly, without following standard change control procedures. Management must approve such changes before they are undertaken or implemented. A change control management process typically covers the following:

■ Change request form
■ Impact assessment
■ Controls
■ Emergency changes
■ Change documentation
■ Maintenance changes
■ Software releases
■ Software distribution

Change Request Form

A change request form ensures that only authorized changes are implemented. It requires that a record is kept of all changes to the system, appropriate resources are allocated, and changes are prioritized. Changes should be prioritized in terms of benefit, urgency, and effort required, as well as possible impact on existing operations. The form should also manage coordination between changes to account for any interdependencies that may exist. Change request procedures should be documented and require:

■ Record of change requests to be kept for each system
■ Definition of the authority and responsibility of the IT department, as well as the user.
■ Approval by management after all the related information is reviewed
■ A schedule for changes, which also allows for changes outside of the schedule (e.g., emergency changes, etc.)
■ Management approval for all changes made outside of the schedule
■ A notification process so that requesters are kept informed regarding the status of their requests

Exhibit 10.1 illustrates an example of a standard Change Request Form.

Exhibit 10.1   Example of a Standard Change Request Form

Change Request Form Form Number: XYZ-WI-CCM-002 Department: Information Technology	Version: 1.0 Effective Date: XX/XX/20XX Revision Date: XX/XX/20XX
CHANGE REQUESTER	
Name and Phone:	
Department:	
Request Date:	
Expected Completion Date:	
Description of Change and Reasons:	
Impacted Department(s):	
Signature:	
CHANGE INFORMATION	
Change Request Number:	
Classification: (Mark with an "X")	Planned Change: _____ Emergency Change: _____ Maintenance Change: _____ Configuration Change: _____
Type: (Mark with an "X")	Application: _____ Database: _____ Operating System: _____ Network: _____ Other: _____ Specify: _____
Requirements:	
Impact Assessment:	
Fall Back Plan/Backup Procedures:	
BUSINESS UNIT MANAGER	
Name and Phone:	
Review Procedures Performed:	
Approval to Start Working with Change:	
TEST PROCEDURES PERFORMED AND RESULTS	

FINAL APPROVALS	Name	Signature	Date
User Acceptance Testing			
Management Review			
Scheduling, Implementation into Production, and Notification to Users			
Documentation Update			

Impact Assessment

Each change requires an impact assessment to ensure that potential negative consequences (resulting from the implementation of the change) are identified and planned for. Changes can introduce risk to the availability, integrity, confidentiality, and performance of a system. Each change request needs to include supporting evidence of the change's impact assessment. The impact analysis should include specific measures compared with prescribed limits. This enables the extent of the impact to be evaluated. Changes should also be reviewed to determine the effect on compliance with existing policy, procedures, and processes.

Controls

Controls via processes and automated tools are needed to ensure the integration of change requests, software changes, and software distribution. This integration is also consistent with the requirements of the Sarbanes-Oxley Act of 2002, which relate to financial reporting, particularly data integrity, completeness, and monitoring. Change controls also ensure that not just authorized changes were made, but also the detection of unauthorized changes, reduction of the errors due to system changes, and an increase in the reliability of changes. Examples of controls over the change control process include independent verifications of the success or failure of implemented changes, as well as updating the infrastructure or system configuration to detect unauthorized changes. Other relevant controls to assess the change control process include ensuring that:

- Business risks and the impact of proposed system changes are assessed by management before implementation into production environments. Assessment results are used when designing, staffing, and scheduling implementation of changes to minimize disruptions to operations.
- Requests for system changes (e.g., upgrades, fixes, emergency changes, etc.) are documented and approved by management before any change-related work is done.
- Documentation related to the change implementation is adequate and complete.
- Change documentation includes the date and time at which changes were (or will be) installed.
- Documentation related to the change implementation has been released and communicated to system users.
- System changes are tested before implementation into the production environment consistent with test plans and cases.
- Test plans and cases involving complete and representative test data (instead of production data) are approved by application owners and development management.

Additional controls over the change control process are shown in Appendix 3 from Chapter 3. The appendix lists controls applicable to most organizations and considered as guiding procedures for both, IT management and IT auditors.

Other sources of controls related to change management are offered by the well-known ISO/IEC 20000, COBIT, and the ITIL frameworks. The purpose of the ISO/IEC 20000 IT service management standard is to ensure that all changes to hardware, communications equipment and software, as well as to system software are evaluated, approved, installed, and monitored in an effective, efficient, and controlled manner. COBIT 5 defines a set of enablers or processes to help achieve the objectives of the organization. Enablers specific to change management are part of one of COBIT 5's main domains: Build, Acquire, and Implement (BAI). Examples of these are BAI05 Manage Organizational Change Enablement, BAI06 Manage Changes, and BAI07 Manage Change Acceptance and Transitioning. These enablers (processes or controls) assist organizations to ensure effective implementation of changes in the IT environment (i.e., minimizing the likelihood of disruption and unapproved changes as well as errors). ITIL provides a set of best practices to prioritize and implement changes in the IT environment in an effective and efficient manner, without negatively affecting customers or agreed-upon service levels. That is, the ITIL change management practices ensure that appropriate methods and procedures are used when implementing changes while minimizing the possibility of disruption to current services. Many organizations supplement ITIL best practices with their own controls, policies, and/or procedures.

Emergency Changes

Emergency changes are changes that are required outside of the prescribed schedule. Normally, emergency changes are required to fix errors in functionality that adversely affect system performance or business processes. Emergency changes may also be required to fix discovered imminent vulnerabilities to availability, integrity, or confidentiality. Conversely, emergency changes should not compromise the integrity, availability, reliability, security, confidentiality, or accuracy of the system. Because of the consequences that can occur with emergency changes, they should only be implemented in declared emergencies.

Emergency change procedures should not only describe the process for implementing emergency changes, but should also include description of what constitutes an emergency change. The definitive parameters and characteristics of an emergency change need to be clearly described. Emergency changes should be documented like regular changes but the documentation may not occur until after the change is made due to the nature of the emergency. These changes do require formal authorization by those responsible for the system and by management before implementation. In some cases, backups before and after the change are retained for later review.

Emergency changes, by their nature, pose increased risk as they bypass some of the formal analyses and processes of the traditional change control process. As a result, audits of change control procedures should pay particular attention to emergency changes.

Change Documentation

In most cases, changes to production environments will require that existing documentation and procedures be updated to reflect the nature of the change. Current documentation ensures the maintainability of the system by any assigned staff member and minimizes reliance on individual staff.

Change control procedures should include a task for updating documentation, operational procedures, help desk resources, and training materials. Changes to business processes should also be considered. Documentation, procedures, and business processes should actually receive the same consideration and testing as other components impacted by the change.

Maintenance Changes

Maintenance updates are also considered changes and should be accounted for and authorized in the change control procedures. Maintenance tasks should be described to the level of detail necessary to ensure appropriate controls. Maintenance actions should be logged and the log reviewed to ensure appropriateness. Access controls should be used to limit the actions of personnel performing the maintenance to only the access required. An example of a routine maintenance task is defragmenting a hard disk to remove fragmented files or lost clusters.

Software Releases

Like any change, new software releases require management approval to ensure that the change is authorized, tested, and documented before the new software release is implemented in the production environment. The following controls address the implementation of new software releases:

- Appropriate backups of the system's data and programs should be made before the change.
- **Version control** should be accounted for in the process. An example of a version control system (VCS) is called git. Git is a free, open-source code VCS tool designed to track modifications in computer files. Git also coordinates work on such modified files among multiple people, and it does it effectively and efficiently.
- Software releases should only be considered received from the prescribed central repository.
- A formal handover process is also required so that authorized personnel are involved in the process, the implemented software is unchanged from what was tested, and software media is prepared by the appropriate function based on the formal build instructions.

Software Distribution

The purpose of software distribution is to ensure that all copies of the software are distributed in accordance with their license agreements. Software distribution minimizes the risk of multiple versions of the software being installed at the same time. Multiple versions of a software package increase support costs as users and support staff need to be trained and skilled in the features, functionality, and issues with each version.

Software distribution should also account for a verification of the software's integrity, as well as verification for compliance with software license agreements. License agreements normally grant permission to use the specified software based on limitations, number of users, location, type of use, and so on. Software licenses can be for unlimited use by a specifically named person, for concurrent use by an unlimited number of simultaneous users, a site license for unlimited use on one site, or an enterprise license for unlimited use by the enterprise.

Violating software agreements has legal ramifications for companies, including costs for installed copies not licensed, damages and legal fees, and loss of corporate reputation. News stories about software piracy mainly focus on court cases related to someone's setting up a Web site distributing software illegally. However, the unprinted stories are those settled out of court with companies. The Software & Information Industry Association (SIIA) is an organization composed of companies of the software and information industry. One of their objectives is to protect the intellectual property of their members. SIIA is instrumental in influencing laws to protect intellectual property and taking action to combat software piracy. SIIA's Corporate Anti-Piracy Program identifies, investigates, and resolves software piracy cases on behalf of its members. Software distribution practices should include the following controls:

- Distribution is made in a timely manner only to those authorized.
- A means is in place for ensuring verification of integrity, and this is incorporated into the installation.

■ A formal record exists of who has received software and where it has been implemented. This record should also match with the number of purchased licenses.

Change Control Management Procedures

Change control management procedures ensure that all members of an organization are following the same process for initiating, approving, and implementing changes to systems and applications. The following are areas to consider when developing change control management procedures.

Objectives

Potential objectives for change control management procedures include:

■ Document reason(s) for the change
■ Identify personnel requesting the change
■ Formalize who will make the change
■ Define how the change will be made
■ Assess the risk of failure and impact of the change
■ Document fall back plans and back up procedures should the need arise
■ Aid in communicating with those affected by the change
■ Identify disaster recovery considerations
■ Identify conflicts between multiple changes
■ Enhance management's awareness of the change management process

Scope

The scope for change control management procedures can include:

■ Hardware
■ Operating system software
■ Database instances
■ Application software
■ Third-party tools
■ Telecommunications
■ Firewalls
■ Network (e.g., local area network, wide area network, routers, servers, etc.)
■ Facilities environment (e.g., uninterruptible power supply, electrical, etc.)

Change Control Management Boards or Committees

Change Control Management Boards or Committees are common entities to deal with coordinating the communication of changes within an organization. Following are possible sources for members of a change control management board or committee:

■ Application development/support teams (e.g., finance, human resources, etc.) who can provide leads

- Data center operations
- Networks/telecommunications
- Help desk
- Key user representatives

The individuals on the change control management board or committee should be selected based on their in-depth perspective and broad knowledge of the areas they represent, as well as their awareness for other functional areas involved. The goal is to ensure that the decisions made are objective as these changes have the potential of affecting the entire organization. Change control management boards or committees should meet daily or weekly. The following types of changes are often considered during daily change review meetings:

- Emergency releases and fixes normally related to circumstances in which the production is down
- Database clones, restores, links, or new instances
- "Fast-tracked" requests for new functionality that cannot wait for the normally scheduled dates for updates
- Application or system maintenance
- Upgrades to development tools

The following topics are often covered during weekly change review meetings:

- Migrations of new releases
- Upgrades to production or development tools
- Environment setup for migrations or tools
- Third-party upgrades
- Configuration changes
- Hardware changes

Criteria for Approving Changes

Approval of changes can be based on the following criteria:

- *State of the production environment.* Before determining if a change should be approved, the change control management board or committee should evaluate the performance and availability of each system in the production environment during the previous week. In general, if the production environment has performed well and has been available to the users, the change control management board or committee is more likely to approve changes that provide new functionality or changes that might have a higher risk of failure. Conversely, if the state of the production environment during the previous week has been poor, the change board or committee is more likely to approve only those changes designed to correct problems.
- *Change level.* As part of the approval process, the change level is examined along with the detailed information and instructions attached to the change request. The attachments should detail the associated risk and impact of the change. Particularly important are the subjective comments provided by the change requester indicating the reasons for the

assigned change level. The change level can be based on six factors: risk, impact, communication requirements, install time, documentation requirements, and education or training requirements.

- *Cumulative effect of all proposed changes.* The change control management board or committee is one place where all of the changes requested come together. The board or committee has the ability to examine several changes, each of which may appear to carry a reasonable risk if taken independently, but when all changes proposed are considered as a group, the composite effect may result in too much change activity—hence, risk. When the board or committee reaches this conclusion, its responsibility is to prioritize the various changes, approve the most significant ones, and recommend that the others be rescheduled.
- *Resource availability.* The change control management board or committee evaluates the availability of people, time, and system resources when considering the scheduling and approval of changes.
- *Criticality.* There are issues that may affect the impact of the change as viewed by the requester. For example, the change requester may feel that the impact is relatively low because his or her change affects a small percentage of the user community. However, the board or committee may view the criticality to be high because that small percentage of users is a critical client or user. An example might be the finance department during a year-end closing process.

Post-Implementation

Following implementation of a change, evaluations of whether the change procedures were followed and met their objectives must be performed. Post-implementation evaluations should also consider whether implementation and fall back plans or back out procedures were adequate and final status of the change (i.e., complete, not complete, in progress, failed, or cancelled).

Points of Change Origination and Initiation

Change control management procedures should identify the users or groups of users who initiate changes. Vendors, computer operators, application or system programmers, or end users are the ones who typically request changes. In addition, procedures should ensure that all requests are submitted using a standard change request form that contains information, such as:

- Date of request
- Name of requester
- Description and reasons for the change
- Areas that may be affected by the change
- Approvals to work with the change and for implementation into the production environment

The urgency of each request should be determined and all requested changes should be filed by priority and date. Procedures must also provide the means for implementing emergency changes in response to an operational problem. Control points for emergency change processing should be established to record, document, and obtain subsequent approval for changes. Emergency changes should be cross-referenced to operations problem reports to help verify the proper recording and handling of the changes.

Each change request should be numbered sequentially and recorded in a change log at a central control point. Most organizations use online tools to facilitate this process. The change request is entered online and, based on the areas affected, electronically routed to those areas for approval. The responsible area's approval is also recorded online. Any and all discussions regarding that change are documented online. Once the change has been scheduled, the online system will notify users of the upcoming change; it will also provide a confirmation once the change has occurred. Storing this information online allows management to perform a subsequent review of all changes for a variety of reasons. One may be the routine occurrence of the same type of emergency, indicating that the changes are fixing the symptom and not the underlying problem. Also, management could analyze what amounts of their resources are devoted to the routine, nonroutine, and emergency changes. If one application is requiring a significant amount of resources for changes, this may be an indicator that the application is reaching the point where replacing it may be more cost effective.

Approval Points

Approval points should be scheduled throughout the change control process. All key people and departments affected by a change should be notified of its implementation schedule. Those who may require notification include:

- End users of the system
- Vendors
- Application programmers
- IT management
- Operations personnel
- Data control personnel
- Network management
- Auditors
- Third-party service providers

An important part of the approval process often overlooked is testing. Ideally, testing would occur in a mirror of the production system (i.e., test environment). System changes need to be tested to ensure that they do not negatively impact the applications. Application changes can be grouped into system or functionality changes. System changes typically do not impact the end user and usually improve the speed of processing transactions. The programming support group, instead of the end users, usually tests these changes. Functionality changes are obvious to the end user and should be tested and approved by them. Functionality changes should be verified in a test environment. The test environment is a mirror of the production environment, which includes the data, programs, or objects. The data files should be expanded to include unusual or nonroutine transactions to ensure that all transaction types are used in the testing.

Approval levels should be predetermined as to who can approve what changes. Part of the change control process should ensure that the appropriate approval level is obtained before any changes are moved into production.

Change Documentation

As a system ages, the task of keeping track of changes and their impact on the operating system, operations environment, and application programs becomes increasingly difficult. The organization should maintain a record of all the changes made to the system. Without such a record, it is impossible to determine how proposed changes will affect the system. Not only should the change be documented with the change request form but also in the programmer's documentation. Programmer documentation is absolutely necessary for future maintenance.

It is also important to know when and why changes were requested but not implemented. With personnel changes, you may be revisiting an issue already deemed to be undeserving of time but it must be revisited to ensure integrity. The opposite can also be true where a previously undeserving issue could have merit due to the changing business environment.

Review Points

Change control management procedures must be carefully coordinated and reviewed if changes to the system are to be successfully implemented. The following steps should be included in the change review process:

- Pending changes are reviewed with key personnel in operations, application programming, network and data control, and auditing.
- Written change notification is sent to all interested parties, informing them of the nature of the change, scheduling of the change, purpose of the change, individual responsible for implementing the change, and systems affected by the change.
- Sufficient response time is provided for interested parties to examine proposed changes. The change notification should indicate the response deadline and the individual to contact for additional information.
- Periodic change control meetings (e.g., daily, weekly, monthly, etc.) should be held to discuss changes with key personnel.
- Reports are filed on implemented changes to record post-implementation results and successes as well as problems.

The change control management process and procedures should be documented in the form of an organizational policy. Appendix 6 illustrates an example of a standard change control management policy.

Configuration Management

Configuration management is controlling the physical inventory and relationships among components that form a set of "baseline" objects that are subject to change. The National Institute of Standards and Technology defined the process of software configuration management (SCM) in its Publication 500-223, "A Frame Work for the Development and Assurance of High Integrity Software." The major objectives of the SCM process are to track the different versions of the software and ensure that each version of the software contains the exact software outputs generated and approved for that version. It must be established before software development starts and continue throughout the software life cycle. SCM is responsible for ensuring that any changes to software during the development processes are made in a controlled and complete manner. The SCM process produces a software configuration management plan. Listed below are sample activities within such SCM plan.

1. Identification of Software **Configuration Items** (CIs)
   - Select the most significant and critical functions that will require constant attention and control throughout software development. These will become the CIs.
   - Assign a unique identifier/number to each CI.
   - Establish baselines for the CIs, that is, documents that have been formally reviewed and agreed upon, which (1) serve as the basis for further development and (2) can be changed only through formal change control procedures. Baselines to be established include:
     • Functional baseline (the completion and acceptance of the system requirements specification): prerequisite for the development of the software requirements specification (SRS) for each CI.
     • Allocated baseline (the review and acceptance of the SRSs): the prerequisite of the development of the software design description for all components making up a CI.
     • Developmental configuration (developer-controlled "rolling" baseline): all documents and code accepted and committed for configuration control up to the establishment of the product baseline.
     • Product baseline (established with the successful conclusion of a configuration audit): prerequisite to the operation and maintenance of the software.
2. Problem Reporting, Tracking, and Corrective Action—Document when a software development activity does not comply with its plan, output deficiency, or anomalous behavior and give the corrective action taken.
3. Change Control—Document, evaluate, resolve, and approve changes to the software.
4. Change Review—Assess problems and changes, implement approved changes, and provide feedback to processes affected by changes.
5. Traceability Analysis—Trace forward and backward through the current software outputs to establish the scope of impacted software.
6. Configuration Control—Delegate authority for controlling changes to software; determine method for processing change requests.
7. Configuration Status Accounting—Keep records detailing the state of the software product's development, for example, record changes made to the software, status of documents, changes in process, change history, release status, etc.
8. Configuration Audits and Reviews—Audit CIs before release of product baseline or updated version of product baseline; review to determine progress and quality of product. Two types of audits include:
   - Functional configuration audit; proves that a CI's actual performance agrees with its software requirements stated in the SRS.
   - Physical configuration audit; ensures that the documentation to be delivered with the software represents the content of the software product.
9. Archive, Retrieval, and Release—Archive software outputs (with backups) so that they cannot be changed without proper authorization, and can be retrieved if/when necessary. Software being released should be described and authorized.

Organizational Change Management

Organizational change management relates to the organization's ability and methods for adopting, managing, and adapting to change. Factors for evaluating change vary based on the scope of the change (i.e., changes to work habits as opposed to changes to the organization itself). Regardless of how IT is managed, it is still enacted by organizations to realize their expected monetary results. Consequently, an IT project can actually be considered a product of the organization's culture.

There are many studies to support the assertion that many IT projects fail due to the inability of the organization to adapt to the change necessary to take advantage of IT. Organizations find it difficult to change their practices and structures, especially if the application is perceived to be in conflict with the company's culture.

Organizational Culture Defined

Organizational culture is composed of structures for incentives, politics, support for interorganizational relationships, and social repercussions.

Incentives offered by the organization can impact the success of the organization in adapting to change because users do not necessarily see the "natural" advantages of adopting the change. As a point of caution, incentives should not conflict with other rewards or incentives or the culture. For example, if employees are told that they will be offered training to learn a new state-of-the-art system, but they will risk losing their bonus due to their increase in nonbillable time, the employee will see the training as a punishment as opposed to being considered an incentive.

Company politics can have significant effects on the success of a new IT system or change in the organization. Most models of organizational change exclude recognition for the importance of political influences over organizational change. For example, a new IT system can shift the power of information. In one case, it provided the corporate office direct online access to the live sales activity in each sales office. Prior to the new system, the field offices were able to determine how and when to present their sales figures to the corporate office. In some cases, the sales office would modify actual sales data with sales that were not yet realized. This enabled the sales offices to present their "best" sales figures. With direct access to the sales information, the corporate office was empowered to see the data when and how they wanted, allowing them insight into the volatility of the sales for a given office. Because it transferred control over the information to the corporate offices, the system was perceived to transfer decision power to the corporate office away from the sales offices. Subsequent power struggles ensued between the sales offices and the corporate office.

Organizational and technical support is critical for the effective use of IT. A supporting infrastructure includes organizational practices, key support staff, and access to technical and organizational skill sets. In this model, individual and organizational learning are considered a subset of the IT system. Many major IT system implementations are now including business process review sessions within their implementation scope. This enables the organization to review their business process in terms of the change that will be introduced into the system. This fit-gap analysis identifies gaps in functionality, which is integral to business processes. Subsequent plans can be made to accommodate for these gaps in redesigned business process, changes in business services, or modifications to the system itself.

Interorganizational relationships and social networks are supported and impacted through the use of IT. The influence of relationships and social networks are believed to explain why some technologies are supported and others are not. Looking at online communities and electronic market places—is it the influence of technology on relationships and networks that created online communities and electronic market places or vice versa?

An example of a successful online community is eBay. Its community is composed of individual buyers and sellers as well as large and small companies. Member relationships are supported with discussion boards. The sense of community is used to ensure that eBay guidelines are followed with a "neighborhood watch" philosophy.

Managing Organizational Change

Culture and structure change should be managed through the life cycle. It includes people, organization, and culture. A culture that shares values and is open to change contributes to success. To facilitate the change process, users should be involved in the design and implementation of the business process and the system. A communications plan and training and professional development plans also support this.

The business processes associated with a software implementation need to be aligned. In adapting packaged software packages, organizations face the question of whether to adapt the organization to the software or vice versa. To minimize maintenance of the software, the company should consider changing the business process to fit the software and the software should be modified as little as possible. Reducing customization reduces errors and improves the ability to utilize newer releases.

Organizational change relies on effective communication. Expectations need to be communicated. Communication, education, and expectations need to be managed throughout the organization. Input from users should also be managed to ensure that requirements, comments, and approval are obtained. Communication includes the formal promotion of the implemented change as well as the organization's progress with adopting the change. Employees should also know the scope, objectives, activities, and updates and the expectation for change in advance.

Training and professional development plans should be incorporated into any effort to introduce change into an organization. Employees need to be trained in the new processes and procedures. Additionally, the team assigned to implement the change requires special training in the process and procedures. Training in adapting to change is also beneficial to all employees.

Audit Involvement

A change control audit or examination would determine whether system changes are authorized, tested, documented, communicated, and controlled. The following areas are typically covered:

- Authorization
- Testing (unit, system, and user acceptance)
- Documentation
- Communication
- Controls

As seen, change control management is a process that has an impact on the production-processing environment. This includes changes to hardware, application software, and networks. With respect to the application change management review, change control management is defined as any modification to programs that may impact production processing. Production processing includes system availability, reliability, and integrity. Many organizations have migrated from a mainframe-dominated environment to a heterogeneous environment with mainframe, client/server, and minicomputers, among others. These systems require very different procedures, tools, and skill sets to manage change.

As organizations continue to evolve, their control environment over the change process may also increase the risk that production could be adversely impacted. Insufficient change control processes may have the following risks:

■ Existing IT systems (e.g., applications, databases, operating systems, networks, etc.) that do not meet the organization's information processing needs, resulting in incomplete, inaccurate, or invalid data.
■ Financial reporting systems not being able to transfer data between other IT systems and their underlying infrastructure components (e.g., network devices, server hardware).
■ Inappropriate development and implementation of IT systems that can result in inaccurate calculations, unreliable processing, incomplete recording of transactions, and other misstatements, among others.
■ Inappropriate procedures to define databases adequately may result in IT systems not capable of processing accurate data.
■ Inappropriate procedures to implement databases effectively may result in data being either unavailable and/or difficult to access.
■ Data loss or system outages resulting from errors, omissions, or malicious intent.
■ Fraud or abuse of company systems and/or data resulting from unauthorized changes.

The objective of a change control management audit is to ensure that changes implemented in production systems and applications do not adversely affect system, application, or data availability or integrity. To that end, auditors need to verify that all changes made to the production systems and applications are appropriately authorized and documented. A critical success factor for change control management is the culture of the organization. Does management understand the critical role of change control management, and does it recognize the impact change control management has on the success of the organization? Are change control management policies, procedures, and processes in place over client/server, mainframe, and desktop environments, for instance? Are controls in place to ensure that changes made to not adversely impact system stability or data integrity? The audit should include procedures such as:

■ Obtaining copies of policies and/or procedures related to change control management.
■ Interviewing application support staff to determine formal, informal, and emergency procedures used to implement changes to production systems and applications.
■ Obtaining copies of change control request forms and logs.
■ Selecting a sample of changes from the changes logged, and determining compliance with policies, procedures, and best practices.
■ Obtaining copies of help desk call logs to determine adverse impacts from changes.
■ Determining if new software, and changes to existing software are properly and formally authorized by the appropriate manager.
■ Determining that all new software and software changes are properly tested, using test files and directories, before they are moved into production.
■ Determining that all test results are effective, and reviewed by someone other than the originator.
■ Determining that program and file names are properly controlled to avoid duplicate names.
■ Determining that all supporting documentation is updated before an application or a change to an application is placed into production.

■ Determining that production applications are stored in a directory protected from unauthorized changes.
■ Determining the means of tracking production application changes. Is there a check-out/check-in system to prevent unauthorized changes?

Appendix 3 (discussed in Chapter 3) provides a sample IT audit program for the change control management general control IT area, which includes a complete list of audit control objectives and activities to be followed and performed when conducting a change control management examination. Depending on the size and complexity of the organization, these control objectives and activities may need to be revised or expanded to obtain adequate audit coverage of the change control management function.

The auditor obtains the necessary background information, determines the key controls, performs limited substantive testing to assess the reliability and effectiveness of the process controls, and evaluates the process. The auditor must take the time to become thoroughly familiar with and understand the change control process. He or she should develop a flowchart documenting the process in which points of origination and initiation, approval points, changes to documentation, and review points are all identified. Furthermore, auditors' developed flowcharts should document procedures for emergency and nonemergency changes. Exhibit 10.2 illustrates a flowchart example of a change control management process carried out in an organization. Notice the different roles involved in the process (e.g., Change Requester, Project Manager, Change Control Board, etc.), and the various activities they all perform.

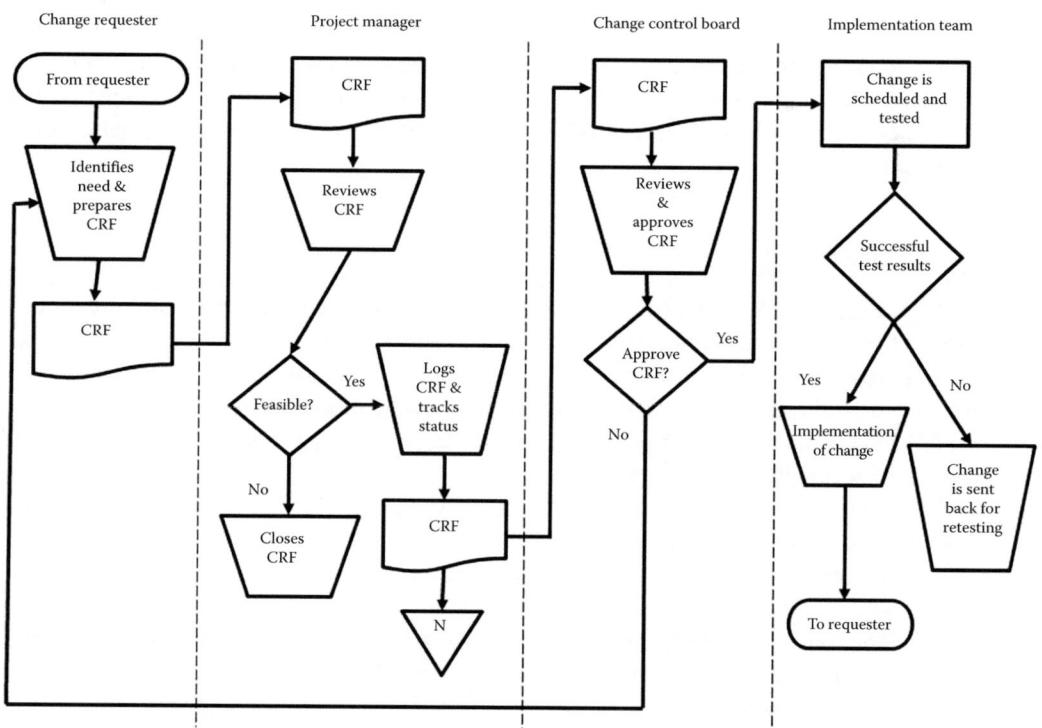

Exhibit 10.2   Flowchart depicting a standard change control management process.

Conclusion

Changes to systems are frequent, and can be introduced to fix bugs, implement new functionality and software releases, or from configuration management and business process redesigns. A change control management process is, therefore, one of the single most important controls to have. An effective change control management process reduces the risk of disruption of IT services. The process also ensures the integrity, availability, reliability, security, confidentiality, and accuracy of an organization or IT system supporting the organization. Moreover, the process not only monitors all changes made to the systems, but ensures adequate segregation of duties between who initiates the change, who approves the change, and who implements the change into the production environment.

To support the process and ensure it is carried out effectively, change control management procedures are put in place. Change control management procedures make certain that all members of the organization follow a standard process for initiating, approving, and implementing changes to systems and applications.

Organizations also need to implement good configuration management practices. Configuration management refers to the process used to monitor the installation of updates to system hardware, operating system software, and firmware to ensure that hardware and software function as expected, and that a historical record of changes is maintained.

Organizational change management also deserves consideration due to its potential impact on the organization and the increased relationships with changes in the IT environment. Organizational change is impacted by limitations introduced by the technology and the organization's culture. An organization's culture is composed of its structures for incentives, politics, and relationships with other organizations and the community.

A change control audit or examination determines whether system changes are authorized, tested, documented, communicated, and controlled. The objective of a change control management audit is to ensure that changes implemented in production systems and applications do not adversely affect system, application, or data availability or integrity. To that end, auditors need to verify that all changes made to the production systems and applications are appropriately authorized and documented.

Review Questions

1. Implementing policies, procedures, and techniques assist changes and modifications to systems (e.g., programs, applications, etc.) to be properly authorized, tested, approved, and carefully distributed or controlled. Without proper controls like the one above, there is a risk that unauthorized changes or modifications could introduce processing irregularities or malicious code, among many others. List examples of other risks that could potentially impact the security of the systems.
2. Explain the benefits for organizations of implementing a well-defined and structured change control management process.
3. Discuss the three types of changes typically implemented in systems and applications.
4. Name and describe four risks associated with the software release process.
5. Describe the controls typically included when following good software distribution practices.
6. Name and summarize the five criteria for approving changes.
7. Once approved, changes should be scheduled for implementation. At this point, all key people and departments affected by a change should be notified of the upcoming implementation. List those who may require such notification.

8. Summarize how the National Institute of Standards and Technology defines the process of software configuration management.
9. Describe the interdependencies between IT change management and organizational change management.
10. What is the objective of a change control management audit? List at least seven procedures in a change control management audit.

Exercises

1. Discuss what emergency changes are and why do they require "special" attention from management.
2. Discuss why revising documentation is an important part of change management.
3. Explain the purpose of a change request form. Why should change request procedures be documented?
4. Using an Internet Web browser, search and examine two recent (within the last 5 years) situations where the implementation of changes and/or modifications to existing financial application systems have failed. Your task: Summarize why such implementations failed. Then, identify solutions or controls that, if implemented, may have avoided these failed implementations. Support your reasons and justifications with IT literature and/or any other valid external source. Include examples as appropriate to evidence your case point. Submit a word file with a cover page, responses to the tasks above, and a reference section at the end. The submitted file should be between 6 and 8 pages long (double line spacing), including cover page and references. Be ready to present your work to the class.
5. Following your recommendation, your organization just created a Change Control Management Board or Committee (Board) to oversee the recently-implemented change control management process. As the Chair of the Board, prepare (using a memorandum format) a document to discuss and present to all members during the first Board meeting. The document should include: (1) description of the responsibilities of the newly-formed Board and (2) the types of changes and/or topics that will be considered and discussed during the meetings.

CASE—CHANGE CONTROL MANAGEMENT AUDIT

NARRATIVE—IT ENVIRONMENT AND CHANGE CONTROL MANAGEMENT PROCESS

The IT audit of client ABC Company is for the fiscal year ended 12/31/20XX. In various planning meetings with client/entity personnel, you gathered the following information pertaining to the client's relevant financial applications. These are the applications that will form the scope of this year's audit:

1. SAP—Handles all accounting and general ledger transactions. SAP's server is SapAppServ1, and its operating system (o/s) version is UNIX AIX VX. SAP is classified as a purchased software requiring significant customization. The database supporting SAP is Oracle, with server name SapDbServ1.

2. Bill-Inv System—Handles billing and inventory management. The Bill-Inv System's server is InfAppServ1, and its o/s version is OS 400 VX. Bill-Inv System is a purchased software with little customization. The database is IBM DB2, with server name InfDbServ1.

3. APS2—Handles all treasury and investment transactions. The server hosting the application is Aps2AppServ1, and its o/s version is OS400 VX. APS2 is a purchased software requiring significant customization. The database is Oracle, with server name Aps2DbServ1.

4. Legacy System—Handles transactions related to the entity's fixed assets. The Legacy System's server is LSAppServ1, and its o/s version is Windows VX. The Legacy System is classified as an in-house developed software. The database is proprietary, with server name LSDbServ1.

5. Kronos—Handles employee time and attendance transactions. The Kronos's server is KAppServ1, and its o/s version is Linux VX. Kronos is also classified as an in-house developed software. The database is proprietary, with server name KDbServ1.

6. HR&P—Handles transactions related to human resources and payroll processing and disbursement. This application is outsourced to a service organization called ADP.

The o/s hosting the applications are the ones hosting the databases. All of the above relevant applications, except for HR&P, are located in the entity's headquarters premises, computer room, first floor, in Melbourne, FL. HR&P is outsourced, meaning that the application, its database and o/s, as well as all related servers are located outside the client's premises. For description of the processes and procedures regarding this outsourced application, refer to the auditor's service organization report.

During the fiscal year, there were no significant modifications performed for the Bill-Inv System, Legacy System, Kronos, and HR&P relevant applications. SAP and APS2, however, were significantly upgraded to their current versions in March 1st and November 15th, respectively.

Client ABC Company has three IT-related departments, all reporting to the IT Director. The IT director reports directly to the entity's Chief Financial Officer. The IT departments and their supporting personnel are listed below:

■ Infrastructure & Operations (I&O)—I&O Manager, Help Desk Supervisor, Trainer & End-User Support, and Network/LAN Administrator
■ Application Systems Programs & Support (ASP&S)—ASP&S Manager, Software Programmer, Analyst Programmer, and SAP Administrator
■ Database and Operating Systems Support (D&OSS)—Database Support Specialist and Operating Systems Support Specialist

Purchased Applications (SAP, Bill-Inv System, and APS2)

The client's process for selecting and purchasing new applications is conducted taking into consideration the economic and operational impact. Whenever the need for acquiring software is identified, both management personnel and representatives of the end user community (users) establish design requirements and compatibilities for the soon-to-be acquired application. According to the IT Director and the I&O Manager, once a need has been identified and requirements have been established, IT management personnel are responsible for evaluating at least three alternatives considering cost–benefit relationships and the impact to the IT environment. Once evaluated, an alternative is preliminary selected and discussed with business managers to ensure alignment between information systems and business initiatives. Applications exceeding $100,000, and/or having the potential of impacting IT risks are subject to risk assessments and business impact analyses. Upon completion of such analyses, the selected alternative (supported with analysis documentation) is submitted to the Chief Financial Officer for final approval. Once a selection of the application is made, IT personnel perform full backups of the old application. IT personnel then prepare a separate environment for users to start testing whether the new application runs as expected, and whether the data are accurate. They also compare the new application against the old application (i.e., parallel testing). After testing is done, users provide acceptance and support for the installation of the new application in the production environment. IT personnel is then responsible for advising all affected users of the installation dates. The above is part of the entity's policy for selecting and purchasing new applications; however, such policy has not been revised and/or updated during the past 5 years. A policy should be reviewed and updated at least once a year to reflect changes in the entity's processing environment.

All changes or upgrades to purchased applications are performed by application vendor personnel and sent back to the entity for installation. Per the ASP&S Manager, changes or upgrades received from the vendors are currently not tracked or logged. The client acknowledges there are available Web-based tools and techniques that can track or log these types of changes, but feel their costs may not be justified. Because these changes or upgrades are from the vendor, the entity trusts that they have been adequately tested (at the vendor site) before the vendor forwards them to the client. Therefore, the entity does not perform full backups of the existing application before the implementation of the changes or upgrades. IT personnel do install the changes/upgrades received in a separate test environment. Tests are performed thoroughly to ensure the new changes are consistent and will conform with current business needs. Test results, if successful, are communicated verbally to the manager in charge. In addition to the IT personnel approval, there are no additional approvals required before implementation of the changes/upgrade into production. IT personnel are responsible for advising all affected users of installation dates in the production environment. The above procedures have been formalized into a policy. The policy is updated annually.

Once implementation of changes or upgrades to purchased applications is performed, IT personnel, led by the ASP&S Manager, perform various tests to validate the integrity, accuracy, and completeness of the information.

In-House Developed Applications (Legacy System and Kronos)

As discussed with both, I&O Manager and ASP&S Manager, the process of developing in-house applications or implementing changes to in-house applications is a standard and common process. The process may result from (1) users identifying system needs; (2) errors being identified and requiring fix; and/or (3) applications themselves forcing implementation of new patches/upgrades. Requests for developing in-house applications or implementing their changes are submitted by users who complete an online "System Modification Request Form" (SMRF). The SMRF is a Web-based tool the entity uses to track and control requests, and includes information, such as the name of the application or system, requester's name, date, department(s) affected, and a description of the requested change. Additionally, the tool provides information on the programmer who will work with the change and the estimated completion date. Once the SMRF is completed and requirements are established, the ASP&S Manager assesses the impact of the change. If the in-house application or change is deemed significant, it is considered as a project and additional resources are allocated. On the other hand, if the in-house application or change is considered to have a minor impact or maintenance, it is assigned to either the Software Programmer, Analyst Programmer, or SAP Administrator, and performed directly in the production environment. Therefore, there is no evidence like test methodologies, test plans and results, and project implementation schedules, maintained for these types of changes. There is also no separate environment established for developing or testing these "minor impact" in-house application or changes.

When determined to be significant, in-house applications or their changes are worked in a development environment separate from production. Full application and data back-ups are not performed prior to developing the in-house application or implementing the changes. Nonetheless, test procedures are documented, and successful results support final implementation. Testing is performed by selected users from the IT and business area. Test procedures performed consist of recreating normal operation transactions and verifying/monitoring the results for accuracy. Test procedures also validate the integrity and completeness of the information. After testing is done and in order to ensure proper segregation of duties, programmers are not allowed to migrate their own changes into the production environment. Instead, they turn in their work to independent, non-programmer personnel (quality assurance team, for example) for migration into the live environment.

Both, the I&O Manager and the ASP&S Manager, indicate that in order to manage and maintain version control, a Software Version Control Configuration Manager (SVCCM) tool is used. This tool allows the identification of changes, labeling them by "revision number," "revision letter," "revision level," or simply just "revision." Change revisions are associated with a timestamp and the name of the user making the change. The SVCCM tool also allows revisions to be compared and restored, as well as combined with other types of files. The above process has not been formally documented in the form of a policy or procedure, nor establishes how it prevents unauthorized changes to the in-house based applications. Additionally, there is currently no version control or management system process used or in place for the entity's purchased applications. The above information was also corroborated with the client's Software Programmer, Analyst Programmer, and the SAP Administrator.

Databases

The entity's process for acquiring and implementing databases is similar to the one conducted for purchased applications, according to the IT Director and the I&O Manager. The process for maintaining databases differs. Changes to the databases supporting all relevant applications, except for Legacy System and Kronos, are mostly dependent on application changes and, thus, are subject to the same application maintenance procedures previously described. Both proprietary databases supporting the Legacy System and Kronos applications are administered and maintained by application programmers (i.e., Software Programmer, Analyst Programmer, and the SAP Administrator). Per conversation with these programmers, whenever a report or particular information is needed from any of these databases, programmers access the live databases to submit a job or generate appropriate queries to produce the desired report.

The database architecture for SAP, Bill-Inv System, and APS2 is a separate database, accessed by the relevant financial application. On the other hand, the database architecture for the Legacy System and Kronos is integrated or individual database used by the particular relevant application.

As discussed with the Database Support Specialist, data dictionaries contain definitions and representations of the data elements stored in the entity's databases. Examples of these would be precise definitions of data elements, integrity constraints, stored procedures, general database structure, and space allocations, among others. Definition of the format of a telephone number field would be an example of one of the uses of a data dictionary. If such format is defined in the data dictionary, the field will be consistent throughout the database even if several different tables hold telephone numbers. Data dictionaries for all purchased applications are only maintained with the purpose of defining data relationships between entities, data/field structures, and to represent data elements. These tasks are used consistently throughout the entity's data dictionaries. Data dictionaries for the in-house applications have been defined within the application code itself.

Networks

Except for HR&P (outsourced to the ADP service organization), all relevant applications are supported by the local network. As discussed with the Network/LAN Administrator, the network consists of a single domain in which all users are grouped and divided using a hierarchical structure. Client computers and peripherals at the entity are connected through switches to the servers located in the computer room. These servers are protected using two firewalls. A network infrastructure diagram, including the locations that are networked together, the equipment used, the activities that are supported by the networked relevant applications, and the interrelationships within the network is not available.

Per the Network/LAN Administrator, whenever configuration changes, upgrades, and/or new network software changes are required (frequently firewall-related changes), these are requested not by regular users, but by IT personnel. Users often do not have the technical background or expertise to define network solution requirements. Network-related change requests are reviewed and approved by the I&O Manager. Once requirements are approved, the needed changes are worked by in-house personnel (typically the Network/LAN Administrator) with support from vendors or external networking consultants, as appropriate. Changes are then tested prior to the implementation into production. Installations of changes are performed during off-peak hours in order to minimize services disruption. The process of requesting, testing, and implementing network changes is done verbally and no documentation is maintained supporting the procedures performed. If documented, this information would be retained to provide insight for the general support of the network, particularly during maintenance activities or if any disruptions occur.

Operating Systems

As discussed with the Operating Systems Support Specialist, the process for acquiring o/s is similar to the process followed for applications (refer to above). Procedures in place to implement and maintain (test) the entity's o/s supporting the relevant financial applications follow:

- UNIX AIX VX (SAP)—Changes or upgrades to UNIX AIX VX are provided by the vendor, IBM. All updates are installed directly into the production environment, because there are no separate environments available to test UNIX AIX VX changes. As a mitigating control, prior to the implementation of UNIX AIX changes or upgrades into production, copy of the existing o/s as well as its application data are backed up, allowing local IT personnel to restore the system to its previous state, should any disruptions in processing occur as a result of the operating system change.

- OS 400 VX (Bill-Inv System and APS2)—All testing related to changes to OS 400 VX is performed in a separate test environment in order to assess any impact that required changes may have to the o/s itself or to the production environment. Testing also ensures the integrity, accuracy, and completeness of the hosted information. Prior to implementation of the changes, the o/s is backed up, allowing restoration of the system to its previous state, should any disruptions in processing arise from the change. All changes are performed during off-peak hours (normally during weekends), allowing the entity to minimize system downtime. Once tested, changes are implemented into production.

- Windows VX (Legacy System)—Updates to Windows are installed directly into the production environment because there are no separate environments available. As a mitigating control, updates that are deemed necessary are deployed first into the less critical servers to undergo compatibility testing with existing applications. Documentation supporting test plans and procedures performed as well as results of the tests related to the Windows updates is not kept.

- Linux VX (Kronos)—Changes or upgrades to Linux are not frequent and are dependent on the application and database it hosts. Whenever changes/upgrades are required, these are provided by the software vendor. Per the Operating Systems Support Specialist, the software vendor performs testing on those changes and provides supporting documentation evidencing results. Testing performed ensures the integrity, accuracy, and completeness of the hosted application information. Upon receipt, changes or upgrades are installed into the Linux o/s's production environment.

Application Controls

For all relevant applications, default application controls validate for mandatory fields, format type, and size of data input. These application controls also number events and transactions (individually and sequentially) thereby verifying accountability among users. Further, default application controls run warning messages to IT personnel when data processing fails, or when calculations are not performed accurately nor completely. The messages, with description of the failures, are stored in a log file for further review. Standard queries generate daily reports with control totals and statistics for output reviews, and in order to detect exceptions and inconsistencies. These reports are forwarded to responsible individuals or departments for further review and to correct any exceptions noted.

Future Plans

In terms of the entity's plans to upgrade or replace existing applications, databases, network, and/or operating systems in the near future, the IT Director along with the I&O Manager and ASP&S Manager, indicate that they will evaluate the possibility of upgrading the current version of the OS/400 VX to a newest version. Update to such current version is expected to take place during the next year. No other plans to upgrade or replace existing systems are set for the immediate future.

TASKS

1. Read the Narrative "IT Environment and Change Control Management Process" and complete Appendix 2—Understanding the IT Environment.
2. From the Narrative, identify potential findings and list them using a table format. Column names are: "Description of Potential Finding," "Area and/or Application Affected," and "Risk Associated with Finding."
3. Support the rationale (the "why") for each potential finding identified and document in the table from #2. This would also be the risk(s) associated with each finding. IT poses specific risks to an entity's internal control, including, for example, unauthorized disclosure of confidential data; unauthorized processing of information; inappropriate manual intervention; system crashes; unauthorized modification of sensitive information; theft or damage to hardware; and loss/theft of information, among many others.
4. Prepare formal communication to management in the form of a Management Letter. Use the format of Exhibit 3.9—Management Letter to prepare your communication. Complete the FINDING, IT RISK, and RECOMMENDATION sections of the Management Letter, but do not include the MANAGEMENT RESPONSE section. Note: For the recommendations, consider including potential IT controls you believe management should implement to address the risk and finding. You may use Appendix 3—Sample IT Audit Programs for General Control IT Areas, as reference, to identify IT controls. Assume the letter will be submitted to the IT Director and to the Chief Financial Officer, and that a preliminary meeting with the IT Director to discuss these findings occurred a month after the Company's fiscal year ended. Lastly, there are no findings repeated from prior years to be included in the letter.

Further Reading

1. Burgess, A. (2009). *Easy Version Control with Git*. Envato Pty Ltd. https://code.tutsplus.com/tutorials/easy-version-control-with-git–net–7449
2. Common Weakness Enumeration (Version 2.10), http://cwe.mitre.org/ (accessed April 6, 2017).
3. Corporate Executive Board, *Change Management Models*, Working Council for Chief Information Officers, January 2003.
4. Deloitte LLP. (2014). *IT Audit Planning Work Papers*. Unpublished internal document.
5. Gallegos, F. and Yin, J. (2000). *Auditing Oracle*, EDP Auditing Series, #74-15-37, Auerbach Publishers, Boca Raton, FL, pp. 1–12.
6. Gallegos, F. and Carlin, A. (2000). *Key Review Points for Auditing Systems Development*, EDP Auditing Series, #74-30-37, Auerbach Publishers, Boca Raton, FL, pp. 1–24.
7. Getting Started—About Version Control. https://git-scm.com/book/en/v2/Getting-Started-About-Version-Control (accessed June 8, 2017).
8. Institute of Internal Auditors. Global Technology Audit Guide (GTAG) 8: Auditing Application Controls. https://na.theiia.org/standards-guidance/recommended-guidance/practice-guides/Pages/GTAG8.aspx (accessed July 2017).

9. IS Audit Basics. The process of auditing information systems. www.isaca.org/knowledge-center/itaf-is-assurance-audit-/pages/is-audit-basics.aspx (accessed July 2017).

10. Information Systems Audit and Control Foundation, IT control practice statement: AI6 Manage changes, Information Systems Audit and Control Foundation, www.isaca.org/popup/Pages/AI6-Manage-Changes.aspx

11. ISACA. (2017). Web application security: Business and risk considerations. http://www.isaca.org/Knowledge-Center/Research/ResearchDeliverables/Pages/Web-Application-Security-Business-and-Risk-Considerations.aspx

12. ISO/IEC 20000-1:2011. Information technology -- Service management -- Part 1: Service management system requirements. www.iso.org/standard/51986.html (accessed June 8, 2017).

13. BMC Software, Inc. (2016). ITIL change management. www.bmc.com/guides/itil-change-management.html

14. Kling, R. and Lamb, R. (2000). IT and organizational change in digital economies: A sociotechnical approach, In *Understanding the Digital Economy: Data Tools and Research*, Brynjolfrson, E. and Kahin, B. Eds., MIT Press, Cambridge, MA, pp. 295–324.

15. Melançon, D. (2006). Beyond checklists: A socratic approach to building a sustainable change auditing practices, Information Systems Audit and Control Association, *Journal Online*.

16. Morella, R. (August 2015). Auditing web applications. IT audit strategies for web applications. ISACA Geek Week. www.isaca.org/chapters3/Atlanta/AboutOurChapter/Documents/GW2015/081115-10AM-WebAppSecurity.pdf

17. National Institute of Standards and Technology. Special Publication 800-53A, Revision 4, Assessing security and privacy controls in federal information systems and organizations, December 2014. http://nvlpubs.nist.gov/nistpubs/SpecialPublications/NIST.SP.800-53Ar4.pdf

18. Oseni, E. (2007). Change management in process change, Information Systems Audit and Control Association, *JournalOnline*.

19. Otero, A. R. (2015). An information security control assessment methodology for organizations' financial information. *Int. J. Acc. Inform. Syst.*, 18(1), 26–45.

20. Otero, A. R. (2015). Impact of IT auditors' involvement in financial audits. *Int. J. Res. Bus. Technol.*, 6(3), 841–849.

21. Otero, A. R., Tejay, G., Otero, L. D., and Ruiz, A. (2012). A fuzzy logic-based information security control assessment for organizations, IEEE Conference on Open Systems.

22. Otero, A. R., Otero, C. E., and Qureshi, A. (2010). A multi-criteria evaluation of information security controls using Boolean features. *Int. J. Network Secur. Appl.*, 2(4), 1–11.

23. Richardson, V. J., Chang, C. J., and Smith, R. (2014). *Accounting Information Systems*, McGraw Hill, New York.

24. Romney, M. B., and Steinbart, P. J. (2015). *Accounting Information Systems*, 13th Edition, Pearson Education, Upper Saddle River, NJ.

25. Senft, S., Gallegos, F., and Davis, A. (2012). *Information Technology Control and Audit*, CRC Press/Taylor & Francis, Boca Raton.

26. Software & Information Industry Association (SIIA). About SIIA, http://www.siia.net/About/About-SIIA.

27. U.S. General Accounting Office. *Federal Information System Controls Audit Manual:* Vol. 1, Financial Statement Audits, AIMD-12.19.6, June 2001.

28. Wallace, D. R., and Ippolito, L. M. *A Framework for the Development and Assurance of High Integrity Software*. NIST Special Publication 500-223, December 1994, Section 3.4 "Software Configuration Management Process". https://www.nist.gov/publications/framework-development-and-assurance-high-integrity-software (accessed March 28, 2017).

Chapter 11

Information Systems Operations

LEARNING OBJECTIVES

1. Describe the importance of policies and procedures regarding information systems operations for both the organization and auditors.
2. Explain how data processing and output controls play a significant role in the completeness, accuracy, and validity of information.
3. Discuss guidelines and controls to protect data files and programs.
4. Discuss controls and procedures related to physical security access.
5. Discuss controls and procedures related to environmental controls.
6. Discuss controls and procedures regarding storage and archival of information.
7. Describe what a business continuity plan is and its significance to the organization.
8. Explain what a disaster recovery plan is and its components. Discuss objectives and procedures when auditing such plan.
9. Describe the importance of end-user computing groups and the steps performed when auditing such groups.
10. Describe the audit involvement in an information systems operations examination.

Within an IT environment, controls related to information systems (IS) operations provide a structure for the day-to-day management of operations and maintenance of existing systems. IS operations is also one of the three major general computer controls used to assess organizations' policies and procedures related to application systems in order to support the effective functioning of application controls. Examples of general controls within IS operations address activities such as job monitoring and tracking of exceptions to completion, access to the job scheduler, and data backups and offsite storage, among others.

This chapter presents an overview of IS operations as a relevant component of the IT infrastructure. Key objectives and controls assessed by both, organizations and IT auditors (consistent with Appendix 3 from Chapter 3), relate to:

- Operating policies and procedures
- Data processing
- Protection of data files and programs
- Physical security and access controls
- Environmental controls
- Program and data backups
- Business continuity plan
- Disaster recovery plan

The chapter discusses the aforementioned, and provides sample objectives and control activities the IT auditor should focus on when examining IS operations. The chapter also describes end-user computing (EUC) groups, and provides guidelines when auditing such groups. Lastly, IT audit involvement and procedures are described when examining IS operations.

Operating Policy and Procedures

Every IT environment should have specific policies and procedures in place covering IS operations to ensure that, at a minimum:

- IS operations support the scheduling, execution, monitoring, and continuity of IT programs.
- IS operations promote the complete, accurate, and valid processing and recording of business transactions.
- Existing facilities protect the integrity of business information.
- IS operations are adequate to safeguard the storage of data files and programs.

Managers should regard the creation, review, and update of operating policies and procedures as a highly important control. Updates and reviews to these policies and procedures should be performed periodically. This can be done through observing execution to determine if such existing policies and procedures are actually being followed in day-to-day IS operations.

Audit objectives when assessing policies and procedures require organizations to provide standards for preparing documentation and ensuring the maintenance of such documentation. The IT operations manager must set documentation standards so that when employees change jobs, become ill, or leave the organization, replacement personnel can adequately perform the task of that employee. The IT operations manager must also periodically test the documentation for clarity, completeness, appropriateness, and accuracy.

Process and procedures related to an organization's IS operations should be documented in the form of an organizational policy. Appendix 7 illustrates an example of an IS operations policy.

Data Processing

An example of data processing, specifically financial data processing, is the daily posting of accounting journal entries to the organization's general ledger. The processing (or posting) of these journal entries typically starts with an unposted, approved batch of journal entries that is scheduled to run at night, or during off-peak hours. If processed successfully, the status of the journal entry batch is changed to indicate that the journal entries have been posted. In other words, journal entries have updated the general ledger. If, on the contrary, errors were identified prohibiting successful posting of the entries, reports should be generated detailing these errors or exceptions. Effective data processing controls detect these as they occur and should prompt IS operators for correction and resolution.

As seen, data processing controls play interdependent roles in the completeness, accuracy, and validity of information. If one of these controls is not functioning properly or an unauthorized intervention overrides complementary control processes, the data processing system becomes vulnerable.

These controls can be oriented toward prevention, detection, or correction of errors and abuse. Preventive controls ensure that events proceed as intended. Detective controls signal an alert or terminate a function and stop further processing when the system is violated or an error occurs. Corrective controls may perform an alert or terminate a function, but they also restore or repair part of the system to its proper state.

Errors in data processing usually relate to **job scheduling** and actual monitoring of the job processing. In fact, an important element of any set of policies and procedures should be the requirement that IS operators maintain logs on which any unusual events or failures resulting from the processing of data are recorded, according to time and in detail. These logs can be used to identify unfavorable trends, detect unauthorized access, and provide a data source for determining the root cause of system failures. Further, to address unusual events, failures, or errors, managers should ask the following key questions:

1. Are there appropriate controls configured to reduce data processing errors and maintain the integrity of data processed?
2. Is there an automated tool used to execute regularly scheduled jobs related to applications, databases, and operating systems, such as scheduled interfaces of data, data purges, table updates, etc.?
3. What are the types of jobs scheduled?
4. How are changes, such as adding, modifying, and deleting jobs and schedules made, and who can make those changes?
5. Does the system use processing checks to detect errors or erroneous data during data processing? If so, which checks?
6. What is the process used to monitor the successful completion of job processing?
7. How the monitoring and review process ensures that exceptions or failures identified during job processing are timely resolved?
8. Are techniques available for detecting erroneous reprocessing of data?
9. Who is responsible for the review and exception tracking of erroneous reprocessing of data?
10. Which reports are reviewed, and what notification systems and mechanisms are currently in place?

Controls that follow data processing and that are also critical in ensuring the accuracy, completeness, and delivery of information are called output controls.

In today's automated environment, most output is posted online or printed and processed by machines. It is important to have completeness and accuracy controls from the time the output is processed until it is posted online or delivered. In addition, security, confidentiality, and privacy need to be maintained from the time the output is created until delivered to the appropriate party. Whether output of information processing is displayed online or on paper form, traditional output controls are needed to ensure the accuracy and completeness of the information. Output controls include balancing and completeness checks, for instance, to confirm that the number of pages processed are created for online or paper printing. This can be accomplished by creating a page total before and after posting or printing the output for comparison. Accuracy can be confirmed by selecting key data fields for comparison before and after output processing. The output controls should also be able to detect where information is missing. It would be difficult to determine the problem from just page count when thousands of pages are printed and there is no way to determine where the failure occurred. In addition, a process needs to be in place to recreate all or a subset of documents where output errors are discovered. Additional controls are needed for sensitive information (e.g., checks, customer lists, trade secrets, payroll data, proprietary data, etc.) as the original documents may need to be destroyed and this needs to be carefully controlled to verify the destruction of such sensitive information.

In an IT audit, a common objective within this area would be to determine whether IS operations support the adequate scheduling, execution, monitoring, and continuity of systems, programs, and processes to ensure the complete, accurate, and valid processing and recording of financial transactions. Some of the control activities the IT auditor can evaluate would relate to whether (1) batch and/or online processing has been defined, timely executed, and monitored for successful completion and whether (2) exceptions identified on batch and/or online processing are timely reviewed and corrected to ensure accurate, complete, and authorized processing of the financial information. Addressing these will ensure that data is validly processed, and that any exceptions noted while processing have been detected and corrected.

Protection of Data Files and Programs

Each IT environment should have a data library that control access to data files, programs, and documentation. An important data library control centers on assurance that all file media are clearly and accurately labeled. That is, external labels should be affixed to or marked upon the data media themselves. On **tape cartridges** and **disk packs**, pressure-sensitive labels are usually affixed to identify both the volume and the file content. Procedures should be in place to assure that all labels are current and that all information they contain is accurate.

The data library should assure that only authorized persons receive files, programs, or documents, and that these persons acknowledge their responsibility at the time of each issuance. Each time a file is removed for processing, controls over data files should assure that a new file would be generated and returned to the library. If appropriate to the backup system in place, both issued and new files should be returned together with the prior version serving as backup.

Control is enhanced by maintaining an inventory of file media within the data library. In other words, an inventory record should exist for each tape cartridge or disk pack. The record should note any utilization or activity. After a given number of users, the file medium or device is cleaned and recertified. Further, if any troubles are encountered in reading or writing to the device, maintenance steps are taken and noted.

Ideally, a full-time person independent of IS operations will be assigned as the data librarian. In smaller IT environments, however, such assignment might not be economically feasible. When an environment cannot afford a full-time data librarian, this custodial duty should be segregated from operations. That is, for adequacy of control, the function of a librarian should be assigned as a specific responsibility to someone who does not have access to the system.

Physical Security and Access Controls

The objective of physical security and access controls is to prevent or deter theft, damage, and unauthorized access to data and software, and control movement of servers, network-related equipment, and attached devices.

Physical security and access controls protect and restrict access to data centers (computer rooms) and EUC areas where intruders could access information resources (i.e., office and network equipment). Physical security and access controls usually include:

- Traditional locks
- Personnel badge-entry systems
- Magnetic doors with security code for the server room
- Closed-circuit television and video surveillance equipment
- Biometric authentication (e.g., retinal scans, fingerprints, etc.)
- Security alarms
- Visitors logs
- Security guards and receptionists to screen visitors

The authority to change the above physical security access controls should be adequately controlled, and limited to appropriate personnel (e.g., Human Resources Management, etc.)

Other controls involve the placement of office and network equipment for further security. For example, network equipment should be placed in areas where the office traffic is light. If possible, the servers, printers, and other equipment should be placed behind locked office doors. Data center operations managers may want to use combination locks to prevent the duplication of keys; another alternative is to use a locking device that operates on magnetic strips or plastic cards—a convenient device when employees regularly carry picture identification badges.

Network equipment should be attached to heavy immovable office equipment, permanent office fixtures, special enclosures, or special microcomputer workstations. The attachment can be achieved with lockdown devices, which consist of a base attached to permanent fixtures and a second interlocking base attached to the microcomputer equipment. The bases lock together, and a key, combination, or extreme force is required to remove the equipment. All network equipment should be locked down to prevent unauthorized movement, installation, or attachment.

Many microcomputers and other equipment attached to the network may contain expensive hardware and security-sensitive devices. The removal of these devices not only incurs replacement costs but could also cause software to fail and allow for unauthorized disclosure of company-sensitive information. Internal equipment can be protected by lockdown devices, as previously discussed, and special locks that replace one or more screws and secure the top of the equipment.

Cabling is also a source of exposure to accidental or intentional damage or loss. Cabling enables users and peripheral equipment to communicate. In many networks, if the cable is severed or damaged, the entire system will be impaired. Cabling should not be accessible to either the environment or individuals. The communications manager may want to route and enclose cabling in an electrical conduit. If possible and if the exposure warrants the cost, cabling can also be encased in concrete tubing. When the cable is encased, unauthorized access through attachment is lessened. In addition, unauthorized movement of the cabling will not occur easily, and this situation will enable the network manager to more efficiently monitor and control the network and access to it. To alleviate potential downtime, cable may be laid in pairs. In this arrangement, if one set is damaged, the alternate set can be readily attached. The second pair is usually protected in the same manner as the original but is not encased in the same tubing, thus preventing a similar type of accident from damaging both cables.

Notebook computers and mobile devices that are used for work purposes (e.g., tablets, smartphones, etc.) should also receive the same care and attention as cited earlier. These are even more vulnerable in that they can be taken and used off-site by employees and then brought back into the office and attached to the network. Off-site vulnerability to theft and sabotage, such as viruses or theft of programs and data is reduced when protected in a secure off-site storage location.

Environmental Controls

All IT and network equipment operates under daily office conditions (e.g., humidity, temperature, electrical flow, etc.). However, a specific office environment may not be suited to a microcomputer because of geographical location, industrial facilities, or employee habits. A primary problem is the sensitivity of microcomputer equipment to dust, water, food, and other contaminants. Water and other substances can not only damage computer equipment, but also may cause electrocution or a fire. To prevent such occurrences, the IS operations manager should adhere to a policy of prohibiting food, liquids, and the like at or near the servers and network equipment.

Although most offices are air-conditioned and temperatures and humidity are usually controlled, these conditions must nonetheless be evaluated by the IS operations manager. If for any reason the environment is not controlled, the IS operations manager should take periodic readings of the temperature and humidity. If the temperature or humidity is excessively high or low, the server equipment and the network should be shut down to prevent loss of equipment, software, and data. When server or network equipment is transported, either within the building or especially outdoors to a new location, the equipment should be left idle at its new location for a short time to allow it to adjust to the new environmental conditions.

Airborne contaminants can enter the equipment and damage the circuitry. Hard disks are susceptible to damage by dust, pollen, air sprays, and gas fumes. Excessive dust between the read/write head and the disk platter can damage the platter or head or cause damage to the data or programs. If there is excessive smoke or dust, the servers should be moved to another location. Static electricity is another air contaminant. Using antistatic carpeting can reduce static electricity as well as pads placed around the server area, antistatic chair and keyboard pads, and special sprays that can be applied to the bottoms of shoes. Machines can also be used to control static electricity in an entire room or building.

Major causes of damage to servers or network equipment are power surges, blackouts, and brownouts. Power surges, or spikes, are sudden fluctuations in voltage or frequency in the electrical supply that originates in the public utility. They are more frequent when the data center is located near an electrical generating plant or power substation. The sudden surge or drop in power supply can damage the electronic boards and chips as well as cause a loss of data or software. If power supply problems occur frequently, special electrical cords and devices can be attached to prevent damage. These devices are commonly referred to as power surge protectors.

Blackouts are caused by a total loss of electrical power and can last seconds, hours, or days. Brownouts occur when the electrical supply is diminished to below-normal levels for several hours or days. Although blackouts and brownouts occur infrequently, they are disruptive to continuing operations. If servers are essential and the organization's normal backup power is limited to necessary functions, special uninterruptible power supply (UPS) equipment can be purchased specifically for the server or network equipment. UPS equipment can be either battery packs or gas-powered generators. Battery packs are typically used for short-term tasks only (i.e., completing a job in progress or supporting operations during a transition to generator power). Gas-powered generators provide long-term power, and conceivably, could be used indefinitely.

To prevent loss of or damage to computer equipment, services, or facilities, organization should implement safeguards or controls such as:

- Avoiding transient surges and outages in power supplies
- Providing alternative sources of power in the event of extended power failures
- Installing devices that stabilize power supplies
- Providing backup generators
- Protecting power cables

Other common and necessary environmental controls to prevent damage to computer equipment include fire suppression equipment and raised floors. Fire suppression systems (e.g., fire sprinkler system, gaseous fire suppression, condensed aerosol fire suppression, etc.) are automatic and do not require human intervention to control and extinguish fires. Raised floors are constructed above the building's original concrete slab floor, leaving the open space created between the two for wiring or cooling infrastructure.

An isolated holding area should be further used for deliveries to and loading from computer rooms supporting critical business activities. All computer and network equipment should be physically secured with antitheft devices if located in an open office environment. Servers and network equipment should be placed in locked cabinets, locked closets, or locked computer rooms.

Program and Data Backups

Laws and regulations may require organizations to maintain or archive their information and records for a specified period of time. Such archives, if containing financial or operational information, allow management to execute useful analyses and comparisons on which to base projections of future operations. In an IT environment, these archives or backups consist of copies of significant programs (i.e., operating systems, applications, and databases) and their related data that are retained and stored in secure storage locations. If programs and data are not backed up regularly nor stored in a secure location, they may not be recoverable in the event of a serious system failure.

Depending on the type of data file being backed up, the retention period may vary. For example, laws and regulations may require organizations to keep backups of general ledger data for a specified number of years, while internal policy may allow certain detailed transaction data to be deleted after a shorter period of time. Also, if data retention laws and regulations are violated, organizations could be subject to regulatory penalties and/or fines.

Establishing backup policies, procedures, standards, and/or guidance ensures the availability of data significant to the operation of the organization. The policies, procedures, standards, and/or guidance should cover areas such as:

- Storage and retention of programs and data
- Backup scheduling and rotation
- Protection of backup media
- Backup monitoring, review, and resolution of exceptions

Organizations should store backups on-site (in a tape library, for example) and off-premises. Typically, backups of programs and data files are stored on-site and off-site. The organization's policies and procedures should require that backup copies of programs and data reflect the latest and updated versions. Organizations should use cycle retention systems to provide backup of current data. Master files and transaction files that are sufficient to recreate the current day's master files should be stored both on and off premises. New backup files should be rotated to the off-premises location before the old files are returned back to the data center.

Backups should be scheduled to run automatically during backup cycles (i.e., daily, weekly, monthly, yearly, quarterly, and/or semiannually) depending on the type of data. Data can be classified as sensitive data, operational and financial data, general and public data, etc. For instance, an organization may schedule partial **incremental** or **differential** backups of all financial data on a daily basis, and a **full system backup** of all organization data every Friday and every last day of the month. The same organization may schedule additional partial or full backups of sensitive and confidential data every quarter and every year. Backups should also be rotated (ideally on a daily basis) and stored off-site. Normally, backup tapes that have been stored on-site in a safe or a secure vault for some time are taken to the off-site facility. The organization should maintain information on which tapes are located on-site and off-site.

The protection of backup media (e.g., tape cartridges, disk packs containing data and software, etc.) should be part of the organization's backup policies. Onsite backups should be stored at a computer center vault, which should be restricted to authorized personnel only (e.g., computer operators, librarians, computer center supervisor, security officer, etc.). Unauthorized personnel must sign a visitor's log and be accompanied by authorized personnel before obtaining access. The computer center vault should be protected with adequate physical and access controls. Similarly, off-site backups should be stored in an area that is restricted to authorized personnel only. The off-site location should also have adequate physical and access controls. Backups stored on-site and off-site should be frequently checked for premature loss due to deterioration of the media. Backup media is susceptible to gradual degradation as the physical material decays. Procedures should be performed to identify possible media degradation or improper creation of backups to prevent loss of data. Periodic scanning of the media, verification of the backup creation, or restoration of the data, will usually indicate whether the data can be read. When media degradation is discovered, the stored data should immediately be transferred to new media. When backups are improperly written, a procedure should exist to correct and reperform the process.

Backups should be monitored frequently and logs should be completed supporting such monitoring and successful completion of the backup. Management should also review these logs per company policy. For example, each morning an IS operator should be responsible for checking his/her computer in order to confirm backup completion, or identify any error messages displayed by the system that prevented the backup from completion. Additionally, system generated logs should be examined by IS operations personnel in order to identify files that might not have been backed up by the system. When exceptions to the backup process are identified, the IS operator should attempt to perform restart procedures in order to resolve them. If the operator is unable to do so, he/she should escalate the problem for resolution. Finally, if unable to correct the exceptions, an external consultant or vendor should be contacted for support. The IT or IS manager must review and maintain control logs of all backups, as well as provide documentation, when necessary, regarding recovery procedures performed and backup results.

Cloud Backups

Cloud backups may offer the perfect and ideal scenario for the future organization. With a cloud backup, files are available everywhere and are no longer dependent on any single computer or server, allowing for a quick and smooth restoring of the data in the event of a disaster. Additional advantages of cloud backups include saving money on storage costs, and the ability to back up more frequently as well as enjoy off-site, redundant storage of critical data. A further advantage is that organizations can outsource cloud backup services from third-party entities that specialize in data backup and protection. Organizations can then eliminate many of the headaches involved in data backup without surrendering control of their most important asset, information. These specialized "outsource" entities also offer the latest advances in security, encryption, disaster recovery, and continuous real-time data protection, among other services.

A research conducted by Forrester Consulting in 2014 concluded that more and more organizations are relying on cloud backups to assist with their continuity and disaster recovery tasks. According to the research, an approximate 44% of the organizations surveyed have already either transferred the majority of their continuity and disaster recovery tasks into the cloud (including backups), or have plans to do so in the near future. Other respondents expressed concern that moving their information in the cloud would still open up opportunities for privacy and security issues, and would therefore remain with their current data environments. All respondents agree that the ultimate goal, whether backing up to the cloud or not, is to have the confidence of knowing that, in the case of catastrophe, the information will be protected and available.

Business Continuity Plan

The objective of a business continuity plan (BCP) is to describe processes, steps, and/or procedures to be carried out in the event of an emergency (i.e., natural disaster or an unplanned interruption to normal business operations) to achieve a timely recovery and availability of all essential business processes, including the information systems. The BCP normally addresses:

- Key computer processing locations
- Application systems and user requirements for key business processes
- End-user activities for key business processes
- Telecommunications and networks
- Key databases, information warehouses, etc.
- Human resources
- Personal safety of employees and others

The plan assists organizations to respond to emergencies while continuing core activities and operating critical business processes at a level acceptable to management.

The lack of a comprehensive BCP in the event of an emergency may translate into delayed restoration of business processes and information systems. This may result in the inability of the organization to continue operations; loss of revenues and incurring in unnecessary expenses; loss of competitive advantage; loss of customer confidence and market share; and fines and sanctions; among others. In the event of an emergency, degraded services may be acceptable for some period of time. Nonetheless, the goal is to restore the affected systems and services to their optimum levels as immediate as possible.

A common control activity tested by IT auditors within this area involves whether the organization's BCP has been prepared and approved by management, based on a business impact assessment. Other controls evaluate if the plan is regularly tested and updated to reflect the results of such tests.

Disaster Recovery Plan

Disasters, whether natural (e.g., earthquakes, tsunamis, hurricanes, tornados, flood, fires, etc.) or unnatural (e.g., cyberattacks, disruption of service, fraud, terrorism, market collapse, etc.) create economic chaos and severe business interruptions. This is why having a Disaster Recovery Plan (DRP) in place is such an important tool for businesses.

A DRP is a survival tool that helps businesses respond to threats and recover in the wake of an event that disrupts normal business operations. Provided the plan is supported by management, updated frequently, and tested and maintained accordingly, it offers the chance for businesses to survive. Should a disaster occur, the payoff is to recover without significant business or operations downtime and loss. Disasters can occur to businesses at any time and can impact them significantly. For instance:

- On September 11, 2001, after the New York Twin Towers disaster, many firms lost connectivity to banks, broker-dealers, and other financial institutions, disrupting their ability to conduct business and determining whether financial transactions like buying and selling stocks, etc., had been executed completely and accurately.
- On August 14, 2003, an enormous power failure blacked out population centers from New York to Cleveland, Detroit, and Toronto, crippling transportation networks and trapping tens of thousands of people in subways, elevators, and trains. Computers became useless to those who did not have battery power.
- One of Japan's major automakers, Honda, suffered a major drop (nearly 90%) in its second quarter profit in 2011 after a massive tsunami and earthquake hammered its production and sales. Small- and medium-sized business would not have been able to stand this type of loss.
- In 2013, part of the Chinese Internet went down in what the government called the largest denial-of-service (DoS) attack it has ever faced. The attack made machines and networks unavailable, and interrupted Internet services. According to the Wall Street Journal, the attack was an indicator of how susceptible the global Internet infrastructure is.

The impact of these and many other related disasters are felt not only by the business, but also by suppliers and customers who relied on that business for their products and sales.

One of the early critical steps in DRP is identifying who is responsible for distributed disaster recovery. Is recovery of all technology the sole responsibility of IT or the business units? The answer depends on who has control over the hardware, software, and data. In most cases, IT and users must work together to identify critical information and resources that will need to be recovered in the event of a disaster.

A DRP should address both partial and total destruction of computing resources. Distributed systems and microcomputer systems should be included within the plan. Critical functions that are performed on these platforms should be identified and procedures established for restoring operations. Microcomputers are an important tool for daily work processing, and the recovery of these tools should not be overlooked. Information on the basic microcomputer configuration, including hardware and software, should be maintained for ease of recreating the processing environment. In addition, a backup of critical data files should be kept off-site along with operating and recovery procedures.

A DRP must be based on the assumption that any computer system is subject to several different types of failures. In particular, procedures must exist and be tested for recovery from failures or losses of equipment, programs, or data files. In the case of equipment failures, each installation might have a contractual agreement covering the use of an alternate site with a comparable computer configuration. Examples of these are cold sites and hot sites. A cold site is an empty building that is prewired for necessary telephone and Internet access, plus a contract with one or more vendors to provide all necessary equipment within a specified period of time. A hot site, on the other hand, refers to a facility that is not only prewired for telephone and Internet access, but also contains all the computing and office equipment the organization needs to perform its essential business activities.

Before assembling a DRP, the assets of the organization (e.g., hardware, software, facilities, personnel, administrative, data, etc.) and their replacement values should be identified. Specific risks that would result in temporary or permanent loss of assets (say from fire, flood, sabotage, viruses, etc.) should also be recognized. Next, the impact of these losses (e.g., modification, destruction, DoS, etc.) must be assessed. Finally, the value of the asset should be compared against the frequency of loss to justify the disaster recovery solution. Following completion of the above, a DRP can be assembled.

DRP Components

The DRP should identify various levels of recovery, from an isolated event to a widespread disaster. The timeliness of recovery will depend on the loss of exposure for the particular program or system. When the plan is completed, it should be tested to identify potential problems. Testing should be conducted on a periodic basis to validate assumptions, and to update the plan based on the constantly changing environment. Testing also provides the opportunity to practice the recovery procedures and identify missing elements that may need to be added. The DRP should address components, such as:

1. Objectives and mission statement
2. Key personnel involved
3. Full and incremental program and data backups
4. Tests and drills
5. Program and data backups stored off-site
6. Disaster recovery chairperson and committee appointed
7. Emergency telephone numbers
8. List of all critical hardware and software applications
9. Insurance coverage
10. Communication plans
11. Up-to-date system and operation documentation
12. Employee relocation plans to alternate work sites

All members of the organization should be familiarized with the DRP. If an emergency occurs, it would be easy for staff members to execute their roles in the plan. Exercising the plan confirms that efforts are not duplicated and all the necessary steps are taken. It is important to have a written DRP with detailed steps as individuals unfamiliar with the process may need to perform the disaster recovery process in a real emergency.

Auditing End-User Computing

EUC groups have grown rapidly in pervasiveness and importance. The knowledge worker's application of technology to help business solve problems has been one of the major forces of change in business today. User dominance will prevail. Auditors, as knowledge workers and users, can assist departments in identifying sensitive or critical PC applications that require special attention. In organizations where controls are inadequate or nonexistent, auditors can play a key role in developing these controls for EUC groups. Once controls are in place, auditors can examine them for adequacy and effectiveness. Auditing EUC groups can encompass the entire spectrum of IS reviews from systems development to disaster recovery. Appendix 8 covers steps performed when auditing EUC groups.

Audit Involvement in Information Systems Operations

An audit of an organization's IS operations, for instance, would provide IT auditors assurance that operations, including processing of data, are adequately designed and ensure the complete, accurate, and valid processing and recording of financial transactions, for instance. Such examination would also provide assurance that financial information and relevant components of the IT infrastructure are appropriately stored and managed.

Insufficient or inadequate IS operations and controls, however, may result in the following risks:

- Incomplete or inaccurate processing of financial transactions whether executed online or through a batch.
- Inability to reconstruct (or restore) financial data from source documentation following an emergency or a serious systems incident or failure.
- Unauthorized personnel being able to access facilities, which may result in loss or substitution of data, programs, and output or malicious damage to the computer facility and equipment.

Common objectives of an IS operations audit include ensuring that:

- IT operations support adequate scheduling, execution, monitoring, and continuity of systems, programs, and processes to ensure the complete, accurate, and valid processing and recording of financial transactions.
- Backups of financial information are appropriately scheduled, managed, and monitored, ensuring information is accurate and complete. Backed up information is also readable and restored effectively without major implications.
- Physical access is appropriately managed to safeguard relevant components of the IT infrastructure and the integrity of financial information.

Without the implementation of appropriate controls, unnecessary damage or disruption to the organization's data processing could occur. Such damage could result in failure of the organization's critical processes. Control activities should be implemented to address risks such as the above. For example, control activities would typically address the completeness of transactions input for processing, including, among others, whether on-line transactions process to normal completion, all necessary batch jobs are processed, processing is performed timely and in the appropriate sequence, and whether inputting and processing transactions is valid and effective. Examples of controls and procedures normally employed by IT auditors when examining data processing include:

- Batch and/or online processing is defined, timely executed, and monitored for successful completion.
- Exceptions identified on batch and/or online processing are timely reviewed and corrected to ensure accurate, complete, and authorized processing of financial information.

To ensure backups are effective and information is accurate, complete, and restored without major implications, IT auditors may evaluate and test control activities such as whether:

- Procedures for the restoration and recovery of financial information from backups have been implemented in the event of processing disruption, shut-down, and restart procedures.
- Automated backup tools have been implemented to manage retention data plans and schedules.
- Backups are controlled, properly labeled, stored in an off-site secured environmentally location, and rotated to such facility on a periodic basis.
- Management plan and schedule (1) backup and retention of data and (2) erasure and release of media when retention is no longer required.
- Management periodically reviews retention and release records.
- Backups are archived off-site to minimize risk that data is lost.
- Management periodically reviews completion of backups to ensure consistency with backup and retention plans and schedules.
- Tests for the readability of backups are performed on a periodic basis. Results support timely and successful restoration of backed up data.
- Procedures for the restoration and recovery of financial information from backups have been implemented in the event of processing disruption, shut-down, and restart procedures consistent with IT policies and procedures.

To ensure whether physical access is appropriately managed to safeguard relevant components of the IT infrastructure and the integrity of financial information, IT auditors may evaluate and test whether:

- Physical access is authorized, monitored, and restricted to individuals who require such access to perform their job duties.
- Users have access to the data center or computer room. If so, which users.
- A physical access control mechanism (e.g., access cards, biometrics, traditional lock and key, security guards, etc.) is used to restrict and record access to the building and to the computer room, and authority to change such mechanism is limited to appropriate personnel.
- Biometrics authentication is employed through fingerprint, palm veins, face recognition, iris recognition, retina scans, voice verification, etc.
- Entry of unauthorized personnel is supervised and logged, and such log is maintained and regularly reviewed by IT management.
- Policies and procedures exist for granting access to the data center.
- Requests and approvals are required and completed before physical access is granted.

■ There is a process in place for changing the access of transferred and/or terminated employees to the data center. Consider (1) naming the personnel involved; (2) how are they being notified to remove such access to the data center; and (3) how timely access is changed to reflect their new status.

■ User access reviews occur frequently to support current physical access granted to the IT environment, and the data center hosting relevant financial applications, databases, operating systems, and other repositories for financial information.

Other typical services provided by IT auditors in the area of IS operations and physical access include examinations of data centers and DRPs.

Audit of Data Centers

Data center audits are performed to evaluate the administrative controls over data center resources and data processing personnel (IS operations, systems analysis, and programming). The scope of the audit may include an evaluation of the planning, staffing, policies/procedures, assignment of responsibilities, budgets, management reports, and performance measures in areas, such as: hardware management, software management, resource protection and recovery, access controls, operations management, and network/communications management. A data center audit may focus on any one of these accountabilities, or may include all of them depending on the size of the data center, operations staff, and time budget. For example, for a large data center with multiple computers and a large number of users, the audit may focus only on access controls and security administration. For a small data center, the audit might include all of the accountabilities.

Common objectives for data center audits relate to the identification of audit risks in the operating environment and the controls in place to mitigate those audit risks in accordance with management's intentions. The IT auditor must evaluate control mechanisms and determine whether objectives have been achieved. Preaudit preparation is required for effective data center audits. These include meeting with IT management to determine possible areas of concern. At this meeting, the following information should be obtained:

■ Current IT organization chart
■ Current job descriptions for IT data center employees
■ List of application software supported and the hardware hosting them
■ IT policies and procedures
■ Systems planning documentation and fiscal budget
■ Business continuity and disaster recovery plans

IT audit personnel should review the preceding information and become familiar with the way the data center provides IT services. In addition, auditors should become familiar with basic terminology and resource definition methodology used in support of the operations environment. Audit engagement personnel should review the audit program and become familiar with the areas assigned for the completion of an audit task.

Audit of a DRP

As stated earlier, a DRP is a plan established to enable organizations and their IT environments to quickly restore operations and resume business in the event of a disaster. The plan must be updated on a regular basis to reduce the likelihood of incorrect decisions being made during the recovery process, and decrease the level of stress that may be placed on the disaster recovery team members during this process.

From an audit standpoint, the DRP to be evaluated and tested by the IT auditor must include a mission statement and objectives. These objectives should be realistic, achievable, and economically feasible. The objectives provide direction in preparing the plan and in continually reevaluating its usefulness. Documentation supporting disaster simulation drills or tests conducted must be available to assess technical and non-technical procedural aspects of the organization's DRP. Tests reduce the opportunity for miscommunication when the plan is implemented during a real disaster. They also offer management an opportunity to spot weaknesses and improve procedures. Some of the control activities the IT auditor can evaluate and test would relate to whether:

- All media (tapes, manuals, guides, etc.) are stored in a secured environmentally-controlled location.
- Adequate insurance coverage has been acquired and maintained.
- On-going readability of backup and retained data is tested periodically through restoration or other methods.
- Removable media are labeled to enable proper identification.

Unfortunately, organizations are often unwilling to carry out a test because of the disruption that occurs to daily operations and the fear that a real disaster may arise as a result of the test procedures. Therefore, a phased approach to testing would be helpful in building up to a full test. A phased test approach would, for example, consider giving personnel prior notice of the test so that they are prepared. The approach would also simulate the disaster with warning (i.e., at a convenient time and during a slow period) and without warning.

Unless a DRP is tested, it seldom remains usable. A practice test of the plan could very well be the difference between its success or failure. The process is parallel to the old adage about the three things it takes for a retail business to be successful: location, location, location. What is needed for an organization's DRP to allow it to continue to stay in business is testing, testing, and more testing.

The audit of a DRP is an important check for both the IT auditor and management. The major elements and areas of the plan should be validated and assessed to ensure that in the event of a disaster, essential business processes and information systems can be recovered timely.

Audit Tools

Exhibit 11.1 illustrates a template of a standard audit checklist that can be used as a starting point when assessing IS operations related to financial applications systems. Appendix 3 (discussed in Chapter 3) also provides a sample IT audit program for the IS operations general control IT area, which includes a complete list of audit control objectives and activities to be followed and performed when conducting such an examination. Depending on the size and complexity of the organization, these control objectives and activities may need to be revised or expanded to obtain adequate audit coverage of the change control management function.

Conclusion

The chapter has provided an overview of IS operations as a relevant component of the IT infrastructure. This overview includes key objectives and controls that relate to the significance

Exhibit 11.1 Sample ISO Audit Checklist

Information Systems Operations—Audit Checklist [Name of] Financial Application System		
Task	*Yes, No, N/A*	*Comments*
OBJECTIVE 1: **IT operations support adequate scheduling, execution, monitoring, and continuity of systems, programs, and processes to ensure the complete, accurate, and valid processing and recording of financial transactions.**		
1. Interview users who are familiar with the control objective and control activities listed below, and ask them to describe the steps involved in achieving and performing such control objective and activities, including but not limited to: • reports used, and how they are used • procedures performed when exceptions or unusual items (e.g., unexpected changes in personnel, etc.) prevent the control objective and activity from being addressed • how the control objective and activity are achieved in their absence		
2. Verify that automated job scheduling tools are implemented to ensure completeness of the data flow processing.		
3. Examine documentation supporting changes to the job schedule. Obtain management's authorization for those changes.		
4. Observe whether logging of changes to the job schedule has been enabled to confirm that such changes are adequately monitored.		
5. Review access of users that can define or modify production schedules. Reassess the reasonableness of such access privileges.		
6. Ensure existing documentation defines batch and online processing procedures.		
7. Ensure that documentation is available supporting the scheduling and timely execution of batch and/or online processing procedures.		
8. Ensure that batch and online processing is managed in accordance with established policies and procedures.		
9. Ensure batch and/or online processing procedures are monitored for successful completion.		
10. Examine entity documentation, such as completed processing logs and access control listings, indicating that the processing is monitored in accordance with established policies and procedures.		

(Continued)

Exhibit 11.1 (*Continued*)   **Sample ISO Audit Checklist**

Information Systems Operations—Audit Checklist *[Name of] Financial Application System*		
Task	*Yes, No,* *N/A*	*Comments*
11. Observe the execution of scheduled processing to confirm that exceptions, if any, are properly recorded in logs.		
12. Observe procedures performed to confirm that exceptions identified on batch and/or online processing are timely reviewed and corrected to ensure accurate, complete, and authorized processing of financial information.		
13. Ensure access to automated scheduling tools and executable programs (i.e., execute, modify, delete, or create) is granted to users consistent with their job tasks and responsibilities.		
14. Sample documentation to be obtained to support the audit procedures above may include: • Operations schedules or task lists • Sample of completed processing log • Policies and procedures regarding job scheduling tools, as well as detection and correction of processing exceptions • Exception, error, or problem logs and reports • Restart/recovery procedures • Organization chart and access listings (e.g., job scheduler function, master scheduler file, etc.)		
OBJECTIVE 2: **Storage of financial information is appropriately managed, accurate, and complete.**		
1. Interview users who are familiar with the control objective and control activities listed below, and ask them to describe the steps involved in achieving and performing such control objective and activities, including but not limited to: • reports used, and how they are used • procedures performed when exceptions or unusual items (e.g., unexpected changes in personnel, etc.) prevent the control objective and activity from being addressed • how the control objective and activity are achieved in their absence		
2. Procedures for the restoration and recovery of financial information from backups have been implemented in the event of processing disruption, shutdown, and restart procedures consistent with IT policies and procedures.		
3. Automated data retention tools (backups) have been implemented to manage retention data plans and schedules.		

(Continued)

Exhibit 11.1 (*Continued*)   Sample ISO Audit Checklist

Information Systems Operations—Audit Checklist *[Name of] Financial Application System*		
Task	*Yes, No, N/A*	*Comments*
4. Backup tools and online schedules have been reviewed and approved by management.		
5. Observe implementation and execution of backup tools.		
6. For errors resulting from backups, examine evidence supporting that such errors have been identified and timely resolved.		
7. Observe any on-site storage location, and ensure it is secured and adequately controlled.		
8. For off-site backups, ensure they are stored in a secured environmentally location.		
9. Verify the adequacy of the off-site facility location, including physical security systems and environmental controls.		
10. Ensure that backups are properly labeled and rotated to the off-site facility on a periodic basis.		
11. Make certain that tests for the readability of backups are performed on a periodic basis. Results must support timely and successful restoration of backed up data.		
12. Examine data in storage and schedule erasure or disposal of such data when no longer required.		
13. Sample documentation to be obtained to support the audit procedures above may include: • Automated data retention tool documentation, including configuration and parameter reports • Examples of management reports generated from the automated data retention tools • Policies and procedures on automated backups, labeling, erasure, retention, and disposal • Job descriptions and responsibilities of records custodian • Business impact analysis on availability of data • Samples of backup logs and rotation schedules • Inventory of on-site and off-site backups		
OBJECTIVE 3: **Physical access is appropriately managed to safeguard relevant components of the IT infrastructure and the integrity of financial information.**		

(*Continued*)

Exhibit 11.1 (*Continued*) Sample ISO Audit Checklist

Information Systems Operations—Audit Checklist *[Name of] Financial Application System*		
Task	*Yes, No, N/A*	*Comments*
1. Interview users who are familiar with the control objective and control activities listed below, and ask them to describe the steps involved in achieving and performing such control objective and activities, including but not limited to: • reports used, and how they are used • procedures performed when exceptions or unusual items (e.g., unexpected changes in personnel, etc.) prevent the control objective and activity from being addressed • how the control objective and activity are achieved in their absence		
2. Physical access control mechanisms are used to restrict and record access to the building and to the computer room (i.e., data center).		
3. Authority to change physical access control mechanisms is limited to appropriate personnel.		
4. Physical access is authorized and granted appropriately consistent with job responsibilities.		
5. Physical access is monitored and restricted to users who require such access to perform their job duties.		
6. Entry of unauthorized personnel is supervised and logged. The log is maintained and regularly reviewed by IT management.		
7. Observe (on an unannounced basis whenever possible) personnel accessing the facilities through access control mechanisms.		
8. Ensure management periodically performs a review of access listings of personnel with authority to access IT resources/facilities and change physical access mechanisms. Corroborate that such access is authorized and granted consistent with job responsibilities, and that unauthorized personnel are removed immediately.		
9. Sample documentation to be obtained to support the audit procedures above may include: • Restricted area access policies and procedures • Access control mechanism monitoring logs • Policies and procedures related to granting/removing access to IT resources and restricted areas, as well as to access to change physical access mechanisms • Listing of users who have access to IT resources and restricted areas, and can change physical access mechanisms • Evidence that violations in access have been timely corrected		

of implementing effective policies and procedures; data processing; physical security; environmental controls; storage of information; and continuity and recovery of operations. These operational controls form an underlying foundation for the availability and security of the entire system, and are extremely important in protecting the applications and support systems. Any breakdown in their effectiveness can have a catastrophic impact to the programs and applications.

Review Questions

1. Policies and procedures related to IS operations are considered essential for every IT environment, why?
2. Data processing controls help ensure that data is validly processed, and that any exceptions noted while processing will be detected and corrected. What are some of the key questions managers ask in order to address unusual events, failures, or errors resulting from data being processed?
3. Why are physical security and access controls important to organizations? List at least six examples of physical security and access controls.
4. Explain the purpose of data center audits.
5. Differentiate between blackouts and brownouts. Research the Internet and provide one example where a blackout took place during the last five years. Do the same for a brownout.
6. List potential areas that backup policies, procedures, standards, and/or guidance should cover to ensure the availability of data significant to the operation of the organization.
7. What is the risk to organizations of not having a comprehensive business continuity plan in place in the event of an emergency?
8. As the Senior IT auditor, you are having a planning meeting with the client's IT management. The IT manager is in the process of creating a disaster recovery plan (DRP) to put the organization in a better position when responding to (and recovering from) threats that may disrupt normal business operations. The IT manager asks you about the components that should be included in a DRP. Provide your response.
9. List control activities the IT auditor can perform to evaluate and test an organization's DRP.
10. Mention potential areas a company policy related to End-user Computing groups should cover.

Exercises

1. List information that the IT auditor should request or obtain at the preaudit meeting in order to conduct a data center audit. Why is this information important for the IT auditor?
2. Document common audit objectives the IT auditor should focus on when auditing storage or archival of information. Also, list control activities that the IT auditor would need to test in order to meet the audit objectives just listed.
3. One of the recommendations you made during last year's IT audit was the implementation of a disaster recovery plan. In performing the IT audit for this year, you find that although a plan was in place, it has not been tested. Document your reasons why the disaster recovery plan should be tested.

4. You are the Senior IT auditor conducting a planning audit meeting with your two IT staff auditors. The main topic discussed at this planning meeting is the upcoming audit of a company's End-User Computing (EUC) groups. One of the staff IT auditors, recently hired from college, is not sure about the specific objectives to include when auditing EUC groups. Summarize and document these objectives to your staff IT auditor.

CASE—BUSINESS CONTINUITY AND DISASTER RECOVERY

SCENARIO: Business continuity and disaster recovery plans are required to counteract interruptions to business activities and to protect critical business processes from the effects of major failures or disasters. The Payroll Department ("Department") of ISO Company, Inc. is classified as a critical business process because of the sensitive, private, and confidential information it hosts. It would be disastrous for the Department if information gets lost or if its business systems go off-line, even for a day. During planning meetings, IT auditors kept the following objectives in mind:

- Are the Department's business systems adequately backed up?
- Are backup copies of the Department's data held in a secure and remote media store?
- Is there evidence that the current backup strategy works in practice?
- Is there an appropriate disaster recovery plan established as part of the company's business continuity plan?
- Is the disaster recovery plan based on a thorough risk assessment?

OBSERVATIONS: As part of the IT audit of ISO Company, Inc.'s Payroll Department, IT auditors uncovered a number of problems with the company's business continuity and disaster recovery plans and practices. While conducting the audit, IT auditors observed that the organization's business continuity and disaster recovery plans, both established 10 years ago, have not been updated to reflect continuity and disaster recovery practices for the current environment. For example, although backup copies were made of the Department's information, upon inspection, IT auditors discovered that those backups were not maintained at the off-site location where they were supposed to be stored. Moreover, when IT auditors asked for documentation supporting the tests performed of the Department's business continuity and disaster recovery plans, they discovered that the Department had never tested the plans. The Department also had not conducted any risk assessment in support of the plans.

The Department's information systems, Payroll System Application (PSA), is open to external attacks since it is interconnected through the network. A collapse of the PSA would bring dire consequences for the Department. In fact, in the event of a crash, switching over to a manual system would not be an option. Manual handling of the company's payroll sensitive, private, and confidential information by staff personnel has resulted in previous loss of such information. Hence, the PSA must operate online at all times. The auditors agree that, based on the above observations, in the event of interruptions due to natural disasters, accidents, equipment failures, and deliberate actions, the Department may not be able to cope with the pressure.

TASK: List the risks the ISO Company, Inc.'s Payroll Department is exposed to as a result of the observations. Also, document audit recommendations you would communicate to ISO

Company, Inc.'s management related to the lack of continuity and disaster recovery procedures observed. Support your reasons and justifications with IT audit literature and/or any other valid external source. Include examples, if appropriate, to evidence your case point. Submit a word file with a cover page, responses to the tasks above, and a reference section at the end. The submitted file should be at least five pages long (double line spacing), including the cover page and the references page. Be ready to present your work to the class.

Further Reading

1. Barron, J. (August 15, 2003). The blackout of 2003: The overview; power surge blacks out Northeast, hitting cities in 8 states and Canada; midday shutdowns disrupt millions. *The New York Times*. Source: http://www.nytimes.com/2003/08/15/nyregion/blackout-2003-overview-power-surge-blacks-northeast-hitting-cities-8-states.html
2. Bartholomew, D. (2014). Northridge earthquake: 1994 quake still fresh in Los Angeles minds after 20 years. *Los Angeles Daily News*. http://www.dailynews.com/general-news/20140111/northridge-earthquake-1994-disaster-still-fresh-in-los-angeles-minds-after-20-years
3. Forrester Research, Inc. (March 2014). Cloud backup and disaster recovery meets next-generation database demands public cloud can lower cost, improve SLAs and deliver on-demand scale. http://scribd-download.com/cloud-backup-and-disaster-recovery-meets-next-generation-database-demands_58c8d228ee34353a2ee07a3e_txt.html
4. Collins, T. (October 2015). Six reasons businesses should choose cloud backup. Atlantech Online, Inc. Source: https://www.atlantech.net/blog/6-reasons-businesses-should-choose-cloud-backup
5. Cox, R. (2013). 5 notorious DDoS attacks in 2013: Big problem for the internet of things. SiliconANGLE Media, Inc. http://siliconangle.com/blog/2013/08/26/5-notorious-ddos-attacks-in-2013-big-problem-for-the-internet-of-things/
6. Deloitte LLP. (2014). *IT Audit Work Papers*. Unpublished internal document.
7. Dobson Technologies. (2013). Whitepaper: 7 reasons why businesses are shifting to cloud backup. Source: http://www.dobson.net/wp-content/uploads/2013/04/7-Reasons-Businesses-are-Shifting-to-Cloud-Backup-Dobson.pdf
8. Full, incremental or differential: How to choose the correct backup type. (August 2008). TechTarget. Source: http://searchdatabackup.techtarget.com/feature/Full-incremental-or-differential-How-to-choose-the-correct-backup-type
9. Govekar, M., Scott, D., Colville, R. J., Curtis, D., Cappelli, W., Adams, P., Brittain, K. et al. (July 7, 2006). *Hype Cycle for IT Operations Management, 2006*, Gartner Research G00141081, Stamford, CT.
10. How long must you keep your data? *Strategic Finance Magazine*. January 2017 edition.
11. Kageyama, Y. (August 1, 2011). Honda's quarterly profit plunges on disaster. The San Diego Union-Tribune. Source: http://www.sandiegouniontribune.com/sdut-hondas-quarterly-profit-plunges-on-disaster-2011aug01-story,amp.html
12. Microsoft's Information Platform. (May 2014). Forrester Consulting study finds cost, business continuity benefits from cloud backup and disaster recovery. Source: https://blogs.technet.microsoft.com/dataplatforminsider/2014/05/02/forrester-consulting-study-finds-cost-business-continuity-benefitsfrom-cloud-backup-and-disaster-recovery/
13. Otero, A. R., (2015). An information security control assessment methodology for organizations' financial information. *International Journal of Accounting Information Systems*, 18(1), 26–45.
14. Otero, A. R. (2015). Impact of IT auditors' involvement in financial audits. *International Journal of Research in Business and Technology*, 6(3), 841–849.
15. Otero, A. R., Tejay, G., Otero, L. D., and Ruiz, A. (2012). A fuzzy logic-based information security control assessment for organizations, IEEE Conference on Open Systems, Kuala Lumpur, Malaysia.

16. Otero, A. R., Otero, C. E., and Qureshi, A. (2010). A multi-criteria evaluation of information security controls using Boolean features. *International Journal of Network Security & Applications*, 2(4), 1–11.
17. Paquet, R. (September 5, 2002). *The Best Approach to Improving IT Management Processes*, Gartner Research TU-17–3745, Stamford, CT.
18. Senft, S., Gallegos, F., and Davis, A. (2012). *Information Technology Control and Audit*. CRC Press/Taylor & Francis, Boca Raton.
19. Summary of "lessons learned" from events of September 11 and implications for business continuity. February 13, 2002. Securities and Exchange Commission. Source: https://www.sec.gov/divisions/marketreg/lessonslearned.htm

Chapter 12

Information Security

LEARNING OBJECTIVES

1. Describe the importance of information security to organizations, and how information represents a critical asset in today's business organizations.
2. Discuss recent technologies that are revolutionizing organizations' IT environments and the significance of implementing adequate security to protect the information.
3. Discuss information security threats and risks, and how they represent a constant challenge to information systems.
4. Describe relevant information security standards and guidelines available for organizations and auditors.
5. Explain what an information security policy is and illustrate examples of its content.
6. Discuss roles and responsibilities of various information system groups within information security.
7. Explain what information security controls are, and their importance in safeguarding the information.
8. Describe the significance of selecting, implementing, and testing information security controls.
9. Describe audit involvement in an information security control examination, and provide reference information on tools and best practices to assist such audits.

Throughout the years, organizations have experienced numerous losses, which have had a direct impact on their most valuable asset, information. One of the major recent attacks against information occurred in September 2017, where hackers gained access to Equifax[*] data on as many as 145 million Americans (nearly half the population of the United States (U.S.) as of the last census). Hackers gained access and exploited a vulnerability on one of the company's U.S.-based web servers. Files hacked included personal information, such as names, dates of birth, social security numbers, and addresses. It was definitely a major score for cybercriminals which, according to the Director of Security and Architecture at Keeper Security, capitalized well by selling such personal information as much as $20 a piece.

---

[*] Equifax is one of the largest credit reporting agencies in North America.

Another common example where information is directly hit is through computer viruses. Based on the McAfee Labs Threats Report for December 2016, the number of malware attacks approximates 650 million. For mobile devices, the number for 2016 is also significant, almost approaching the 13.5 million mark. Further, in its 2017 Threats Predictions report, McAfee Labs predicts the following, among others:

- Attackers will continue to look for opportunities to break traditional (non-mobile) computer systems, and exploit vulnerabilities. Attackers are well able to exploit information systems whose firmware (permanent software programmed into a read-only memory) either controls input and output operations, or constitute solid-state drives, network cards, and Wi-Fi devices. These types of exploits are probable to show in common malware attacks.
- Ransomware on mobile devices will continue its growth even though attackers will likely combine these mobile device lock attacks with others, such as credential theft, which allow access to bank accounts, credit cards, etc.

Other examples of information losses suffered by organizations result from fraud and economic crimes (also known as **white-collar crime**). According to the Federal Bureau of Investigation's (FBI) 2017 White-Collar Crime Overview, corporate fraud continues to be one of the FBI's highest criminal priorities. Corporate fraud results in significant financial losses to companies and investors, and continue causing immeasurable damage to the U.S. economy and investor confidence. The FBI states that the majority of corporate fraud cases pursued mostly involve:

- Accounting schemes:
  - False accounting entries and/or misrepresentations of financial condition;
  - Fraudulent trades designed to inflate profits or hide losses; and
  - Illicit transactions designed to evade regulatory oversight.
- Self-dealing by corporate executives and insiders:
  - Insider trading (trading based on material, non-public information);
  - Kickbacks;
  - Misuse of corporate property for personal gain; and
  - Individual tax violations related to self-dealing.

These fraud cases are designed to deceive investors, auditors, and analysts about the true financial condition of a corporation or business entity. Through the manipulation of financial data, share price, or other valuation measurements, financial performance of a corporation may remain artificially inflated based on fictitious performance indicators provided to the investing public.

To add to the above, in a Global Economic Crime Survey performed by PricewaterhouseCoopers LLP in 2014, the views of more than 5,000 participants from over 100 countries were featured on the prevalence and direction of economic crime since 2011. The survey revealed that 54% of U.S. participants reported their companies experienced fraud in excess of $100,000 with 8% reporting fraud in excess of $5 million.

This chapter talks about the importance of information security to organizations, and how information represents a critical asset in today's business environment. As you may recall, information security is one of the three major general computer controls used to assess organization's policies and procedures related to application systems in order to support the effective functioning of application controls. Examples of general controls within information security address activities such as access requests and user account administration; access terminations; and physical security. This chapter also discusses recent technologies that are revolutionizing organizations and, specifically, the need for adequate security to protect their information. Information security threats and risks, and how they continue to affect information systems are also described. Relevant information security standards and guidelines available for organizations and auditors will then be discussed, along with the significance of an information security policy. This chapter continues with a discussion of roles and responsibilities of information security personnel. This chapter ends with explanations of information security controls, the significance of selecting, implementing, and testing such controls, and the IT audit involvement in an information security assessment.

Information Security

Information represents a critical asset in many organizations today. Without reliable and properly secured information, organizations would most likely go out of business. Surprisingly, although investing in security can bring them many benefits in that it helps protect valuable information assets and prevent other devastating consequences; many organizations do not spend enough on security expenditures. According to many chief security officers, it is very hard to prove the value of investments in security unless a catastrophe occurs.

The preservation and enhancement of an organization's reputation is directly linked to the way in which information is managed. Maintaining an adequate level of security is one of the several important aspects of managing information and information systems. Systems should be designed with security to integrate with the existing security architecture. The security architecture is not a set of products. Security architecture is a model that specifies what services, such as authentication, authorization, auditing, and intrusion detection, need to be addressed by technologies. It provides a model to which applications can be compared to answer questions such as "How are users authenticated?" In addition, security architecture helps developers recognize that the same security services are needed by many different applications and those applications should be designed to the same security model.

Effective implementation of information security helps ensure that the organization's strategic business objectives are met. The three fundamental objectives for information are confidentiality, integrity, and availability. These objectives are explained below along with the associated risks that would prevent achieving them.

- *Confidentiality* is the protection of information from unauthorized access. This is important in maintaining the organization image and complying with privacy laws. A possible risk associated with confidentiality includes information security breaches allowing for unauthorized access or disclosure of sensitive or valuable company data (e.g., policyholder information or corporate strategic plans to competitors or public, etc.).
- *Integrity* is the correctness and completeness of information. This is important in maintaining the quality of information for decision-making. A potential risk associated with information integrity includes unauthorized access to information systems, resulting in corrupted information and fraud or misuse of company information or systems.
- *Availability* refers to maintaining information systems in support of business processes. This is important in keeping operational efficiency and effectiveness. Possible risks associated with availability include disruption or failure of information systems, loss of the ability to process

business transactions, and crash of information systems due to sources like catastrophes, viruses, or sabotage.

As an example of the importance of protecting the confidentiality, integrity, and availability of information, the U.S. Federal Government has published Federal Information Processing Standards (FIPS) 200. FIPS 200 includes minimum security requirements to protect federal information systems and the information processed, stored, and transmitted by those systems. These requirements or security-related areas include (1) access control; (2) awareness and training; (3) audit and accountability; (4) certification, accreditation, and security assessments; (5) configuration management; (6) contingency planning; (7) identification and authentication; (8) incident response; (9) maintenance; (10) media protection; (11) physical and environmental protection; (12) planning; (13) personnel security; (14) risk assessment; (15) systems and services acquisition; (16) system and communications protection; and (17) system and information integrity. The 17 areas represent a broad-based, balanced information security program that addresses the management, operational, and technical aspects of protecting not only federal information and information systems, but also all information from never-ending threats and risks.

Information Security in the Current IT Environment

Technology is constantly evolving and finding ways to shape today's IT environment in the organization. The following sections briefly describe recent technologies that have already started to revolutionize organizations, how business is done, and the dynamics of the workplace. With these technologies, adequate security must need to be in place in order to mitigate risks and protect the information.

Enterprise Resource Planning (ERP)

According to the June 2016 edition of Apps Run the World, a technology market-research company devoted to the applications space, the worldwide market of ERP systems will reach $84.1 billion by 2020 compared to $82.1 billion in 2015. ERP is a software that provides standard business functionality in an integrated IT environment system (e.g., procurement, inventory, accounting, and human resources). In essence, ERP systems allow multiple functions to access a common database—reducing storage costs and increasing consistency and accuracy of data from a single source. Some of the primary ERP suppliers today include SAP, FIS Global, Oracle, Fiserv, Intuit, Inc., Cerner Corporation, Microsoft, Ericsson, Infor, and McKesson.

Despite the many advantages of ERPs, they are not much different than purchased or packaged systems, and may therefore require extensive modifications to new or existing business processes. ERP modifications (i.e., software releases) require considerable programming to retrofit all of the organization-specific code. Because packaged systems are generic by nature, organizations may need to modify their business operations to match the vendor's method of processing, for instance. Changes in business operations may not fit well into the organization's culture or other processes, and may also be costly due to training. Additionally, as ERPs are offered by a single vendor, risks associated with having a single supplier apply (e.g., depending on a single supplier for maintenance and support, specific hardware or software requirements, etc.).

Cloud Computing

Cloud computing continues to have an increasing impact on the IT environment. Cloud computing has shaped business across the globe, with some organizations utilizing it to perform business critical processes. Based on the July 2015's ISACA Innovation Insights report, cloud computing is considered one of the key trends driving business strategy. The International Data Corporation, in its 2015 publication, also predicts that cloud computing will grow at 19.4% annually over the next 5 years. Moreover, Deloitte's 2016 Perspective's Cloud Computing report indicates that for private companies, cloud computing will continue to be a dominant factor.

Cloud computing, as defined by PC Magazine, refers to the use of the Internet (versus one's computer's hard drive) to store and access data and programs. In a more formal way, the NIST defines cloud computing as a "model for enabling ubiquitous, convenient, on-demand network access to a shared pool of configurable computing resources (e.g., networks, servers, storage, applications, and services) that can be rapidly provisioned and released with minimal management effort or service provider interaction." NIST also stress that availability is significantly promoted by this particular (cloud) model.

The highly flexible services that can be managed in the virtual environment makes cloud computing very attractive for business organizations. Nonetheless, organizations do not yet feel fully comfortable when storing their information and applications on systems residing outside of their on-site premises. Migrating information into a shared infrastructure (such as a cloud environment) exposes organizations' sensitive/critical information to risks of potential unauthorized access and exposure, among others. Deloitte, one of the major global accounting and auditing firms, also supports the significance of security and privacy above, and added, based on its 2016 Perspective's Cloud Computing report, that cloud-stored information related to patient data, banking details, and personnel records, to name a few, is vulnerable and susceptible to misuse if fallen into the wrong hands.

Mobile Device Management (MDM)

MDM, also known as Enterprise Mobility Management, is also shaping the IT environment in organizations. MDM is responsible for managing and administering mobile devices (e.g., smartphones, laptops, tablets, mobile printers, etc.) provided to employees as part of their work responsibilities. Specifically, and according to PC Magazine, MDM ensures these mobile devices:

- integrate well within the organization and are implemented to comply with organization policies and procedures
- protect corporate information (e.g., emails, corporate documents, etc.) and configuration settings for all mobile devices within the organization

Mobile devices, also used by employees for personal reasons, can be brought to the organization. In other words, employees can bring their own mobile device to the organization (also referred to as bring-your-own-device or BYOD) to perform their work. Allowing employees to use organization-provided mobile devices for work and personal reasons has proved to appeal to the average employee. Nevertheless, organizations should monitor and control the tasks performed by employees when using mobile devices, and ensure employees remain focused and productive. It does represent a risk to the organization's security and a distraction to employees when mobile devices are used for personal and work purposes. Additionally, allowing direct access to corporate information always represents an ongoing risk, as well as raises security and compliance concerns to the organization.

Other Technology Systems Impacting the IT Environment

The Internet of Things (IoT) has a potential transformational effect on IT environments, data centers, technology providers, etc. A 2016 Business Insider report stated that there will be 34 billion devices connected to the Internet by 2020, up from 10 billion in 2015. Further, IoT devices will account for 24 billion of them, while traditional computing devices (e.g., smartphones, tablets, smartwatches, etc.) will comprise 10 billion. And, nearly US $6 trillion will be spent on IoT solutions over the next 5 years.

IoT, as defined by Gartner, Inc., is a system that allows remote assets from "things" (e.g., stationary or mobile devices, sensors, objects, etc.) to interact and communicate among them and with other network systems. Assets, for example, communicate information on their actual status, location, and functionality, among others. This information not only provides a more accurate understanding of the assets, but maximizes their utilization and productivity, resulting in an enhanced decision-making process. The huge volumes of raw data or data sets (also referred to as Big Data) generated as a result of these massive interactions between devices and systems need to be processed and analyzed effectively in order to generate information that is meaningful and useful in the decision-making process.

Industry is changing fast and new IoT use cases are maturing. More and more functionality is being added to IoT systems for first-to-market advantages and functional benefits, while security of IoT system devices is often ignored during design. This is evident from recent hacks:

- The US Food and Drug Administration issued safety advice for cardiac devices over hacking threat, and St. Jude Children's Research Hospital patched vulnerable medical IoT devices.
- Hackers demonstrated a wireless attack on the Tesla Model S automobile.
- Researchers hacked Vizio Smart TVs to access a home network.

Other sources of missed security opportunities occur during IoT installation and post-installation configuration. A ForeScout IoT security survey stated that "Respondents, who initially thought they had no IoT devices on their networks, actually had eight IoT device types (when asked to choose from a list of devices) and only 44% of respondents had a known security policy for IoT." Only 30% are confident they really know what IoT devices are on their network. These hacks and the implications of the ForeScout survey results indicate that IoT security needs to be implemented holistically.

Big Data, as defined by the TechAmerica Foundation's Federal Big Data Commission (2012), "describes large volumes of high velocity, complex and variable data that require advanced techniques and technologies to enable the capture, storage, distribution, management, and analysis of the information." Gartner, Inc. further defines it as "… high-volume, high-velocity and/or high-variety information assets that demand cost-effective, innovative forms of information processing that enable enhanced insight, decision making, and process automation."

Even though accurate Big Data may lead to more confident decision-making process, and better decisions often result in greater operational efficiency, cost reduction, and reduced risk, many challenges currently exist and must be addressed. Challenges of Big Data include, for instance, analysis, capture, data curation, search, sharing, storage, transfer, visualization, querying, as well as updating. Ernst and Young, on their EY Center for Board Matters' September 2015 publication, states that challenges for auditors include the limited access to audit relevant data; scarcity of available and qualified personnel to process and analyze such particular data; and the timely integration of analytics into the audit.

Other recent emerging technologies that are currently impacting IT environments include wearables (e.g., smartwatches, etc.), consumer 3D printing, autonomous vehicles, cryptocurrencies, **blockchain**, and speech-to-speech translation, among others.

Information Security Threats and Risks

Expanding computer use has resulted in serious abuses of data communications systems. Computer hackers and sometimes employees use an organization's data communications system to tamper with the organization's data, destroying information, introducing fraudulent records, and stealing assets with the touch of a few keys. First occurrences of this vulnerability appeared in 1981. A grand jury in Pennsylvania charged nine students (aged 17–22 years) with using computers and private telephone services to make illegal long-distance calls and have merchandise delivered to three mail drops in the Philadelphia area without getting billed. Over a 6 month period, the group was responsible for $212,000 in theft of services and $100,000 in stolen merchandise.

Several decades later, the methods have become more sophisticated and the vulnerabilities continue to exist. For example, on November 8, 2008, a series of theft occurred across the globe nearly simultaneously. Over 2,100 money machines in at least 280 cities on three continents in countries, such as the U.S., Canada, Italy, Hong Kong, Japan, Estonia, Russia, and Ukraine were compromised. Within 12 hours, the thieves—lead by four hackers—stole a total of more than $9 million in cash. The only reason the theft ceased was because the ATMs were out of money. This was one of the most sophisticated and organized computer fraud attacks ever conducted.

According to the 2016 Internet Crime Report, the FBI's Internet Crime Complaint Center (IC3) received a total of 298,728 complaints with reported losses in excess of $1.3 billion. In 2015, the FBI received 127,145 complaints from a total of 288,012 concerning suspected Internet-facilitated criminal activity which actually reported having experienced a loss. Total losses reported on 2015 amounted to $1,070,711,522 (or almost a 134% increase from 2014's total reported loss of $800,492,073). In 2014, there were 123,684 complaints received (from a total of 269,422) by the FBI that actually reported a loss from online criminal activity. In 2015, most of the continuing complaints received by the FBI involved criminals hosting fraudulent government services Websites in order to acquire personally identifiable information and to collect fraudulent fees from consumers. Other notable ones within from 2014 through 2016 involved "non-payment" (i.e., goods/services shipped or provided, but payment never rendered); "non-delivery" (i.e., payment sent, but goods/services never received); identity theft; personal data breach; extortion; impersonation; and others. Some of the most frequently reported Internet crimes from 2014 through 2016 are listed in Chapter 2's Exhibit 2.1. Well-known techniques, according to Malware Labs' Cybercrime Tactics and Techniques for Q1 2017, to commit cybercrimes include malware, ransomware, social media scams, and tech support scams. Other commonly-used techniques to commit these cybercrimes are shown on Exhibit 12.1.

The FBI has also identified multiple sources of threats to our nation's critical infrastructures, including foreign nation states engaged in information warfare, domestic criminals, hackers, and terrorists, along with disgruntled employees working within an organization. These are shown in Exhibit 12.2. The Computer Emergency Readiness Team (CERT) Center has further been reporting increasing activity in vulnerabilities.

Exhibit 12.1 Techniques Used to Commit Cybercrimes

Technique	Description
Spamming	Disruptive online messages, especially commercial messages posted on a computer network or sent as email.
Phishing	A high-tech scam that frequently uses spam or pop-up messages to deceive people into disclosing their personal information (i.e., credit card numbers, bank account information, social security numbers, passwords, or other sensitive information). Internet scammers use e-mail bait to "phish" for passwords and financial data from the sea of Internet users.
Spoofing	Creating a fraudulent Website to mimic an actual, well-known Website run by another party. E-mail spoofing occurs when the sender address and other parts of an e-mail header are altered to appear as though the e-mail originated from a different source. Spoofing hides the origin of an e-mail message.
Pharming	A method used by phishers to deceive users into believing that they are communicating with a legitimate Website. Pharming uses a variety of technical methods to redirect a user to a fraudulent or spoofed Website when the user types in a legitimate Web address. For example, one pharming technique is to redirect users—without their knowledge—to a different Website from the one they intended to access. Also, software vulnerabilities may be exploited or malware employed to redirect the user to a fraudulent Website when the user types in a legitimate address.
Denial-of-service attack	Attack designed to disable a network by flooding it with useless traffic.
Distributed denial-of-service	A variant of the denial-of-service attack that uses a coordinated attack from a distributed system of computers rather than from a single source. It often makes use of worms to spread to multiple computers that can then attack the target.
Viruses	Piece of program code that contains self-reproducing logic, which piggybacks onto other programs and cannot survive by itself.
Trojan horse	Piece of code inside a program that causes damage by destroying data or obtaining information.
Worm	Independent program code that replicates itself and eats away at data, uses up memory, and slows down processing.
Malware	Malicious code that infiltrates a computer. It is intrusive software with the purpose of damaging or disabling computers and computer systems.
Spyware	Malware installed without the user's knowledge to surreptitiously track and/or transmit data to an unauthorized third party.

(Continued)

Exhibit 12.1 (*Continued*)   Techniques Used to Commit Cybercrimes

Technique	Description
Botnet	A network of remotely controlled systems used to coordinate attacks and distribute malware, spam, and phishing scams. Bots (short for "robots") are programs that are covertly installed on a targeted system allowing an unauthorized user to remotely control the compromised computer for a variety of malicious purposes.

Adapted from United States General Accounting Office, CYBERCRIME—Public and Private Entities Face Challenges in Addressing Cyber Threats, GAO-07-705, June 22, 2007.

Exhibit 12.2   Sources of Cyber Threats to the U.S. Critical Infrastructure Observed by the FBI

Threat Source	Description
Criminal groups	Groups of individuals or entities that attack information systems for monetary gain. There is an increased use of cyber intrusions by criminal groups.
Foreign nation states	Foreign intelligence services use cyber tools as part of their information gathering and espionage activities. Also, several nations are aggressively working to develop information warfare doctrine, programs, and capabilities. Such capabilities enable a single entity to have a significant and serious impact by disrupting the supply, communications, and economic infrastructures that support military power—impacts that, according to the director of the Central Intelligence Agency, can affect the daily lives of Americans across the country.
Hackers	Hackers sometimes crack into networks for the thrill of the challenge or for bragging rights in the hacker community. While remote cracking once required a fair amount of skill or computer knowledge, hackers can now download attack scripts and protocols from the Internet and launch them against victim sites. Thus, attack tools have become more sophisticated and easier to use.
Hacktivists	Hacktivism refers to politically motivated attacks on publicly accessible Web pages or e-mail servers. These groups and individuals overload e-mail servers and hack into Websites to send a political message.
Disgruntle insiders	The disgruntled insider, working from within an organization, is a principal source of computer crimes. Insiders may not need a great deal of knowledge about computer intrusions because their knowledge of a victim system often allows them to gain unrestricted access to cause damage to the system or to steal system data. The insider threat also includes contractor personnel.

(*Continued*)

Exhibit 12.2 (*Continued*)   Sources of Cyber Threats to the U.S. Critical Infrastructure Observed by the FBI

Threat Source	Description
Terrorists	Terrorists seek to destroy, incapacitate, or exploit critical infrastructures to threaten national security, cause mass casualties, weaken the U.S. economy, and damage public morale and confidence. However, terrorist adversaries of the U.S. are less developed in their computer network capabilities than other adversaries. Terrorists likely pose a limited cyber threat. The Central Intelligence Agency believes terrorists will stay focused on traditional attack methods, but it anticipates growing cyber threats as a more technically competent generation enters the ranks.

Adapted from United States General Accounting Office, *Information Security: TVA Needs to Address Weaknesses in Control Systems and Networks*, GAO-08-526, May 21, 2008.

Along with these increasing threats, the number of computer security vulnerabilities reported to NIST's National Vulnerability Database has reached over 89,700 vulnerabilities by August 2017. According to a Government Accounting Office (GAO) report, the director of the CERT Center stated that as much as 80% of actual security incidents go unreported in most cases because the organization (1) was unable to recognize that its systems had been penetrated because there was no indication of penetration or attack or (2) was reluctant to report the incidents.

As both governments and businesses worldwide place increasing reliance on interconnected systems and electronic data, a corresponding rise is occurring of risks in fraud, inappropriate disclosure of sensitive data, and disruption of critical operations and services, among others. The same factors that benefit business and government operations also make it possible for individuals and organizations to inexpensively interfere with, or eavesdrop on, these operations from remote locations for purposes of fraud or sabotage, or other mischievous or malicious purposes.

Information Security Standards

Information security standards and guidelines provide a framework for implementing comprehensive security processes and controls. Three widely recognized and best practice information security standards include: ISACA's COBIT, the British Standard International Organization for Standardization (ISO)/International Electro technical Commission 27002 (ISO/IEC 27002), and the National Institute of Standards and Technology (NIST). These standards provide organizations with the means to address different angles within the information security arena.

COBIT

COBIT helps organizations meet today's business challenges in the areas of information security, regulatory compliance, risk management, and alignment of the IT strategy with organizational goals, among others. COBIT is an authoritative, international set of generally accepted IT standards, practices, or control objectives designed to help employees, managers, executives, and auditors in: understanding IT systems, discharging fiduciary responsibilities, and deciding adequate levels of security and controls.

COBIT supports the need to research, develop, publicize, and promote up-to-date internationally accepted IT control objectives. The primary emphasis of COBIT is to ensure that technology provides businesses with relevant, timely, and quality information for decision-making purposes.

COBIT, now on its fifth edition (COBIT 5), allows management to benchmark its IT/security environment and compare it to other organizations. IT auditors can also use COBIT to substantiate their internal control assessments and opinions. Because the standard is comprehensive, it provides assurances that IT security and controls exist.

COBIT 5 helps organizations create optimal value from IT by maintaining a balance between realizing benefits and optimizing risk levels and resource use. COBIT 5 is based on five principles. It considers the IT/security needs of internal and external stakeholders (Principle 1), while fully covering the organization's governance and management of information and related technology (Principle 2). COBIT 5 provides an integrated framework that aligns and integrates easily with other frameworks (e.g., Committee of Sponsoring Organizations of the Treadway Commission-Enterprise Risk Management (COSO-ERM), etc.), standards, and best practices used (Principle 3). COBIT 5 enables IT to be governed and managed in a holistic manner for the entire organization (Principle 4). Lastly, it assists organizations in adequately separating governance from management objectives (Principle 5).

COBIT is valuable for all size types organizations, including commercial, not-for-profit, or in the public sector. The comprehensive standard provides a set of IT and security control objectives that not only helps IT management and governance professionals manage their IT operations, but IT auditors in their quests for examining those objectives.

ISO/IEC 27002

ISO/IEC 27002 is a global standard (used together with ISO/IEC 27001) that provides best practice recommendations related to the management of information security. The standard applies to those in charge of initiating, implementing, and/or maintaining information security management systems. This standard also assists in implementing commonly accepted information security controls (ISC) and procedures.

ISO/IEC 27002 is the rename of the ISO 17799 standard, and is a code of practice for information security. It outlines hundreds of potential controls and control mechanisms in major sections such as risk assessment; security policy; asset management; human resources security; physical and environmental security; access control; information systems acquisition, development, and maintenance; information security incident management; business continuity management; and compliance. The basis of the standard was originally a document published by the U.K. government, which became a standard "proper" in 1995, when it was republished by BSI as BS7799. In 2000, it was again republished, this time by ISO/IEC, as ISO/IEC 17799. A new version of this appeared in 2005, along with a new publication, ISO/IEC 27001. These two documents are intended to be used together, with one complementing the other.

The ISO/IEC 27000 Toolkit is the major support resource for the ISO/IEC 27001 and ISO/IEC 27002 standards. It contains a number of items specifically engineered to assist with implementation and audit. These include:

- Copy of ISO/IEC 27001 itself
- Copy of ISO/IEC 27002 itself
- Full business impact assessment (BIA) questionnaire
- Certification guide/roadmap

- Network audit checklist
- Complete set of ISO/IEC 27002 aligned information security policies
- Disaster recovery kit, including checklist and questionnaire
- Management presentation to frame the context of the standards
- Glossary of IS and IT terms and phrases

An enterprise that is ISO/IEC 27002 certified could win business over competitors who are not certified. If a potential customer is choosing between two different services, and security is a concern, they will usually go with the certified choice. In addition, a certified enterprise will realize:

- Improved enterprise security
- More effective security planning and management
- More secure partnerships and e-commerce
- Enhanced customer confidence
- More accurate and reliable security audits
- Reduced liability

The ISO/IEC 27002 framework promotes sound information security in organizations as it:

- includes ISC to help organizations comply with legal, statutory, regulatory, and contractual requirements, as well as with organizations' established policies, principles, standards, and/ or objectives (ISO/IEC 27002, 2005).
- is designed to address the confidentiality, integrity, and availability aspects of IT systems within organizations.
- defines the fundamental guidelines to ensure adequate and sound information security in the organization. ISO/IEC 27002 common best practices offer procedures and methods, proven in practice, which could be adapted to specific company requirements.
- covers all types of organizations (e.g., commercial, governmental, not-for-profit, etc.).
- is based on a management systems approach and represents a viable choice of many organizations for developing information security programs.

The ISO/IEC 27000 family of standards includes techniques that help organizations secure their information assets. Some standards, in addition to the ones mentioned above, involve IT security techniques related to:

- Requirements for establishing, implementing, maintaining, assessing, and continually improving an information security management system within the context of the organization. These requirements are generic and are intended to be applicable to all organizations, regardless of type, size, or nature. (ISO/IEC 27001:2013)
- Guidance for information security management system implementation. (ISO/IEC DIS 27003)
- Guidelines for implementing information security management (i.e., initiating, implementing, maintaining, and improving information security) for intersector and interorganizational communications. (ISO/IEC 27010:2015)
- Guidance on the integrated implementation of an information security management system, as specified in ISO/IEC 27001, and a service management system, as specified in ISO/IEC 20000-1 (ISO/IEC 27013:2015).

The family of standards assists organizations in managing the security of assets, including, but not limited to, financial information, intellectual property, employee details or information entrusted by third parties.

NIST

A major focus of NIST activities in IT is providing measurement criteria to support the development of pivotal, forward-looking technology. NIST standards and guidelines are issued as Federal Information Processing Standards (FIPS) for government-wide use. NIST develops FIPS when there are compelling federal government requirements for IT standards related to security and interoperability, and there are no acceptable industry standards or solutions.

One of the first of several federal standards issued by NIST in 1974 was FIPS 31, "Guidelines for Automatic Data Processing Physical Security and Risk Management." This standard provided the initial guidance to federal organizations in developing physical security and risk management programs for information system facilities. Then, in March 2006, NIST issued FIPS 200 "Minimum Security Requirements for Federal Information and Information Systems," where federal agencies were responsible for including within their information "policies and procedures that ensure compliance with minimally acceptable system configuration requirements, as determined by the agency."

Managing system configurations is also a minimum security requirement identified in FIPS 200 and NIST SP 800–53, "Security and Privacy Controls for Federal Information Systems and Organizations," came to define security and privacy controls that supported this requirement. In August 2011, NIST issued SP 800–128, "Guide for Security-Focused Configuration Management of IS." Configuration management concepts and principles described in this special publication provided supporting information for NIST SP 800–53, and complied with the Risk Management Framework (RMF) that is discussed in NIST SP 800–37, "Guide for Applying the Risk Management Framework to Federal Information Systems: A Security Life Cycle Approach," as amended. More specific guidelines on the implementation of the monitor step of the RMF are provided in NIST SP 800–137, "Information Security Continuous Monitoring for Federal IS and Organizations." The purpose of the NIST SP 800–137 in the RMF is to continuously monitor the effectiveness of all security controls selected, implemented, and authorized for protecting organizational information and information systems, which includes the configuration management security controls identified in SP 800–53. These documents are a very good starting point for understanding the basis and many approaches one can use in assessing risk in IT today.

When assessing risks related to IT, particular attention should be provided to the NIST SP 800–30 guide, "Guide for Conducting Risk Assessments."[*] The NIST SP 800–30 guide provides a common foundation for organizations' personnel with or without experience, who either use or support the risk management process for their IT systems. Organizations' personnel include: senior management, IT security managers, technical support personnel, IT consultants, and IT auditors, among others. The NIST SP 800–30's risk assessment standard can be implemented in single or multiple interrelated systems, from small to large organizations.

NIST guidelines have assisted federal agencies and organizations in significantly improving their overall IT security quality by:

■ providing a standard framework for managing and assessing organizations' information systems risks, while supporting organizational missions and business functions;

---

[*] http://nvlpubs.nist.gov/nistpubs/Legacy/SP/nistspecialpublication800-30r1.pdf.

- allowing for making risk-based determinations, while ensuring cost-effective implementations;
- describing a more flexible and dynamic approach that can be used for monitoring the information security status of organizations' information systems;
- supporting a bottom-up approach in regards to information security, centering on individual information systems that support the organization; and
- promoting a top-down approach related to information security, focusing on specific IT-related issues from a corporate perspective.

Organizations within the private sector use NIST guidelines to promote secured critical business functions, including customers' confidence in organizations' abilities to protect their personal and sensitive information. Furthermore, the flexibility of implementing NIST guidelines provides organizations appropriate tools to demonstrate compliance with regulations.

Other sources of **information security standards include** the well-known Information Technology Infrastructure Library (ITIL), Payment Card Industry Data Security Standard (PCI DSS), and Cloud Security Alliance (CSA) frameworks. The purpose of the ITIL standard, for instance, is to focus on aligning IT services with the needs of business organizations. Specifically, ITIL defines the organizational structure, skill requirements, and a set of standard operational management procedures and practices to allow the organization to establish a baseline from which it can plan, implement, manage, and measure an IT operation and the associated infrastructure. ITIL is mainly used to demonstrate compliance and to measure improvement.

PCI DSS refers to technical and operational requirements applicable to entities that store, process, or transmit cardholder data, with the intention of protecting such data in order to reduce credit card fraud. PCI DSS are maintained, managed, and promoted by the PCI Security Standards Council (Council) worldwide to protect cardholder data. The Council was founded in 2006 by major credit card companies, such as American Express, Discover, JCB International, MasterCard, and Visa, Inc. These companies share equally in governance, execution, and compliance of the Council's work.

Cloud Security Alliance is defined as "the world's leading organization dedicated to defining and raising awareness of best practices to help ensure a secure cloud computing environment."[*] Among others, CSA offers:

- cloud security-specific research, education, certification, and products.
- networking activities for all community impacted by cloud, including providers, customers, governments, entrepreneurs, and the assurance industry.
- forums through which diverse parties can work together to create and maintain a trusted cloud ecosystem.
- a cloud security provider certification program (i.e., CSA Security, Trust & Assurance Registry (STAR)). The CSA-STAR is a three-tiered provider assurance program of self-assessment, third party audit, and continuous monitoring.
- high quality educational events around the world and online.

Information Security Policy

According to the SANS Institute, the term policy typically refers to a document (or set of documents) that summarizes rules and requirements that are usually point-specific, must be met, and

---

[*] https://cloudsecurityalliance.org/about/.

cover a single area. An "Acceptable Use" information system policy, for instance, cover rules for appropriate use of the information system and computing facilities. A policy differs from a standard or a guideline. A standard refers to a collection of system-specific or procedural-specific requirements that enforce a given policy. Standards are established by authority, custom, or general consent as a model or example, and their compliance is mandatory. An example of a standard would be the specification of minimum password requirements (e.g., passwords must be at least eight characters in length, and require at least one number and one special character, etc.) that should be configured by all users in order to improve computer security. The aforementioned standard helps enforce compliance with a "Password Policy," for instance. A standard can also be in the form of a technology selection (e.g., selection of a particular technology for continuous security monitoring, etc.) that complies with a specific policy. Guidelines, on the other hand, are also system-specific or procedural-specific; however, they are "suggestions" for best practice. In other words, they do not refer to rules or requirements to be met, but strong recommendations to consider. An effective information security policy makes frequent references to standards and guidelines that exist within an organization.

An information security policy defines the security practices that align to the strategic objectives of the organization. It describes ways to prevent and respond to a variety of threats to information and information systems including unauthorized access, disclosure, duplication, modification, appropriation, destruction, loss, misuse, and denial of use. The information security policy is intended to guide management, users, and system designers in making decisions about information security. It provides high-level statements of information security goals, objectives, beliefs, ethics, controls, and responsibilities.

An important factor to implement an information security policy in an organization is to do an assessment of security needs. This is achieved by first understanding the organization's business needs and second by establishing security goals. There are some common questions to be answered:

- What information is critical to the business?
- Who creates that critical information?
- Who uses that information?
- What would happen if critical information is stolen, corrupted, or lost?
- How long can the company operate without access to critical information?

Information security crosses multiple areas. The information security policy must be coordinated with systems development, change control, disaster recovery, compliance, and human resource policies to ensure consistency. An information security policy should state Web and e-mail usage ethics and discuss access limitations, confidentiality policy, and any other security issue. Good policies give employees exact instructions as to how events are handled and recovery escalated if necessary. The policy should be available and distributed to all users within the organization.

The SANS Institute has several information security policy templates available for over 25 relevant security requirements.* These serve as a great starting point for rapid development and implementation of information security policies. The Institute has developed information security policy templates and classified them under the following categories: General, Network Security, Server Security, and Application Security. Below are some of the policy template areas offered by each category:

- General—Includes information security policy templates covering the areas of: Acceptable Encryption Policy, Acceptable Use Policy, Clean Desk Policy, Data Breach Response Policy,

---

* https://www.sans.org/security-resources/policies/#template.

Disaster Recovery Plan Policy, Digital Signature Acceptance Policy, Email Policy, Ethics Policy, Pandemic Response Planning Policy, Password Construction Guidelines, Password Protection Policy, Security Response Plan Policy, and End User Encryption Key Protection Policy.

■ Network Security—Includes information security policy templates covering the areas of: Acquisition Assessment Policy, Bluetooth Baseline Requirements Policy, Remote Access Policy, Remote Access Tools Policy, Router and Switch Security Policy, Wireless Communication Policy, and Wireless Communication Standard.

■ Server Security—Includes information security policy templates covering the areas of: Database Credentials Policy, Technology Equipment Disposal Policy, Information Logging Standard, Lab Security Policy, Server Security Policy, Software Installation Policy, and Workstation Security (for HIPAA) Policy.

■ Application Security—Includes information security policy templates covering the area of Web Application Security Policy.

Information Classification Designations

Information security policies are also helpful for organizations when documenting information classification designation requirements. Organizations need to establish an information classification system that categorizes information into groupings. Information groupings help determine how information is to be protected. In the private sector, there may be legal or regulatory reasons to classify information into public, internal, or confidential. In the government sector, there may also be national security reasons to classify information into various categories (e.g., top secret, etc.).

If information is sensitive, from the time it is created until the time it is destroyed or declassified, it must be labeled (marked) with an appropriate information classification designation. Such markings must appear on all manifestations of the information (hard copies, external media, etc.). The vast majority of information falls into the Internal Use Only category. For this reason, it is not necessary to apply a label to Internal Use Only information. Information without a label is therefore by default classified as Internal Use Only.

Access to information in the possession of, or under the control of, the organization must be provided based on the need to know. In other words, information must be disclosed only to people who have a legitimate need for the information. At the same time, users must not withhold access to information when the owner of the information in question instructs that it be shared. To implement the need-to-know concept, organizations should adopt an access request and owner approval process. Users must not attempt to access sensitive information unless granted access rights by the relevant owner.

Organization information, or information that has been entrusted to the organization, must be protected in a manner commensurate with its sensitivity and criticality. Security measures must be employed regardless of the media on which information is stored (hardcopy or electronic), the systems that process it (e.g., personal computers, mobile devices, etc.), or the methods by which it is moved (e.g., electronic mail, instant messaging, face-to-face conversation, etc.). Information must also be consistently protected no matter what its stage is in the life cycle from origination to destruction.

Information Security Roles and Responsibilities

Information security is achieved through a team effort involving the participation and support of every user who deals with information and information systems. An information security department typically has the primary responsibility for establishing guidelines, direction, and authority over information security activities. However, all groups have a role and specific responsibilities in protecting the organization's information, as described in the following sections.

Information Owner Responsibilities

Information owners are the department managers, senior management, or their designees within the organization who bear the responsibility for the acquisition, development, and maintenance of production applications that process information. Production applications are computer programs that regularly provide reports in support of decision making and other organization activities. All production application system information must have a designated owner. For each type of information, owners designate the relevant sensitivity classification, designate the appropriate level of criticality, define which users will be granted access, as well as approve requests for various ways in which the information will be used.

Information Custodian Responsibilities

Custodians are in physical or logical possession of either organization information or information that has been entrusted to the organization. Whereas IT staff members clearly are custodians, local system administrators are also custodians. Whenever information is maintained only on a personal computer, the user is necessarily also the custodian. Each type of application system information must have one or more designated custodians. Custodians are responsible for safeguarding the information, including implementing access control systems to prevent inappropriate disclosure and making backups so that critical information will not be lost. Custodians are also required to implement, operate, and maintain the security measures defined by information owners.

User Responsibilities

Users are responsible for familiarizing themselves (and complying) with all policies, procedures, and standards dealing with information security. Questions about the appropriate handling of a specific type of information should be directed to either the custodian or the owner of the involved information. As information systems become increasingly distributed (e.g., through mobile computing, desktop computing, etc.), users are increasingly placed in a position where they must handle information security matters that they did not handle in days gone past. These new distributed systems force users to play security roles that they had not played previously.

Third-Party Responsibilities

Access to information from third parties needs to be formally controlled. With the use of contractors and outsourcing, third parties will have the need to access the organization's information. There must be a process in place to grant the required access while complying with rules and regulations. This process should include a nondisclosure agreement signed by the third party that defines responsibility for use of that information. A similar process should be in place when individuals in the organization have access to third-party information.

Information Security Controls

In today's organizational culture, most information security challenges are being addressed with security tools and technologies, such as encryption, firewalls, access management, etc. Although tools and technologies are certainly an integral part of organizations' information security plans, the literature argues that they alone are not sufficient to address information security problems. To improve overall information security, organizations must implement ISC that satisfy their specific security requirements.

According to ISO/IEC 27002, "information security is achieved by implementing a suitable set of controls, including policies, processes, procedures, organizational structures and software and hardware functions." These controls need to be designed and implemented effectively to ensure that specific organization security and business objectives are actually met. NIST defines ISC or security controls as "a safeguard or countermeasure prescribed for an information system or an organization designed to protect the confidentiality, integrity, and availability of its information and to meet a set of defined security requirements."*

A very common example of an ISC in organizations involves the review and monitoring of computer security relevant events. Computer systems handling sensitive, valuable, or critical information must securely log all significant computer security relevant events. Examples of computer security relevant events include password-guessing attempts, attempts to use privileges that have not been authorized, modifications to production application software, and modifications to system software. Logs of computer security relevant events must provide sufficient data to support comprehensive audits of the effectiveness of and compliance with security measures. All commands issued by computer system operators must be traceable to specific individuals through the use of comprehensive logs. Logs containing computer security relevant events must be retained per established archiving procedures. During this period, such logs must be secured such that they cannot be modified, and such that they can be read only by authorized persons. These logs are important for error correction, forensic auditing, security breach recovery, and related efforts. To assure that users are held accountable for their actions on computer systems, one or more records tracing security relevant activities to specific users must be securely maintained for a reasonable period. Computerized records reflecting the access privileges of each user of multiuser systems and networks must be securely maintained for a reasonable period.

Other common ISC that assist organizations are included as part of the vulnerability, threat, trust, identity, and incident management processes described below.

Vulnerability Management

Vulnerabilities refer to "weakness or exposures in IT assets or processes that may lead to a business risk or a security risk." A vulnerability management process is needed to combat these specific risks. The process includes the identification, evaluation, and remediation of vulnerabilities. Prerequisites for responding to vulnerabilities include asset management processes to determine the software installed on organization hardware, as well as change management processes to manage the testing of patches. Patches should be reviewed and tested before implementation to verify that the system continues to work as intended, and that no new vulnerabilities are introduced. With these processes in place, the information security group can identify the vulnerabilities that apply to the organization. Once identified, the vulnerabilities need to be prioritized and implemented based on the risk of the particular issue.

---

* http://nvlpubs.nist.gov/nistpubs/SpecialPublications/NIST.SP.800-53Ar4.pdf.

Threat Management

Threat management includes virus protection, spam control, intrusion detection, and security event management. Virus protection software should be loaded on all workstations and the servers to regularly scan the system for new infections. Sooner or later a virus will find its way into a system. Even some of the largest software vendors have sent out products with viruses by mistake. Policies regarding virus protection should be implemented to prevent, detect, and correct viruses. Virus software must be continuously updated with virus definitions as new viruses are introduced daily. User awareness training is another important control for making users aware of the danger to the system of infected software that is downloaded from any source.

Trust Management

Trust management includes encryption and access controls. To ensure cryptography is applied in conformance with sound disciplines, there has to be a formal policy on the use of cryptography that applies across the organization. A formal policy should be supported by comprehensive standards/ guidelines (e.g., for selection of algorithms, cryptographic key management, etc.) and take into account cross-border restrictions. Many encryption routines require that the user provide a seed or a key as input. Users must protect these security parameters from unauthorized disclosure, just as they would protect passwords from unauthorized disclosure. Rules for choosing strong seeds or keys should likewise follow the rules for choosing strong passwords.

Encryption technologies electronically store information in an encoded form that can only be decoded by an authorized individual who has the appropriate decryption technology and authorization to decrypt. Encryption provides a number of important security components to protect electronic information such as:

- *Identification.* Who are you?
- *Authentication.* Can you prove who you are?
- *Authorization.* What can you do?
- *Auditing.* What did you do?
- *Integrity.* Is it tamperproof?
- *Privacy.* Who can see it?
- *Nonrepudiation.* Can I prove that you said what you said?

When information is encoded, it is first translated into a numerical form and then encrypted using a mathematical algorithm. The algorithm requires a number or message, called a key, to encode or decode the information. The algorithm cannot decode the encrypted information without a decode key.

Identity Management

Identity management is the process used to determine who has access to what in an organization. It is also one of the most difficult areas to manage due to the number of functions that must work together to implement proper controls. Identity management must be a collaborative effort between information security, applications development, operations, human resources, contracts/procurement, and business groups to implement. There are many reasons for implementing an identity management solution: regulatory compliance, risk management, and expense reduction to mention a few.

Automating identity management into a single application that manages access to systems speeds the development of applications and reduces operating costs. Organizations have developed systems and applications over time with stand-alone user identity programs. With the number of applications and systems increasing, users have a difficult time remembering the number of user IDs and passwords. This causes users to create easy-to-guess passwords, write down passwords, not change them, or change a single digit.

Implementing identity management can result in savings for the help desk with reduced call volume and operations from fewer password changes, and to users with increased productivity from reduced log-on time and password resets. Implementing common process for administering access rights provides a consistent level and security and accountability across applications. Automating identity management can enable implementation of security access rights based on business roles and improve the turnaround time for adding, changing, and removing access. Additional benefits include:

- Reduced manual processes and potential for human error
- Improved management reporting of user access rights
- Ability to enforce segregation of duties according to business rules
- Automatically revoked access rights of inactive employees
- Audit trail of requests and approvals

Integrating all these systems with a common identity management program can be costly and time consuming. The Gartner Group recommends implementing identity management over time by first proving success with a single function or application.

Incident Management

Security incidents, such as malfunctions, loss of power or communications services, overloads, mistakes by users or personnel running installations, access violations, and so forth, have to be managed and dealt with in accordance with a formal process. The process has to apply to all forms of security incident. Incidents have to be:

- Identified and recorded
- Reported to a focal point
- Prioritized for action
- Analyzed and acted upon

Each incident has to be dealt with by a person equipped to understand the full implications of the incident as well as the consequences for the organization and initiate appropriate action. Significant incidents, and the pattern of incidents over time, have to be reported to and reviewed by the person in charge and by user representatives, so that appropriate action can be initiated and properly documented. Incidents have to be reported to management.

Selection and Testing of Information Security Controls

Due to a variety of organizational-specific constraints (e.g., costs, availability of resources, etc.), organizations do not have the luxury of selecting all required ISC. Adequate selection of ISC is crucial to organizations in maintaining sound information security as well as in protecting their financial information assets. The literature points out several issues, gaps, and/or weaknesses that prevent an effective selection of ISC in organizations. These weaknesses also impact the overall protection of the information's confidentiality, integrity, and availability. In other words, the lack of adequate information security over valuable, sensitive, or critical financial information may allow for: (1) fraud, manipulation, and/or misuse of data; (2) security-related deficiencies and findings; (3) bogus trades to inflate profits or hide losses; (4) false accounting journal entries; (5) computer security breaches; and (6) false transactions to evade regulators, among many others.

ISO/IEC 27002 states that following the identification of security risks, appropriate controls should be selected (and implemented) to ensure that those risks are reduced to acceptable levels. Commonly-used ISC have been included in Chapter 3's Appendix 3. Additional ISC are offered by the ISO/IEC 27002 standard. The main point here is that the selection of ISC is dependent upon organizational decisions and specific needs. Such selection is also based on the criteria for risk acceptance, risk treatment options, and the general risk management approach applied to the organization. The selection of ISC should also be subject to all relevant national and international legislation and regulations. ISC included in Chapter 3's Appendix 3 and in ISO/IEC 27002 are applicable to most organizations and considered as guiding principles for both, information security management and IT auditors.

Testing ISC is essential to maintain adequate and secured information systems. Nevertheless, most organizations either do not perform effective (and thorough) testing of the information systems, or lack the mechanisms and controls to perform the required testing. Nothing can substitute for assessing ISC. Some of the reasons for the lack of testing involve:

- Leadership not providing clear expectations for assessing controls and/or testing schedules
- Inadequate oversight of the risk management program
- Lack of skilled test managers and testers/security assessors
- Leadership pressure to condense the testing cycle due to the schedule having a higher priority than the security of a system

The only way to know whether an ISC works or not, or passes or fails, is to test it. Testing ISC cannot be achieved through a vulnerability-scanning tool, which only checks a small number of security controls. A vulnerability scan often tests a fraction, approximately five percent, of the security controls.

When testing ISC, the NIST RMF recommends the development and execution of a test plan. The test plan should include all controls applicable to the specific information system. Testers should execute the test plan with the information system owner and record the results, which per the NIST RMF framework, include:

- A list of applicable security controls
- A test plan encompassing all of the applicable security controls
- A test report (pass/fail)
- Mitigations for any failed controls

Test results provide the risk executive with the information that is required to make a risk decision. The risk executive is often the chief information officer (CIO), deputy CIO, chief information security officer (CISO) or director of risk management. From an IT audit standpoint, test results support the audit work and conclusion, and form the base for the formal exit communication with the organization's management.

Involvement in an Information Security Audit

According to ISACA, an audit of an organization's information security provides management with an assessment of the effectiveness of the information security function over IT systems (e.g., computers, servers, mainframes, network routers, switches, etc.). The information security audit also assist in determining whether essential security processes, controls, and/or procedures are either missing or actually in place (meaning they are adequately designed, implemented, and operate effectively). Such assessments are normally performed manually and/or through automated procedures. Manual assessments may include interviewing organizations staff (e.g., security personnel, etc.), performing security vulnerability scans, reviewing system and application access controls, and analyzing physical access to the systems, among others. Automated assessments typically involve the use of Computer-Assisted Audit Techniques (CAATs) or software to help auditors review and test application controls, as well as select and analyze computerized data for substantive audit tests.

Auditing information security cover topics like: Security Administration Function, Security Policies and Procedures, Security Software Tools and Techniques, Unique User Identifiers and Passwords, Physical Security, Security Administrator and Privilege Access, Information Security Logs, and others. Insufficient or inadequate information security procedures or controls, however, may result in risks to organizations such as the following:

■ If logical security tools and techniques are not implemented, configured, or administered appropriately, control activities within the significant flows of transactions may be ineffective, desired segregation of duties may not be enforced, and significant information resources may be modified inappropriately, disclosed without authorization, become unavailable when needed, and/or deleted without authorization. Furthermore, such security breaches may go undetected.
■ If an organization relies on security features of its application systems to restrict access to sensitive application functions, weaknesses in network or operating system security (e.g., user authentication and overall system access, etc.) may render such application security features ineffective.

Common audit objectives of an information security audit include ensuring that:

■ Security configuration of applications, databases, networks, and operating systems is adequately managed to protect against unauthorized changes to programs and data that may result in incomplete, inaccurate, or invalid processing or recording of financial information.
■ Effective security is implemented to protect against unauthorized access and modifications of systems and information, which may result in the processing or recording of incomplete, inaccurate, or invalid financial information.

Without the implementation of appropriate controls, unnecessary damage or disruption to the organization's information could occur. Such damage could result in failure of the organization's critical processes. Control activities should be implemented to reduce risks and address objectives such as the above. For example, implementation of adequate control activities would ensure security and protection against unauthorized changes to programs and data within applications, databases, networks, and operating systems, which may result in incomplete, inaccurate, or invalid processing or recording of financial information (first objective above). Information security controls and procedures reviewed by IT auditors would normally test and examine that:

■ The security administration function should be separate from the IT function.
■ Formal policies and procedures define the organization's information security objectives and the responsibilities of employees with respect to the protection and disclosure of informational resources. Management monitors compliance with security policies and procedures, and agreement to these are evidenced by the signature of employees.
■ Security-related software tools and techniques are in place to restrict and segregate access to sensitive IT functions (e.g., programming, administrative functions, implementation of changes in production environments, etc.) within the systems. Changes related to access are assessed by management for adequate segregation of duties.
■ Implementation and configuration of security software tools and techniques are reviewed and approved by management.
■ Unique user identifiers have been assigned to users as required on the information security policies and procedures, to distinguish them and to enforce accountability.
■ Local and remote users are required to authenticate to applications, databases, networks, and operating systems via passwords to enhance computer security.
■ Passwords promote acceptable levels of security (consistent with policies and/or best industry practices) by enforcing confidentiality and a strong password format.
■ Vendor-supplied passwords embedded on the applications, databases, networks, and operating systems are modified, eliminated, or disabled to avoid security vulnerabilities from being exploited in the systems.
■ Administrator, privileged, or super-user account access within the systems are limited to appropriate personnel. Changes to these accounts (e.g., system security parameters, security roles, security configuration over systems, etc.) are logged and reviewed by management.
■ Information security logs are configured and activated on applications, databases, networks, and operating systems to record and report security events consistent with information security policies and procedures.
■ Reports generated from information security logs (e.g., security violation reports, unauthorized attempts to access information, etc.) are frequently reviewed and acted upon as necessary.

To ensure effective security is implemented to protect against unauthorized access and modifications of systems and information, which may result in the processing or recording of incomplete, inaccurate, or invalid financial information, control activities would test that:

■ Training programs have been established for all personnel within the following areas:
  – Organizational security policies
  – Disclosure of sensitive data
  – Access privileges to IT resources
  – Reporting of security incidents
  – Naming conventions for user passwords
■ System owners authorize user accounts and the nature and extent of their access privileges.
■ User account access privileges are periodically reviewed by system and application owners to determine whether they are reasonable and/or remain appropriate.
■ Users who have changed roles or tasks within the organization, or that have been transferred, or terminated are immediately informed to the security department for user account access revision in order to reflect the new and/or revised status.
■ Transmission of sensitive information is encrypted consistent with security policies and procedures to protect its confidentiality.

Audit Tools and Best Practices

As stated earlier, the SANS Institute provides several information security policy templates,* which are very practical and serve as a great starting point for rapid development and implementation of information security policies. Information security policy templates developed by the Institute cover common and relevant areas in the information security field such as: General Information Security, Network Security, Server Security, and Application Security, among others.

Appendix 3 (discussed in Chapter 3) also provides a sample IT audit program for the Information Security general control IT area, which includes a complete list of audit control objectives and activities to be followed and performed when conducting such an examination. Depending on the size and complexity of the organization, these control objectives and activities may need to be revised or expanded to obtain adequate audit coverage of the information security function.

Another good source for tools and best practices is ISACA. ISACA currently provides Audit/Assurance Programs† based on COBIT 5 for the following, among others:

- Mobile Computing Audit/Assurance Program
- Data Privacy Audit Program
- Outsourced IT Audit/Assurance Program
- Cybersecurity: Based on the NIST Cybersecurity Framework
- Bring Your Own Device (BYOD) Security Audit/Assurance Program
- Change Management Audit/Assurance Program
- Cloud Computing Management Audit/Assurance Program
- IT Risk Management Audit/Assurance Program
- PCI DSS Compliance Program Audit/Assurance Program and ICQ
- SAP ERP Revenue Business Cycle Audit/Assurance Program and ICQ
- SAP ERP Expenditure Business Cycle Audit/Assurance Program and ICQ
- SAP ERP Inventory Business Cycle Audit/Assurance Program and ICQ
- SAP ERP Financial Accounting (FI) Audit/Assurance Program and ICQ
- SAP ERP Managerial Accounting (CO) Audit/Assurance Program and ICQ
- SAP ERP Human Capital Management Cycle Audit/Assurance Program and ICQ
- SAP ERP BASIS Administration and Security Audit/Assurance Program and ICQ
- SAP ERP Control Environment ICQ

Additional references of relevant audit information or good practices when embarking in information security audits particularly related to recent technologies impacting the IT environment are shown in Exhibit 12.3.

Conclusion

Information represents a critical asset in most organizations today. Without reliable and properly secured information, organizations would not last long and, instead, may quickly go out of business. Information security helps ensure that the organization's strategic business objectives are met. Three

---

* https://www.sans.org/security-resources/policies/#template.
† http://www.isaca.org/knowledge-center/research/pages/audit-assurance-programs.aspx?cid=1003563&appeal=pr.

Exhibit 12.3   Additional References of Relevant Audit Information or Good Practices When Embarking in Information Security Audits Related to Recent Technologies Impacting the IT Environment

Technology/Source	Description
Cloud Computing	
Cloud Security Alliance	CloudAudit Working Group
National Institute of Standards and Technology, Special Publication 800-144	Guidelines on Security and Privacy in Public Cloud Computing
Deloitte	Cloud Computing—The Non-IT Auditor's Guide to Auditing the Cloud
PWC	A Guide to Cloud Audits
ISACA	Auditing Cloud Computing: A Security and Privacy Guide
The SANS Institute	Cloud Security Framework Audit Methods
Big Data	
IIA	Global Technology Audit Guide (GTAG): Understanding and Auditing Big Data
AICPA	Audit Analytics and Continuous Audit—Looking Toward the Future
ISACA	What is Big Data and What Does it have to do with IT Audit?
Mobile Devices	
National Institute of Standards and Technology, Special Publication 800-124 Revision 1	Guidelines for Managing and Securing Mobile Devices in the Enterprise
Internet of Things	
Deloitte	Auditing the Internet of Things
ISACA	Internet of Things: Risk and Value Considerations
OWASP	IoT Testing Guides
EY	Cybersecurity and the Internet of Things
Blockchain	
Deloitte	Blockchain & Cyber Security
EY	Blockchain and the future of audit
PWC	Auditing blockchain: A new frontier
ISACA	How to Navigate Blockchain—The Technology That Could Change Everything
ISO/TC 307	Blockchain and distributed ledger technologies

fundamental objectives for information and information systems that are essential to maintain competitive edge, cash flow, profitability, legal compliance, and respected organization image are confidentiality, integrity, and availability. The meeting of these objectives is also fundamental when protecting against the increasingly number of information security threats and risks.

Technology is constantly evolving and finding ways to shape today's IT environment in the organization. Recent technologies (e.g., ERPs, Cloud Computing, MDM, IoT, Big Data, Blockchain, Wearables, etc.) have already started to revolutionize organizations, how business is done, and the dynamics of the workplace. Effective implementation of information security within these technologies is paramount in order to mitigate risks and protect the information.

Information security standards and guidelines provide a framework for implementing comprehensive security processes and controls. Three widely recognized and best practice information security standards are COBIT, the ISO/IEC 27002, and NIST. They provide organizations with the means to address different angles within the information security arena.

An information security policy is intended to guide organizations in making decisions about information security. An information security policy provides high-level statements of information security goals, objectives, beliefs, ethics, controls, and responsibilities. Standards and guidelines that define specific implementation of the policies are documented separately. The organization, management, and staff and the IT audit, control, and security professionals must work together to establish, maintain, and monitor the organization's information security policy.

To be effective, information security must be achieved through a team effort involving the participation and support of every user who deals with information and information systems. An information security department typically has the primary responsibility for establishing guidelines, direction, and authority over information security activities. However, all groups should have a role and specific responsibilities in protecting the organization's information.

ISC assist organization to achieve adequate levels of security. They include implemented policies, processes, procedures, organizational structures, and software and hardware functions, among others, that enforce security in the organization. ISC need to be designed and implemented effectively to ensure that the organization security and business objectives are achieved. They must also be monitored, reviewed, and improved, when necessary. Organizations must select and test ISC to satisfy specific security requirements and improve overall information security.

Review Questions

1. Explain each of the three organization's strategic business objectives attained through implementation of information security. What are the associated risks that would prevent achieving them?
2. Pick two of the recent technologies discussed in this chapter that have already started to revolutionize organizations, how business is done, and the dynamics of the workplace. Describe the technology and provide examples of three risks each technology would likely add to the organization.
3. Briefly describe six commonly-used techniques used to commit cybercrimes according to this chapter.
4. Define COBIT. Describe the COBIT 5 principles that help organizations create optimal value from IT by maintaining a balance between realizing benefits and optimizing risk levels and resource use.
5. What is the purpose of an information security policy?

6. List and describe typical roles within information security, and their responsibilities in protecting the organization's information.
7. Provide two or three examples of information security controls within the following management processes:
   a. Vulnerability
   b. Threat
   c. Trust
   d. Identity
   e. Incident
8. Information security test results should be recorded and, according to NIST, those test results should include?
9. The Company you work for is in the process of determining whether to have an information security audit (ISA) performed. Even though the Company is not (yet) required to have an ISA for compliance purposes with laws, rules, and/or regulations, they are very aware of the benefits such audit can provide. However, they also know how pricy these specialized audits are. Would you be inclined to advise your Company go through such type of audit, yes or no? Explain your position.
10. List 10 sources for audit tools, best practices, and/or relevant audit information when performing information security audits that were discussed in this chapter.

Exercises

1. Exhibit 12.1 lists common techniques used to commit cybercrimes. For each of these techniques, research the Internet and provide the names of one or two entities that have been impacted by such technique in the last 5–7 years. Briefly describe how the technique was used in the attack.
2. List information, screenshots, reports, etc. that the IT auditor would likely request from a client in order to conduct an information security audit. Why is this information important for the IT auditor?
3. A potential client asks you to provide a draft of the IT audit program (objectives and control procedures) you would use and follow in order to audit information security at her organization. Provide your response in memo format, documenting (a) audit objectives the audit program will focus on, and (b) the control activities that would need to be assessed in order to meet the audit objectives just listed.

CASE—INFORMATION SECURITY AUDIT PROGRAM

INSTRUCTIONS: Research the Internet and identify two recent major cyberattacks within the last 3 years.

TASK: For each attack, (1) briefly describe the nature and operations of the organization that suffered the attack, as well as provide a brief summary description of the attack itself. Then, (2) put together a list of information security audit controls and procedures that, had they been in place, would have helped mitigate or reduce the impact of the attack. Lastly, (3) explain how each information security audit control you listed would have helped mitigate or reduce the attack. You are required to search beyond the chapter (i.e., IT literature and/ or any other valid external source) to back up your response. Include examples, as appropriate, to evidence your case point. Submit a word file with a cover page, responses to the task above, and a reference section at the end. The submitted file should be between 8 and 10 pages long (double line spacing), including cover page and references. Be ready to present your work to the class.

Further Reading

1. AICPA. Audit analytics and continuous audit—Looking toward the future. www.aicpa.org/ InterestAreas/FRC/AssuranceAdvisoryServices/DownloadableDocuments/AuditAnalytics_ LookingTowardFuture.pdf (accessed August 2017).
2. PWC. Auditing blockchain: A new frontier. www.pwc.com/us/en/financial-services/research-institute/blog/blockchain-audit-a-michael-smith.html (accessed September 2017).
3. Bacon, M. St. Jude Medical finally patches vulnerable medical IoT devices, *TechTarget*, https://search-security.techtarget.com/news/450410935/St-Jude-Medical-finally-patches-vulnerable-medical-IoT-devices (accessed 13 January 2017).
4. BI Intelligence. Here's how the internet of things will explode by 2020, *Business Insider*, www.businessinsider.com/iot-ecosystem-internet-of-things-forecasts-and-business-opportunities-2016-2 (accessed 31 August 2016).
5. Deloitte. Blockchain & cyber security. www2.deloitte.com/content/dam/Deloitte/ie/Documents/ Technology/IE_C_BlockchainandCyberPOV_0417.pdf (accessed September 2017).
6. ISO/TC 307. Blockchain and distributed ledger technologies. www.iso.org/committee/6266604. html (accessed September 2017).
7. EY. Blockchain and the future of audit. www.ey.com/gl/en/services/assurance/ey-reporting-blockchain-and-the-future-of-audit (accessed September 2017).
8. Cloud computing in 2016- Private company issues and opportunities. Deloitte. www2.deloitte.com/ us/en/pages/deloitte-growth-enterprise-services/articles/private-company-cloud-computing.html
9. Cloud Security Alliance. https://cloudsecurityalliance.org/about/ (accessed August 2017).
10. Cloud Security Alliance's CloudAudit Working Group. https://cloudsecurityalliance.org/group/ cloudaudit/#_overview (accessed June 2017).
11. Cybercrime tactics and techniques for Q1 2017. Malware Labs. www.malwarebytes.com/pdf/labs/ Cybercrime-Tactics-and-Techniques-Q1-2017.pdf
12. Da Veiga, A. and Eloff, J.H.P. (2007). An information security governance framework. *Inform. Sys. Manag.*, 24(4), 361–372.
13. Deloitte LLP. (2014). *IT Audit Work Papers*. Unpublished internal document.
14. Deloitte's auditing the Internet of Things. www2.deloitte.com/gz/en/pages/risk/articles/auditing-theinternet-of-things.html (accessed July 2017).
15. Deloitte's cloud computing, The non-IT auditor's guide to auditing the cloud. www.iia.org.uk/ media/1283828/cloud-computing-20150617.pdf (accessed June 2017).
16. Disterer, G. (2013). ISO/IEC 27000, 27001 and 27002 for information security management. *J. Inform. Sec.*, 4(2), 92–100.
17. Dubsky, L. (2016). Assessing security controls: Keystone of the risk management framework. *ISACA J.*, 6, 2016.
18. EY big data and analytics in the audit process. (2015). EY Center for Board Matters' September 2015. www.ey.com/Publication/vwLUAssets/ey-big-data-and-analytics-in-the-audit-process/$FILE/ ey-big-data-and-analytics-in-the-audit-process.pdf
19. EY's Cybersecurity and the Internet of Things. www.ey.com/Publication/vwLUAssets/ EY-cybersecurity-and-the-internet-of-things/$FILE/EY-cybersecurity-and-the-internet-of-things.pdf (accessed August 2017).

20. ForeScout, IoT security survey results. www.forescout.com/iot-security-survey-results/ (accessed June 2017).

21. Gartner IT Glossary. (n.d.). www.gartner.com/it-glossary/big-data/ (accessed October 2017).

22. Gartner's 2015 Hype cycle for emerging technologies identifies the computing innovations that organizations should monitor. (2015). www.gartner.com/newsroom/id/3114217

23. Gartner says the internet of things will transform the data center. (2014). http://www.gartner.com/newsroom/id/2684616

24. Gikas, C. (2010). A general comparison of FISMA, HIPAA, ISO 27000 and PCI-DSS standards. *Inform. Secur. J.*, 19(3), 132.

25. Golson, J. Car hackers demonstrate wireless attack on Tesla Model S, *The Verge*, www.theverge.com/2016/9/19/12985120/tesla-model-s-hack-vulnerability-keen-labs (accessed 19 September 2016).

26. Gressin, S. (2017). The Equifax data breach: What to do. Federal Trade Commission—Consumer Information. www.consumer.ftc.gov/blog/2017/09/equifax-data-breach-what-do

27. Herath, T. and Rao, H.R. (2009). Encouraging information security behaviors in organizations: Role of penalties, pressures, and perceived effectiveness. *Decis. Support Syst.*, 47(2), 154–165.

28. ISACA. How to navigate blockchain—The technology that could change everything. http://www.isaca.org/About-ISACA/Press-room/News-Releases/2017/Pages/ISACA-Guidance-How-to-Navigate-Blockchain.aspx (accessed June 2017).

29. IDC. Worldwide public cloud services spending forecast to reach $266 billion in 2021, according to IDC, USA, www.idc.com/getdoc.jsp?containerId=prUS42889917 (accessed 18 July 2017).

30. IIA's Global Technology Audit Guide (GTAG): Understanding and Auditing Big Data. https://global.theiia.org/standards-guidance/recommended-guidance/practice-guides/Pages/GTAG-Understanding-and-Auditing-Big-Data.aspx (accessed August 2017).

31. ISACA, *COBIT 5: A Business Framework for the Governance and Management of Enterprise IT* (ISACA, 2012), 94.

32. ISACA, Innovation insights: Top digital trends that affect strategy, 2015. http://www.isaca.org/knowledge-Center/Research/Pages/isaca-innovation-insights.aspx

33. ISACA innovation insights. ISACA. www.isaca.org/knowledge-center/research/pages/cloud.aspx (accessed September 2016).

34. ISACA's Internet of Things: Risk and value considerations. www.isaca.org/knowledge-center/research/researchdeliverables/pages/internet-of-things-risk-and-value-considerations.aspx (accessed August 2017).

35. ISACA's What is big data and what does it have to do with IT audit? www.isaca.org/Journal/archives/2013/Volume-3/Pages/What-Is-Big-Data-and-What-Does-It-Have-to-Do-With-IT-Audit.aspx (accessed August 2017).

36. ISO/IEC 27001- Information security management. www.iso.org/iso/home/standards/management-standards/iso27001.htm (accessed January 2017).

37. Mathews, L. (2017). Equifax data breach impacts 143 million americans. *Forbes*. https://www.forbes.com/sites/leemathews/2017/09/07/equifax-data-breach-impacts-143-million-americans/#3ab97325356f

38. McAfee Labs. (2017). Threats predictions report issued on November 2016. www.mcafee.com/au/resources/reports/rp-threats-predictions-2017.pdf (accessed October 2017).

39. McAfee Labs Threats Report—December 2016. www.mcafee.com/ca/resources/reports/rp-quarterly-threats-dec-2016.pdf (accessed October 2017).

40. National Vulnerability Database. National Institute of Standards and Technology. https://nvd.nist.gov/vuln/search (accessed August 2017).

41. NIST SP 800-144's Guidelines on Security and Privacy in Public Cloud Computing. https://nvlpubs.nist.gov/nistpubs/Legacy/SP/nistspecialpublication800-144.pdf (accessed July 2017).

42. NIST SP 800-124's Guidelines for Managing and Securing Mobile Devices in the Enterprise. https://csrc.nist.gov/csrc/media/publications/sp/800-124/rev-1/final/documents/draft_sp800-124-rev1.pdf (accessed July 2017).

43. Otero, A.R. (2015). An information security control assessment methodology for organizations' financial information. *Int. J. Acc. Inform. Sys.*, 18(1), 26–45.

44. Otero, A.R. (2015). Impact of IT auditors' involvement in financial audits. *Int. J. Res. Bus .Technol.*, 6(3), 841–849.

45. Otero, A.R., Tejay, G., Otero, L.D., and Ruiz, A. (2012). A fuzzy logic-based information security control assessment for organizations, IEEE Conference on Open Systems, 21–24 Oct. 2012, Kuala Lumpur, Malaysia.

46. Otero, A.R., Otero, C.E., and Qureshi, A. (2010). A multi-criteria evaluation of information security controls using boolean features. *Int. J. Network Secur .Appl.*, 2(4), 1–11.

47. OWASP's IoT Testing Guides. www.owasp.org/index.php/IoT_Testing_Guides (accessed August 2017).

48. Payment Card Industry Data Security Standards (PCI DSS)'s Security standards for account data protection. www.pcisecuritystandards.org/ (accessed July 2017).

49. PCI Security. (2016). PCI Security Standards Council. www.pcisecuritystandards.org/pci_security/

50. PWC's A guide to cloud audits. www.pwc.com/us/en/risk-assurance-services/publications/assets/internal-cloud-audit-risk-guide.pdf (accessed June 2017).

51. Ross, R. (2007). Managing enterprise security risk with NIST standards. *IEEE Comp Soc.*, 40(8), 88–91. doi: 10.1109/MC.2007.284.

52. Senft, S., Gallegos, F., and Davis, A. (2012). *Information Technology Control and Audit*. Boca Raton: CRC Press/Taylor & Francis.

53. Singh, A. N., Picot, A., Kranz, J., Gupta, M. P., and Ojha, A. (2013). Information security management (ISM) practices: Lessons from select cases from India and Germany. *Global. J. Flexible Sys. Manag.*, 14(4), 225–239.

54. Srinivasan, M. (2012). Building a secure enterprise model for cloud computing env. *Acad. Inform. Manag. Sci. J.*, 15(1), 127–133.

55. TechAmerica Foundation's Federal big data commission (2012). Demystifying big data: A practical guide to transforming the business of government. www.techamerica.org/Docs/fileManager.cfm?f=techamerica-bigdatareport-final.pdf

56. The best mobile device management (MDM) solutions of 2016. *PC Magazine*. www.pcmag.com/article/342695/the-best-mobile-device-management-mdm-software-of-2016

57. The SANS Institute's Cloud security framework audit methods. https://www.sans.org/reading-room/whitepapers/cloud/cloud-security-framework-audit-methods-36922 (accessed June 2017).

58. Top 10 ERP software vendors and market forecast 2015–2020. (2016). Apps run the world. www.appsruntheworld.com/top-10-erp-software-vendors-and-market-forecast-2015-2020/

59. United States General Accounting Office. *CYBERCRIME—Public and Private Entities Face Challenges in Addressing Cyber Threats*, GAO-07–705, June 22, 2007.

60. United States General Accounting Office. *Information Security: TVA Needs to Address Weaknesses in Control Systems and Networks*, GAO-08–526, May 21, 2008.

61. U.S. Department of Justice, Federal Bureau of Investigation. 2016. Internet Crime Report. https://pdf.ic3.gov/2016_IC3Report.pdf

62. U.S. Department of Justice, Federal Bureau of Investigation. 2015. Internet Crime Report. https://pdf.ic3.gov/2015_IC3Report.pdf

63. U.S. Department of Justice, Federal Bureau of Investigation. 2014. Internet Crime Report. https://pdf.ic3.gov/2014_IC3Report.pdf

64. U.S. Supplement to the 2014 Global Economic Crime Survey, PricewaterhouseCoopers LLP, http://www.pwc.com/gx/en/economic-crime-survey/

65. What is blockchain? *Journal of Accountancy*. (July 2017). www.journalofaccountancy.com/issues/2017/jul/what-is-blockchain.html

66. What is cloud computing? *PC Magazine*. (2016). www.pcmag.com/article2/0,2817,2372163,00.asp

67. White-collar crime overview. FBI major threats & programs. What we investigate. www.fbi.gov/investigate/white-collar-crime (accessed October 2017).

68. Zorz, Z. Researchers hack vizio smart TVs to access home network, *Help Net Security*, www.helpnetsecurity.com/2015/11/12/researchers-hack-vizio-smart-tvs-to-access-home-network/ (accessed 12 November 2015).

Chapter 13

Systems Acquisition, Service Management, and Outsourcing

LEARNING OBJECTIVES

1. Discuss the importance of establishing a system acquisition strategy.
2. Describe the system acquisition process.
3. Explain the service management process and its areas.
4. Describe outsourcing, including services, benefits, and risks. Discuss the process of outsourcing IT systems.
5. Describe audit involvement when auditing both, software acquisitions and service organizations.

An organization's reason for acquiring or outsourcing systems is to effectively and efficiently support one or more business processes. Once the business objectives have been defined for the solution(s) being sought, the acquisition or outsourcing process for systems and/or services can begin. Both acquisition and outsourcing must be managed and monitored adequately to ensure that suppliers live up to their commitments, and provide services consistent with management goals and objectives. IT auditors and management need to be aware of the importance of these areas and the critical control processes that need to be in place to support and protect the organization.

This chapter discusses critical success factors when acquiring systems or services from third parties, including establishing a strategy, following a formal acquisition process, and learning about acquisition contract terms and issues. This chapter also explains service management and expectations for an effective partnership between organizations and suppliers. All of these help ensure that the expected value is indeed delivered from the contract and supplier. Outsourcing IT, discussed next, refers to hiring an outside company to handle all or part of an organization's IT processing activities. It is a viable, strategic, and economic business solution that allows organizations to focus on what they do best (i.e., core competencies) and leave IT processing activities (e.g., data processing, etc.) to qualified computer companies. Lastly, IT audit involvement and procedures when examining system acquisitions as well as outsourced IT services are described.

Systems Acquisition Strategy

It is important to have a strategy in place that assists departments with the selection, purchase and, if applicable, implementation of technology-related systems (i.e., hardware, software, services). The goal of this strategy is to ensure that systems to be acquired (1) align to the objectives and goals of the organization; (2) integrate well and are compatible with existing software and technology infrastructure; and that they (3) result in a positive return on investment.

The key to getting value for organizations is to establish a win–win partnership with the vendor(s). Contracts that do not have a profit for the vendor nor assist organizations to meet business objectives are doomed to failure. Once the partnership is established, both sides should be looking for additional win–win opportunities and a symbiotic relationship. On the flip side, if the relationship is failing and the win–win is not possible due to inefficiencies with the vendor, etc. then there must be an exit strategy from the relationship that will allow value to continue with a new vendor with minimized transition costs.

Systems Acquisition Process

There has been much written in this area. The system acquisition process should include the identification and analysis of alternative solutions that are each compared with the established business requirements. In general, the acquisition process consists of the following:

1. Defining system requirements
2. Identifying alternatives
3. Performing a feasibility analysis
4. Conducting a risk analysis
5. Carrying out the selection process
6. Procuring selected software
7. Completing final acceptance

Defining System Requirements

One of the greater challenges in procuring any information system is to define its requirements. System requirements describe the needs or objectives of the system. They define the problem to be solved, business and system goals, system processes to be accomplished, and the deliverables and expectations for the system. System requirements include defining the information being given to the system to process, the information to be processed within the system, and the information expected out of the system. Each requirement should be clearly defined so that later gaps in expectations are avoided.

System requirements can be captured by interviewing management and those expected to use the information produced by the system to understand their expectations. Gathering requirements can also be accomplished by:

- Inspecting existing systems.
- Reviewing related paper and electronic forms and reports.
- Observing related business processes.
- Meeting with IT management and support staff regarding their expectations and constraints for implementing and supporting the system.
- Researching other companies in a related industry, of similar company size, and with a similar technical environment to identify best practices and lessons learned.

A system requirements document formally records expectations of the system, and typically provides the following information:

- *Intended users.* The users of the system include those who actually interact with it as well as those who use the information that it produces.
- *Scope and objectives.* The scope should be "holistic," incorporating both the technical environment as well as a business perspective.
- *Problem statement.* A description of the problem needing to be solved by the system.
- *System goals.* Be sure to include aims and objectives from a technical as well as a business perspective.
- *Feasibility analysis.* Defines the constraints or limitations for the system from a technical and business stand points. Feasibility should be assessed in the following categories: economic, technical, operational, schedule, legal or contractual, and political. Feasibility can include those things that are tangible, intangible, one-time, or recurring.
- *Other assumptions.* Additional expectations made regarding the system such as compliance with existing business practices.
- *Expected system functions.* Anticipated system functions to be provided such as authorizing payments and providing account status.
- *System attributes.* Attributes such as ease of use, fault tolerance, response time, and integration with existing platforms.
- *Context or environment in which the system is expected to operate.* This includes a description of the system that is expected to fit or interface within the environment (e.g., industry context, company culture, technical environment, etc.).

Identifying Alternatives

Many options exist in procuring system solutions (i.e., software), which include any combination of the following: off-the-shelf product, purchased supplier package, contracting for development or developing the system in-house, or outsourcing from another organization. Refer to Exhibit 13.1 for descriptions of these options.

A sourcing policy defines where IT systems or services are to be acquired. For instance, in a centralized shared-services model, the business units are required to purchase from the central IT group. This approach allows IT to standardize technology solutions and leverage size/scale for procurement. There may be situations where the central IT group cannot provide a requested system or service. In this case, there needs to be a process in place to source solutions through a third party. Although third-party systems/services may be used, it is important that IT manage the procurement and relationship to ensure compliance with internal standards.

Exhibit 13.1   System Acquisition Options

System Acquisition Option	Description
Off-the-Shelf Product	Purchasing commercially available products requires the organization to adapt to the functionality of the system. This business adaptation process means that the organization would need to customize the software product, and subsequently maintain those customizations within the processes that have been modified and changed. The advantages of using off-the-shelf products are shorter implementation time, use of proven technology, availability of technical expertise from outside the company, availability of maintenance and support, and easier-to-define costs. The disadvantages include incompatibility between packaged system capabilities and the company's requirements, long-term reliance on a supplier for maintenance and support, specific hardware or software requirements, and limitations on the use or customization of the software.
Purchased Supplier Package	Suppliers develop packaged systems for wide distribution that satisfy a generic business problem. For example, a payroll system is somewhat generic for most organizations. Often, a purchased package system can satisfy the business needs for much less than developing a system internally. If various package systems are available, organizations will develop a request for proposal that defines the system requirements and asks for the supplier's solution to those requirements. Organizations then weigh how well each package system meets the business requirements and whether they make sense. When selecting a purchased supplier package, organizations should consider the following: • Stability of the supplier company • Volatility of system upgrades • Existing customer base • Supplier's ability to provide support • Required hardware or software in support of the supplier system • Required modifications of the base software Although purchased supplier package systems may appear to be less costly to implement than developing a new system internally, there are risks to consider before selecting this option. A packaged system may not meet the majority of the business needs, resulting in extensive modification or changes to business processes. Also, any future releases of this system may require extensive programming to retrofit all of the company-specific code. Because packaged systems are generic by their nature, the organization may need to modify its business operations to match the supplier's method of processing. Changes in business operations may be costly due to training and the new processes may not fit into the organization's culture or other processes.

(*Continued*)

Exhibit 13.1 (*Continued*) System Acquisition Options

System Acquisition Option	Description
Contracted or In-house Development	Contracted development requires that the organization procure personnel to develop a new system or customize an existing system to the company's specifications. Contracting for systems development can provide increased control over costs and implementation schedules, legal and financial leverage over the contractor, additional technical expertise, and the ability to adhere to company policies, processes, and standards. Disadvantages associated with contracting for development include higher labor costs when compared to in-house staff, turnover of contract staff, business viability of the contracted company, exclusion of maintenance in development costs, and a lack of organizational understanding by the contractor. In-house developed systems (i.e., software) are generated internally by the organization. For instance, organizations develop software "in-house" either because there is no such software (or similar versions) available in the market, or because the organization wants to customize the software based on specific needs. Another advantage of in-house developed software is that it may later become available for commercial use upon discretion of the developing organization.
Outsourcing from Another Organization	Many companies choose to outsource system functionality from another organization. Outsourcing allows the company to cost-effectively remain focused on their core competencies and quickly respond to business needs as well as take advantage of the expertise of another organization. Outsourcing provides increased control over costs without the need to acquire or maintain hardware, software, and related staff. However, outsourcing systems increases reliance on the outsourcer, limits the company to what is provided by the outsourcer, and decreases the ability of the company to acquire related experience and expertise. Outsourcing is discussed in more detail later in the chapter.

Adapted from Senft, S., Gallegos, F., and Davis, A. 2012. *Information Technology Control and Audit*. Boca Raton: CRC Press/Taylor & Francis.

Performing a Feasibility Analysis

A feasibility analysis defines the constraints or limitations for each alternative from a technical as well as a business perspective. Feasibility analysis includes the following categories: economic, technical, operational, legal or contractual, and political.

■ *Economic feasibility* analysis provides a cost–benefit justification. The expenses of a system include procurement, start-up, project-specific issues, and impact to operations. It includes one-time and recurring costs. Sample of costs includes the following: consultants, start-up infrastructure, support staff, application software, maintenance contracts, training, communications, data conversion, and leases. Benefits include cost reduction or avoidance, error reduction, increased speed, improved management decisions, improved response to business needs, timely information, improved organizational flexibility, better efficiency, and better resource utilization.

■ *Technical feasibility* analyzes the technical practicality of the proposed system. It evaluates the consistency of the proposed system with the company's technical strategy, infrastructure, and resources. It answers the question of whether the organization has the resources to install and support the solution. Technical feasibility evaluates whether the company has the necessary hardware, software, and network resources to support the application as well as whether it provides reliability and capacity for growth. It also assesses the technical expertise requirements and compares it with those provided by the organization.

■ *Operational feasibility* examines how well the proposed system solves business problems or provides opportunities to the business. It also evaluates the extent of organizational changes required to accommodate the system. These changes can include personnel, business processes, and products or services offered.

■ *Legal and contractual feasibility* reviews any related legal or contractual obligations associated with the proposed system. Legal constraints include federal or state law as well as industry-related regulations. In addition to any new contract obligations introduced by the new system, existing contracts are also reviewed to ensure that there are no preexisting commitments that regulate the installation or use of the proposed system. Legal counsel should be involved in this process and is one of the critical points for IT auditors to review. Note that the underlying theme is protection of the company and the establishment of the remedy process should the contractor fail to perform or deliver as promised. Organizations looking for assistance in this area should refer to their legal counsel or an organization whose members specialize in this area.

■ *Political feasibility* evaluates how the internal organization will accept the new system. This includes an assessment of the desirability of the system within the organization as well as its fit with the organization's corporate culture.

Conducting a Risk Analysis

A risk analysis reviews the security of the proposed system. It includes an analysis of security threats and potential vulnerabilities and impacts, as well as the feasibility of other controls that can be used to reduce the identified risks. Controls include systematic or automated methods as well as audit trails. Risks, as discussed in other chapters of this book, affect control objectives in the areas of confidentiality and privacy of information, integrity and accuracy of the data, timeliness of the information for decision making, ability to access the system, as well as staff organization and knowledge required to support the system.

Carrying Out the Selection Process

The selection process includes identifying the best match between the available alternatives and the identified requirements. According to KPMG's 2017 State of the Outsourcing, Shared Services, and Operations Industry Survey Report, the top three selection criteria for an external service supplier are: return on investment (21%), business outcomes (16%), and automation (11%) (note that return on investment, not cost, was the most important consideration for buyers). The survey also showed that smaller organizations place more importance on traditional selection criteria like industry experience and service quality than business outcomes and automation.

Activities for selecting an external service supplier start with soliciting information for a specific purpose followed by requesting proposals from interested providers. The process ends with evaluating the proposals in terms of the identified requirements and selecting the best available alternative (supplier). The selection process should be structured to ensure that the process would be completed diligently and in a timely manner. If done correctly, the selection process promotes buy-in for the selected solution. The selection process is described in Exhibit 13.2.

Exhibit 13.2   Description of Supplier Selection Process

Request for information ⇨ Request for proposal ⇨ Evaluation and selection	
Selection Process Activity	*Description*
Request for Information (RFI)	A RFI seeks information from suppliers for a specific purpose. However, neither the company nor the suppliers are obligated by the response to the RFI. The RFI serves as a tool for determining the alternatives or associated alternatives for meeting the organization's needs. A RFI often asks suppliers to respond to questions that will assist the organization in obtaining additional relevant information. Information from the RFI may then be used to prepare a request for proposal.
Request for Proposal (RFP)	A RFP is a document that specifies the minimally acceptable requirements (functional, technical, and contractual), as well as the evaluation criteria used in the selection process. A RFP offers flexibility to respondents to further define or explore the requested requirements. RFPs may lead to a purchase or continued negotiation. Potential suppliers are supplied with copies of the RFP and are requested to submit proposals by a specified date. After a supplier has submitted a proposal, the response cannot be changed. The RFP should be communicated to as many prospective bidders as possible. It should contain the selection criteria that will be used. The criteria should be written with enough detail that it prevents any misunderstanding or misinterpretation. Any specific criterion that will influence the selection must appear in the RFP. The RFP should also describe the members of the selection committee. All questions from bidders should be answered in writing and made available to all bidders. Verbal answers to questions should be avoided. All questions should be received with enough time before the deadline so that answers can be incorporated in the proposal. Public meetings such as "bidder conferences" are used as a means to receive and respond to questions from all prospective bidders. The basic components of a RFP are: • Background information about the company, business problem, and the computing environment. It may also include results of any needs assessment performed.

Exhibit 13.2 (*Continued*)   Description of Supplier Selection Process

Selection Process Activity	Description
	• Schedule of important dates, such as when the supplier's RFP response is due, when the decision is expected, when the actual purchase is expected, and when implementation is expected. • Contact names and sources for answering questions for the RFP. • Instructions for formatting the response to the RFP. Some RFPs include an explicit description of what the supplier should and should not include in their response. • Specific requirements being sought. • Technical requirements for the system, such as specifications for an operating system or a network environment. • List of documents required as attachments, such as sample reports and standard contract language. • Additional requirements for the selection process, such as supplier presentations, supplier demonstrations, or on-site installation and testing. The components listed above must be followed to the letter before the next step can be performed. If it is not followed as the company has specified, a "bid protest" could be filed by one of the potential suppliers. Again, this is another audit, management, or legal counsel review point before the evaluation process can begin.
Evaluation and Selection	A selection committee of one or more key stakeholders evaluates submitted proposals using a list of objective selection criteria. A list of the objective selection criteria is used as a means for identifying the best match between the product's features and functionality and the identified requirements. The basis for the selection criteria is the user and system requirements. Features and functionality are normally the most significant factors in the decision-making process. Selection criteria may also include evaluating the consistency of the proposal with the company's business and IT strategy, the breadth of the supplier's products and services, supplier's relevant experience, supplier's customer support, scalability of the solution, supplier viability, total cost of ownership, integration and growth capabilities, and reliability. Participants in the selection process often include representatives from key stakeholders, such as management and anticipated users as well as the IT department. The selection committee should consist of representatives that are impartial to any one provider or solution, have knowledge of best practices, and knowledge of the market. However, the participants should not have any conflict of interest and should sign a statement before they serve on the selection committee, indicating that they do not. If they do, someone who meets the qualifications warranted and does not have a conflict of interest must replace them.

Adapted from Senft, S., Gallegos, F., and Davis, A. 2012. *Information Technology Control and Audit*. Boca Raton: CRC Press/Taylor & Francis.

Procuring Selected Software

Once a technical solution has been selected, the procurement process helps ensure that the right terms and conditions are negotiated. One of the integral processes in any project is the procurement of services, hardware, and software. In most cases, organizations consider whether to make or buy systems. In either case, the procurement of external services is usually required. Depending on the extent of the service, a formal RFP or other requirements document needs to be prepared to request competitive bids. The requirements should include service levels with contract penalties and tracking metrics/success criteria.

The procurement process sounds simple but is actually the most complicated of all the acquisition steps. It requires that the purchase price and conditions be stipulated and agreed upon. These agreements take the form of contracts.

When procuring software, IT managers complain about unpredictable license fees, pressured sales methods, poor technical support, and unclear pricing for ongoing maintenance fees.

Software Contracts and Licenses Agreements are used to document agreements for development, marketing, distribution, licensing, maintenance, or any combination. Contracts can specify a fixed price or a price based on time and materials. Contracts based on time and materials state that the fees charged are directly attributable to actual expenses of time (hourly) or materials. These contracts place more financial risk on the buyer if the initial definition of pricing, the scope, or desired requirements are unclear or poorly defined. Contract terms and conditions normally include the following:

- A functional definition of the work to be performed.
- Specifications for input or output designs, such as interfaces, screens, or reports.
- Detailed description of the necessary hardware.
- Description of the software systems or tools required for development or implementation.
- Terms or limitations with the use of any related trademark rights or copyrights.
- Requirements for the conversion or transfer of data.
- System performance or capacity such as speed, throughput, or storage.
- Testing procedures used to identify problems and the results expected to define acceptance. Information and system requirements serve as the basis for defining the acceptance tests.
- Supplier staffing and specified qualifications.
- Contact and relationship protocols between the buyer and the supplier.
- Expected schedules for development, implementation, and delivery.
- Methods for providing progress reports, such as meetings or reports.
- Definition of deliverables, which includes a clear description of each item to be delivered or provided by the supplier, when it is to be delivered, and any consequences for missed deliverables.
- Explicit criteria for defining acceptance of each deliverable as well as for final acceptance.
- Requirements and expectations for installation.
- Documentation expected to be provided to the supplier or by the supplier as well as any intellectual property rights needed to maintain or customize the documentation.
- Training expected to be provided as part of the product or service.
- Any applicable warranties or maintenance, including provisions for future versions currently in development.
- Any requirements for indemnity or recovery for losses, such as insurances or bonding requirements.

- A statement of future support that is to be provided as well as anticipated costs.
- Clear definition of ownership or licensing of relevant copyrights and patents.
- Terms and conditions related to confidentiality or trade secrets for either party.
- Terms or limitations related to staff changing employers from one party to another (e.g., raiding staff, etc.).
- Description of payment terms.
- Process for accepting changes to contract definition, such as changes in terms, scope, or deliverables.

Other considerations, specifically for software contracts and licenses, should also address the following:

- Flexibility and choice for upgrades and updates. Some contracts specify required upgrades to receive updates or maintenance.
- Service-level agreements (SLAs) for defining expectations for support and maintenance.
- Annual maintenance costs. Such maintenance costs should be fixed at the time of purchase and should not vary.
- Provisions for protecting the company against unforeseen problems such as software interoperability.
- Intellectual property rights for modifications. Customer may not be granted the rights for modifications.
- Terms and conditions for termination options, such as what transfer process will take place when the license ends, length of the transition period, and the impacts from the termination.
- Assignment clauses requiring consent. Assignment clauses allow the supplier to segregate the customer's payments from that of the service provider. Under an assignment clause, the supplier can transfer the customer's financial obligations to another firm or ongoing service components to a third party. With the payment and service separated, the supplier's motivation to perform to the terms of the contract may be reduced.
- Verification of any export or import restrictions by customers. Specifically, the export of the specified technology should be allowed by U.S. legal restrictions and the import allowed by the foreign government. The license should specify responsibility for any costs or duties.
- Regulatory approvals that may be required. Some governments, such as Japan, require that they approve the license. If the approval is not sought, the license can be considered void. The license should specify which party is responsible for obtaining the approval.
- Review of competition or antitrust laws to ensure compliance with any related legal requirements.
- Consideration of currency exchange rates. Exchange rates should be agreed in the contract to minimize the risk of fluctuating currency rates.
- In the case of outsourcing, specification in the contract for the financial and legal interests in the company's software now being supported by the outsourcer. For example, in some cases, the software license may transfer to the outsourcer, which will require relicensing its own software if the company later chooses to discontinue outsourcing. The other option is to allow the outsourcer to use the software with the company retaining responsibility for all license agreements.

Completing Final Acceptance

An acceptance plan should be agreed upon and defined in the contract. This plan defines the terms and condition for acceptance. Normally, final payment is withheld until all acceptance tests have been completed and the software and equipment meet all specifications in the contract.

Technology acquisition is dependent on other key processes and must integrate to operate effectively. For example, acquired devices must integrate with existing infrastructure architecture and project dependencies tied to the delivery of components. In addition, the life cycle of acquired devices does not end with the acquisition process. Assets must be installed, secured, tracked, maintained, and disposed of properly.

Service Management

The service management process begins once a contract is signed and helps ensure that suppliers live up to their commitments. Using external resources provides flexibility and scalability by leveraging the expertise and staff of third-party suppliers for temporary staffing needs for development projects. There is also an opportunity to drive development costs down by outsourcing programming to offshore suppliers with mature development processes and inexpensive labor rates. There are a variety of sourcing models from internal delivery to full outsourcing. All models require internal processes to manage service levels, costs, and risk.

The service management process for an organization includes defined IT services; SLAs; design services and pricing; service engagement and delivery; and service measurements to track performance.

Defined IT Services

Defining services requires the organization to consider what services are important from a customer perspective. For example, the finance function is dependent on the availability of the accounting systems. Finance users are not really interested in platform availability, although this may be an important aspect of delivering the accounting application. The accounting application only works if all components involved in delivering the application work. The client hardware and software, the networking components, the server hardware and software, and the database must all function well together to deliver the accounting application to the end user. From a customer perspective, the measurement of end-to-end availability of the accounting application is important. From an IT perspective, each of the individual platforms must be available to deliver the service expected from the customer. The processes that go into delivering platform availability include all aspects of IT.

Delivering high-quality IT services is dependent on the delivery of high-quality applications and all of the processes that go into managing IT operations: capacity management, change control, security, disaster recovery, vendor management, and so on. Another complicating factor in measuring and delivering IT services is the dependencies on third-party providers of services, networks, hardware, and software. With all this complexity, delivering high-quality IT services is very challenging.

An organization with a large volume of applications makes the task of measuring service even more difficult. It is probably not cost effective to measure every application in an organization; therefore, it makes more sense to measure selected key applications. These and other factors have to be negotiated with the identified customers, keeping in mind that trade-offs will need to be made on the value delivered compared to the cost.

Service-Level Agreements

An SLA is a formal agreement between a customer requiring services and the organization that is responsible for providing those services. Below are basic defined terms related to SLAs.

- *Service.* A set of deliverables that passes between a service provider and a service consumer.
- *Level.* The measurement of services promised, services delivered, and the delta between the two.
- *Agreement.* Contract between two entities—the one providing the service and the recipient.

SLAs should include a definition of the services, the expected quality level, how service will be measured, the planned capacity, the cost of the service, the roles and responsibilities of the parties, and a recourse process for nonperformance. For an internal service agreement, performance can be tied to compensation (e.g., bonuses, etc.). External agreements may involve payment of a penalty or compensation for poor performance.

It is important to engage the customer in the service-level definition process to help gain agreement and buy-in. Organization management will make trade-offs between service quality and cost. These decisions must be communicated across the organization to set expectations at all levels. Otherwise, customer satisfaction results may reflect unreasonable expectations versus agreed IT service levels.

SLAs can be made between IT and its customers, operations and application groups, and between suppliers and IT. A customer service-level agreement (CSLA) encompasses all of the subservices being provided both internally and externally to deliver the services needed by the customer as defined earlier. An operating-level agreement (OLA) between operations and application groups defines the underlying operating services required to deliver projects and applications to the customer under the CSLA. Supplier SLAs define the services required by operations, applications, or end users to deliver in accordance with the CSLA. These common types of SLAs are explained in Exhibit 13.3.

Design Services and Pricing

Service design can be quite complicated, depending on the breakdown of the organization and the components needed to support an application. Once the services are defined, the internal functions and processes have to be mapped to each of the services. At the lowest level, IT organizations typically separate costs into cost centers aligned to individual managers. These cost centers will need to be aligned to services entirely or as a percentage based on an allocation of the time spent or some other appropriate measurement. For example, an operations group may spend 20% of its time supporting the client environment, 20% supporting the mainframe environment, 30% supporting the distributed environment, and 30% supporting the network environment. For example, developing the pricing for mainframe application hosting will require the aggregation of all the costs that go into supporting the mainframe environment. Mainframe application hosting may include engineering, backup, archive, recovery, disaster recovery, performance monitoring, capacity management, and provisioning.

Exhibit 13.3   Common Types of Service-Level Agreements

Service-Level Agreement	Description
Customer Service–Level Agreement (CSLA)	A CSLA is a formal agreement between IT and the user organization. It encompasses all of the subservices being provided both internally and externally to deliver the services needed by the customer. The customer is a user group defined as the first step in the process and is usually an organization function that is the primary user of a key application (e.g., finance, human resources (HR), etc.). It is better to have the primary owner for each key application to gain agreement on service levels and pricing. The primary user organization can include the IT services and cost in services to its customers. Although it is possible to split the cost of a shared application into separate user groups, it is not possible to provide varying levels of service for the same application. Examples of customer services include: • Client services—desktop, laptop, mobile devices, software, support, etc. • Storage services—direct access storage devices, storage area network, tape, etc. • Networking services—telecom, data, Internet, etc. • ERP services—accounting, HR, procurement, etc. • Business processing services—manufacturing, policy processing, etc. Service levels that are provided to all user groups (e.g., personal computer services, etc.) may have to be agreed at the organization level. It is possible to offer different hardware, software, and support options to end users; however, this can be more costly. Standardizing hardware, software, and support allows an organization to leverage scale in negotiation and gain processing efficiencies. These are decisions that are best made at the organization level.
Operating-Level Agreement (OLA)	OLAs, whether formal or informal, define the underlying services required by application development and maintenance groups to deliver the end-to-end application services to customers. A formal agreement helps set expectations between the application and operations groups on the quality of service; roles and responsibilities; expected capacity; and in some cases, service cost. Examples of operating services include: • Mainframe application hosting services • Distributed application hosting services • Web application hosting services • Storage services, such, as DASD, SAN, tape, etc. • Networking services, such as telecom, data, Internet, etc. Storage and network services may be provided directly to the customer without bundling with other services. They may also be shown on the OLA to have all underlying operating services in one agreement for accountability purposes.

(Continued)

Exhibit 13.3 (*Continued*)   Common Types of Service-Level Agreements

Service-Level Agreement	Description
Supplier Service–Level Agreements	A portion or all of a service may be provided by a third party through outsourcing, cosourcing, or selective sourcing. It is important to align the SLAs from third parties to customer SLAs to prevent a misunderstanding or misinterpretation of the agreement. For example, it would be virtually impossible to be successful if the supplier's SLA promises 95% availability and the customer's SLA promises 99%. To successfully manage services provided by a third party, IT must have a solid understanding of the services it provides, the services expected from the third party, and how the internal and external services work together. Examples of third-party services include: • Service desk, desktop support, etc. • Application development or maintenance • Application hosting • Intrusion detection To be successful, organizations must have processes in place to consistently monitor fulfillment of service levels from all key suppliers. As third-party services and spending represent a significant portion of IT services and costs, it is worth the investment in managing third-party service delivery.

Adapted from Senft, S., Gallegos, F., and Davis, A. 2012. *Information Technology Control and Audit*. Boca Raton: CRC Press/Taylor & Francis.

Behind customer services are one or more IT functions or processes required to deliver those services. For example, designing an ERP service requires adding the ERP application maintenance costs plus the underlying infrastructure service costs plus the overhead costs.

Further complicating service design is providing varying levels of service based on availability requirements. The more options, the more complex the service model, and the more complicated the pricing model and resulting measurement process. Organization and IT management must understand the cost/benefit before implementing a complicated service/pricing model. The author recommends keeping the service structure as simple as possible for the initial implementation. More complexity can be added once the service management processes (e.g., measurement, etc.) are mature.

There are some services that will most likely be required for all customers depending on the organization's service strategy. Required services may include minimum levels of disaster recovery, security, risk management, and project management. Making these services mandatory reduces some of the conflict in pricing projects and services as they are required by the organization and built into the service and pricing model. It is important to explicitly state the mandatory services that are included in the service or pricing model to improve transparency into IT services.

Services Engagement and Delivery

An important aspect of service management is how services are delivered to the user organization. Provisioning service is the first point of contact with users and the point of capture for usage information. For delivering end-user services (e.g., client, telecom, etc.), the first point of contact sets the stage for delivering customer satisfaction. Procurement, asset management, and service desk are processes that are dependent on service provisioning to function properly. Procurement processes for end-user services should be agreed with the user organizations to ensure that only approved hardware, software, and system access are provided to end users. The procurement process can be the trigger for creating user and asset information used by security, finance, asset management, operations, service management, and the service desk to identify users and client configurations.

For application development and maintenance, the entry point into IT is entry into the systems development life cycle, and mature processes in this area can make or break the relationship with the customer. Demand management, requirements definition, and change control were discussed in previous chapters. Again, these processes should be agreed between IT and organization management to ensure alignment on the rules of engagement. These processes are even more critical in an outsourced environment where service requests need to be approved before incurring charges. There are automated solutions for entering, approving, and tracking application requirements and service requests that were discussed in earlier chapters.

Service Measurements

Areas that are measured get the focus and tend to improve as a result. Focusing only on customer service may lead to an improvement in one area at the sacrifice of others (e.g., cost and security, etc.). That is why it is important to measure the right things. According to the Gartner Group, it is also important to have different internal and external measures. Internal measures keep the IT group focused on what they need to do to achieve service levels. External (customer) measures should be focused on things the customer cares about.

Because internal IT functions and processes impact customer services in different ways, it is important to align the internal measures to the external customer services. The internal IT measurements need to be designed to hold individual groups accountable for delivering their portion of the service.

It is also important to coordinate measuring IT services with IT process improvement and governance metrics. Measuring too many processes for too many purposes can create measurement overload. Leveraging the same information for multiple purposes is far more efficient. A little up-front planning on what to measure and the use of the measurements will go a long way toward implementing an effective measurement process. Once the measurements have been decided, the source of the information needs to be identified. The source must be reliable and consistent to ensure the measurement can be delivered. The rules around measurement must also be decided and agreed by all parties. Having a measurement that is not trusted is counterproductive.

What to Measure

What to measure will depend on the services provided and the agreed quality level. It is important to keep any measurement program simple as the cost to deliver more detailed measurements may far outweigh the benefit. However, measurement must be granular enough to help identify the root cause of problems and hold individuals accountable for their share of the service. Consideration also needs to be given to the level of maturity in the IT organization. It may be necessary to start with simple measurements that get expanded as the IT organization matures. The underlying data must be available for more complicated metrics and the acquisition of the right data needs to be planned for in advance.

Application services typically include maintaining existing applications, making changes and enhancements to existing applications, and building new applications (i.e., projects). As discussed earlier, customers care about the availability of their key applications. To determine availability, all components of an application must be measured to determine end-to-end availability.

Infrastructure services include building and maintaining operating platforms. Availability can be measured at the operating platform (e.g., mainframe, UNIX, Windows, Linux, etc.) or the environment level (e.g., online, application, database, etc.). Infrastructure metrics will depend on the services provided.

Determining the required availability percentage will depend on the organization's objectives, function of the application, tolerance for risk, and cost/benefit of higher availability. With the complexity of today's applications and technology, increasing availability from 95% to 100% would be costly and maybe even impossible.

Measuring project success includes delivering to the agreed-upon scope, schedule, budget, and quality. Measuring change could include the average cost per change, time to deliver, and quality. Measuring quality could include issue/problem quantity, severity, backlog, and time to resolve. The challenge with quality metrics is that there are many factors that go into a quality system that may not be within the control of the IT group. Quality problems can arise from a failure on the part of the user group to define requirements, validate acceptance, and train users on how to effectively use the system. It is important to track the underlying cause of issues/problems to be able to identify the root cause and attribute only those problems associated with IT against the SLA.

An important measure for both application and infrastructure services is the number of changes. A high quantity of changes can indicate a failure on the part of user groups to properly define requirements, a failure on the part of the applications group to understand requirements or implement high-quality code, a failure of the operations group to carefully manage change to the infrastructure, or an overall problem with the change control process.

How to Measure

As important as what to measure is how to measure. Whatever method is selected, it should be used consistently and clearly communicated to the user organizations. The following are possible approaches to measuring processor usage:

- *Peak consumption.* This method measures the highest level of consumption. As hardware/software must be purchased to support peak capacity, it more closely aligns costs to consumption. However, the organization needs to decide how to measure and charge for peak versus off-peak consumption.
- *Capacity allocated.* This method measures the processors or capacity allocated to user groups or applications. The challenge comes with measuring shared environments and the capacity allocated to system functions. If system capacity is excluded, only production capacity is measured and allocated. However, what happens during the month when the accounting systems need additional capacity for month-end processing?
- *Average consumption.* This method measures either average capacity or consumption over a period of time. This approach seems to overcome some of the limitations with the other two methods. However, it does not encourage workload optimization to minimize peak workload.
- *End-to-end availability is based on component outages as well as reported problems/outages during the core hours of services.* End-to-end availability can be measured by dividing the application

downtime by the available connect time. However, this does not take into account the number of users impacted. An outage in the early morning may have much less impact than an outage at peak usage during the day. A more meaningful measure from a customer perspective would be the number of users impacted multiplied by the amount of time the system is unavailable. Multiplied by the average cost per user can give an approximate financial impact of an outage. This information can be used when determining the cost/benefit of increasing service levels or considering process improvements.

A common tool used to reflect many of these is graphs, charts, or tables. It is very important that if such data are analyzed and shown, the information about the time period and source of data be clearly provided. Also, any anomalies or assumptions should be communicated clearly, especially if the data are forecasted or projected. Without the proper communication, such tools can easily distort the results and be counterproductive to the organization.

Service Management Tools

There are many tools available to assist organizations in implementing service management processes. Tools are needed to capture performance, usage metrics from the various platforms, and to consolidate and report on all of this information. Automation is required to deliver an efficient measurement and reporting process. Many of the performance management tools used by systems programmers, operations, and network administrators can also be used to measure service delivery. Common examples of service management tools include: Customer Satisfaction Surveys and Benchmarking.

A customer satisfaction survey is a good example of an important tool used to measure the quality of the services provided by IT. There may be multiple customer satisfaction surveys for different services. Senior management may be asked questions on the value of IT, and end users may be asked questions on satisfaction with the service desk and application availability.

Senior management satisfaction should be measured separately as there will be different objectives and questions. Senior management will be more focused on the value delivered by IT. This may include project delivery, IT understanding of business needs, application delivery and support, cost effectiveness, service quality, and overall satisfaction.

Customer satisfaction surveys may include the time to provision requests, time for the service desk to answer a call, satisfaction with issue resolution, and system response time. Using customer satisfaction surveys to measure system response time in addition to using technical measures helps to determine if there is an expectation gap. This information can then be used in discussions with the organization management to either increase the expected service level or improve communication to the user population.

Along with reporting survey results, a follow-up process is needed to respond to issues raised in the survey. This is an important communication tool and can also help boost customer satisfaction. Customer satisfaction is partly a result of expectations and just active listening can boost results. That is another reason why customer satisfaction may have more to do with communication and the IT/organization relationship than with actual service issues.

The biggest challenge with **benchmarking** is finding an organization that is similar in type, size, and structure with similar business needs. This is particularly difficult as every organization is different. One organization may be an early adopter of technology, whereas similar organizations may be late adopters. These differences may impact service and cost comparisons, for instance, in a benchmarking exercise.

Benchmarking services attempt to align costs and services by using standard definitions of service and cost elements. This makes benchmarking a time-consuming and costly task as financial information and service structures must be restated to align to the benchmark. Even with the restatement of services and costs, benchmarking will have limited value as there may be valid reasons for cost difference from the benchmark. For example, an organization may have lower automation and a higher unit cost offset by efficient manual operations. The opposite could be true, high automation with low unit cost combined with inefficient manual operations. The bottom line is no one data point can provide conclusive information on the efficiency or effectiveness of IT services.

Another issue that reduces the value of benchmarking information is that the results will not be the same services that user organizations are familiar with. This makes it difficult to confirm and communicate that the unit cost charged to individual user functions compares favorably to the benchmark. The advantage of using external benchmark information is the independent source of comparison data. The information provided to the benchmark provider must be auditable to ensure the credibility of the results. Having internal audit validate the submission may also be a good way to validate the results.

Benchmarking can be a useful tool in evaluating the design, quality, and cost of IT services. However, because of limitations, benchmarking should be considered as an input into evaluating the underlying cost of IT services rather than the end result. Exhibit 13.4 shows the cyclical nature of the service management process just discussed.

Outsourcing IT Systems

Outsourcing refers to the transfer of service delivery to a third party, allowing companies to concentrate on core competencies. According to KPMG's 2017 State of the Outsourcing, Shared Services and Operations Industry Survey Report, IT continues to be the largest user of outsourcing, with 94% of organizations using at least some IT for both, applications management and infrastructure. Almost half of the responding enterprises have outsourced their IT processes to outside parties.

Outsourcing was initially used for standardized payroll and accounting applications, or by companies to sell their hardware. In an effort to reduce costs, many organizations started outsourcing their payroll and human resource management (HRM) functions to payroll service bureaus (PSB) and professional employer organizations (PEO). A PSB maintains the payroll master data for each of its clients and process payroll for them. In addition to performing the services PSBs do, a PEO also provides HRM services, including employee benefit design and administration.

When organizations outsource payroll processing, they send time and attendance data along with information about personnel changes to the PSB or PEO at the end of each pay period. The PSB or PEO then uses that data to prepare employee paychecks, earnings statements, and a payroll register. The payroll processing service also periodically produces employee W-2 forms and other tax-related reports. PSBs and PEOs represent an attractive option for businesses to reduce costs, such as:

- eliminating the needs of preparing paychecks for a large number of companies;
- developing and maintaining the expertise required to comply with the constantly changing tax laws;
- offering a wider range of benefits across all their clients; and
- freeing up of computer resources (e.g., payroll and benefits management application systems, etc.).

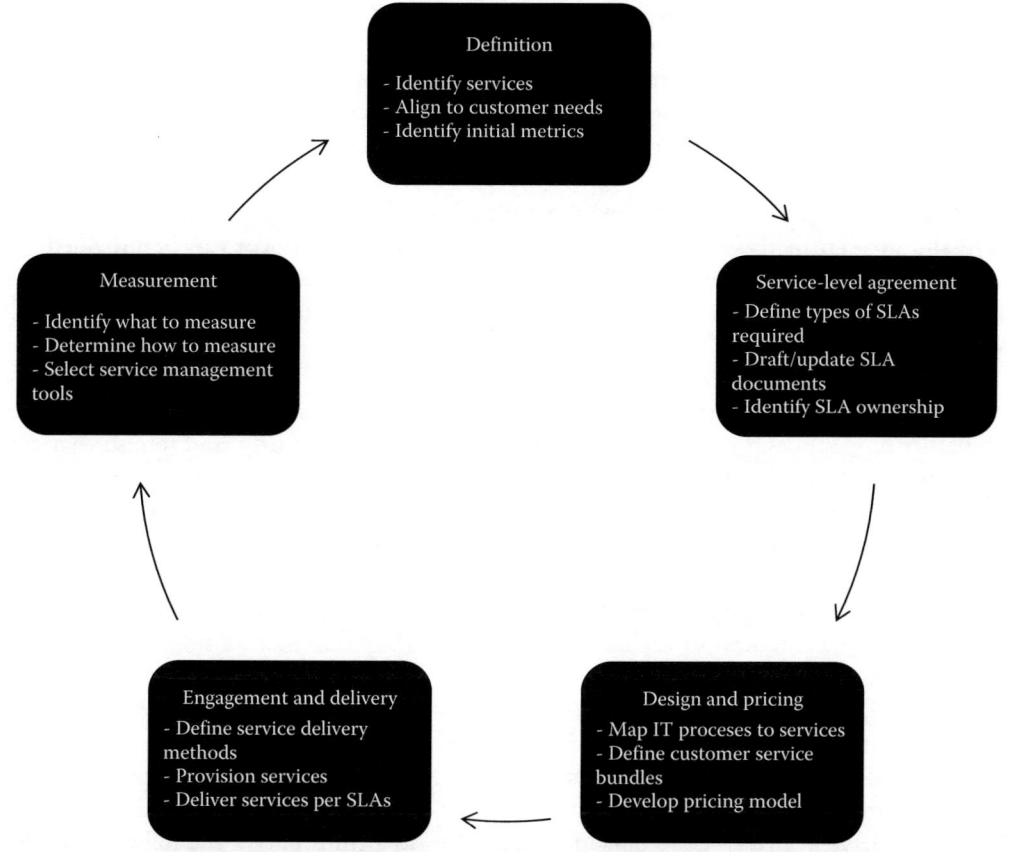

Exhibit 13.4   Cyclical Service Management Process. (Adapted from Senft, S., Gallegos, F., and Davis, A. 2012. *Information Technology Control and Audit*. Boca Raton: CRC Press/Taylor & Francis.)

Off-shoring is the transfer of service delivery to a third party outside the client's country. Based on the KPMG's 2017 Survey Report, off-shoring is most commonly used in IT and Finance and Accounting, both with 38%, followed by customer service with 24%. Company size (i.e., organizations with annual revenues of $5 Billion or more) showed to be the biggest indicator of offshore use. Both, outsourcing and off-shoring, have become common business practice in the various industries for decades as shown by the two examples below.

■ In the late eighties, the first landmark IT outsourcing deal was signed between The Eastman Kodak Company (Kodak) and IBM. Kodak hired IBM to take care of its data processing operations, as well as other companies to run telecommunications functions and PC operations. Results showed a significant decrease in computer operating expenses. Annual savings related to IS were also significant as a result of the outsourcing agreement.

■ During the mid-nineties, Xerox Corporation awarded a 10 year, $3.2 billion contract to Electronic Data Systems (EDS) to operate the Xerox worldwide computer, software management, and telecommunications network. The agreement was believed to be the largest commercial contract of its kind and the first on a global scale.

Although organizations can increase productivity and reduce costs using internal resources, there are additional benefits to outsourcing or using offshore suppliers. Outsourcing can provide increased flexibility by leveraging the resources of a third party to ramp-up or ramp-down resources for a variable workload. In some organizations, it is very difficult to reduce the labor force due to legislation or union agreements. Outsourcing allows an organization to transfer this responsibility to a third party. A global supplier can transfer resources to other engagements eliminating the need for labor force reductions and providing individuals with better career opportunities.

Off-shoring can increase productivity if sequential work or services can be provided 24×7 with resources in different time zones. Help desk support is a good example of a service that can be outsourced to a third party in multiple time zones to provide customers with around the clock service.

Outsourcing to a mature service provider can enable an organization to leverage the technology and processes of the third party. For example, a large desktop support supplier will already have mature processes and technology for deployment, support, security, and refresh that can be implemented in the target organization.

There are also many risks with outsourcing, including reliance on a third party, reduced flexibility (e.g., locked-in systems, long-term contracts difficult or costly to break, etc.), loss of control, and increased costs. Companies that outsource may also often experience a significant reduction in competitive advantage, employee morale, productivity, and service levels; poor service and unfulfilled goals; failure to realize savings; and weak governance and service management processes; among others.

Outsourcing and off-shoring may require significant changes to the way people work. Managers will have to start managing deliverables rather than people. This change may require a paradigm shift as well as a process shift for many managers. Organizations need to be prepared for an off-shoring program by having the processes in place to ensure success. Service management is a critical process to have in place before outsourcing services to a third party. The following decisions and processes need to be considered:

- A clear definition of the work to be outsourced and be performed offshore.
- How work will be initiated, transferred, and received from the third party.
- Educate staff on working with people remotely, with a different culture, different language, and probably a different time zone.
- Are there data privacy issues that need to be considered? In some countries like Germany, some data cannot be taken out of the country or viewed by people outside the country. Work-arounds may be needed to accommodate these requirements.

One of the challenges the sourcing project team has to manage is scope creep. It is very important to clearly define the scope of work that will be outsourced to the supplier and the work to be retained by the organization. Clearly defined roles and responsibilities, processes, and SLAs are also key to ensure a successful outsourcing engagement. A mature organization will be in a better position to manage the work being performed by an outsourcer.

Key to the up-front planning is to develop a statement of work for the transition project that defines the deliverables. Examples of deliverables include documentation, the future SLA, and people with knowledge. Acceptance criteria for each of these are essential and must be documented in advance. Every effort should be made to make the transition fixed price or even totally at vendor expense, and the team should not move to the steady-state outsourced mode until the agreed acceptance criteria are passed.

Effective governance over the process helps ensure that the benefits are realized and the risks are mitigated. Although service delivery is transferred—accountability remains with the client organization. SLAs and benchmarking existing performance are absolutely critical if the outsourcing is to be service based rather than just staff augmentation. The ultimate intention is to "commoditize" the service and be in a position to expect service against an agreed level and move to another provider with total continuity of service if the service is not adequate.

IT Audit Involvement

Without the implementation of appropriate controls related to purchasing or outsourcing systems/software, unnecessary damage or disruption to the organization's information could occur. Such damage could result in failure of the organization's critical processes. Control activities should be implemented to reduce risks and address objectives such as the above. The following sections describe the IT audit involvement when auditing both software acquisitions and service organizations.

Auditing Software Acquisitions

A purchased software solution should effectively and efficiently satisfy user requirements. It is also a situation where IT audit may be called upon to provide an external evaluation of the processes and procedures in place and whether the acquisition was in compliance with institutional processes and operating procedures. IT can also be a place where these procedures are lacking and the IT auditor can offer help and suggestions for improvement.

The most common risks associated with acquiring software include the selected solution not being able to satisfy the intended purpose, or that is not technically feasible. The consequences of a poor software purchase are increased costs, missed deadlines, or neglected requirements. To offset these risks, Appendix 9 lists recommended control areas IT auditors should evaluate when auditing software acquisitions.

Auditing Service Organizations

Service organizations (S.O.) are established to offer services to organizations that decide to outsource, for example, their data processing services. The entities that receive services are referred to as user entities. Organizations typically outsource service to third parties, such as payroll processing, cloud computing services, Internet sales, information security services, and IS services. AICPA AU-C 402 (PCAOB 324) requires the auditors of the user entity (user auditors) to obtain an understanding of how the organization uses the services of the S.O., including the nature and significance of the services as well as the effect on internal control. If the user auditors are unable to obtain a sufficient understanding from the user entity, they should obtain that understanding through performing one or more of the following procedures:

- Contact the S.O., through the user entity, to obtain specific information.
- Visit (or engage another auditor to visit) the S.O. and perform necessary procedures about the relevant controls at the S.O.
- Obtain and consider the report of a service auditor (i.e., auditor who examines and reports on the internal control of service organization's controls) on the S.O.'s controls (reports defined below).

The auditors may find that controls implemented by the user entity are effective to ensure that material errors or fraud in transactions are detected. For example, user entity's personnel may generate input control totals and compare them to the S.O.'s output, noting no significant difference. They may also re-perform computer calculations on a test basis. When the results of such controls are effective, the auditors need test only client controls (also referred to as user-based controls) to obtain appropriate evidence; meaning there is no need to perform audit procedures (e.g., tests of controls, etc.) at the service organization. On the other hand, if auditors determine that the controls implemented at the user entity may not be effective in ensuring detection of material errors or transaction fraud, auditors may contact the S.O. and perform procedures to obtain required understanding. In addition, if the audit plan includes a presumption that certain controls operate effectively, the auditors should obtain evidence of their operating effectiveness regardless of whether those controls are applied by the client or by the S.O.

Most S.O. tend to perform similar processing services for numerous clients. If the auditors of each user entity were to visit the S.O., they would ask similar questions and test similar controls. It is therefore common for the S.O. to hire its own audit firm (service auditor) to examine or review controls that are likely to be relevant to user entities' internal controls over financial reporting. The user auditors may then elect to rely on the audit results from the service auditors (in the form of a report) as an alternative or in addition to performing procedures at the S.O. themselves.

The AICPA's Statements on Standards for Attestation Engagements (SSAE) 16, section AT 801 (PCAOB 324), "Reporting on IT Controls at Service Organizations," addresses examination engagements undertaken by a service auditor to report on controls at organizations that provide services to user entities when those controls are likely to be relevant to user entities' internal control over financial reporting. SSAE 16, section AT 801, describes audit considerations relating to entities using a service organization and also presents two types of reports that service auditors may provide:

- *Type 1 Report*: Report on management's description of a S.O.'s system and the suitability of the design of controls.
- *Type 2 Report*: Report on management's description of a S.O.'s system and the suitability of the design and operating effectiveness of controls throughout the period covered by the service auditor's report. A *Type 2 Report* may provide the user auditor with a basis for assessing control risk below the maximum.

User auditors in determining the sufficiency and appropriateness of the audit evidence provided by the service auditor's report, should consider the professional competence of the service auditors and their independence with respect to the S.O. When the user auditor's risk assessment and/or audit requirements includes an expectation that controls at the S.O. operate effectively, the user auditors should either obtain a *Type 2 Report* (if available) or perform appropriate tests of controls to gain assurance that controls at the service organization operate effectively. If the service auditor's report provides an adequate basis for the user auditors' assessment, usually there is no need for the user auditors to perform their own tests at the S.O.

SSAE 18, effective for report dates on or after May 1, 2017, updates SSAE 16 (among other audit standards) by clarifying and formalizing requirements for performing and reporting on the examination, review, and agreed-upon procedures engagements. The following highlights relevant information about the SSAE 18:

■ Provides attestation standards that establish requirements and provide application guidance to auditors for performing and reporting on examination, review, and agreed-upon procedures engagements, including Service Organization Controls (SOC) attestations.

■ Requires the relationship between service organizations and a subservice organizations to be accurately disclosed. This means that service organizations must now identify all subservice organizations used in providing the services, and describe subservice organization controls (if any) relied by the service organization in providing principal services to customers.

■ Requires service organizations to provide service auditors with risk assessment information regarding key internal risks in the organization.

■ Requires service organizations to continuously monitor the effectiveness of relevant controls at the subservice organization. Service auditors must report on the effectiveness of the procedures used by service organizations to monitor relevant controls at the subservice organization. Monitoring may include one or any combination of the following:

   – Reviewing and reconciling output reports or files
   – Periodic discussion with subservice organization personnel
   – Regular site visits
   – Testing controls at the subservice organization
   – Monitoring external communications
   – Reviewing SOC reports of the subservice organization's system

In addition to processes related to financial statements, SSAE 18 will report on an organization's compliance with certain laws or regulations, contractual arrangements, or another set of defined agreed-upon procedures.

SSAE 18's section on Concepts Common to All Attestation Engagements applies to engagements in which a CPA in the practice of public accounting is engaged to issue, or does issue, a practitioner's examination, review, or agreed-upon procedures report on subject matter or an assertion about subject matter that is the responsibility of another party. This section contains performance and reporting requirements and application guidance for all examination engagements. The requirements and guidance in this section supplement the requirements and guidance in section 105, Concepts Common to All Attestation Engagements.

Conclusion

Software acquisitions are often thought to be faster, easier, and cheaper for companies to meet their business needs. Although acquiring software can be very successful, it can also miss the mark. Purchased software can miss user requirements, exceed implementation goals or implementation costs, as well as introduce delays in business or project schedules.

A service management process helps ensure that expected values are received. To be effective, service management is dependent on all areas of IT. Service management is dependent on well-functioning processes in asset management, financial management, service delivery, service desk, problem management, change management, and relationship management. In addition, the customer has the responsibility for ensuring the successful delivery of IT services by communicating, defining requirements, and complying with processes. It is critical that the negotiated contract clearly defined SLAs, performance measures, pricing terms, and escalation processes to effectively manage a third-party supplier. It is also important to have an effective relationship with the supplier and fair contract terms that enable both parties to be successful.

Outsourcing refers to the transfer of service delivery to a third party, allowing companies to concentrate on core competencies. Outsourcing was initially used for standardized payroll and accounting applications, or by companies to sell their hardware. Off-shoring is the transfer of service delivery to a third party outside the client's country. Both outsourcing and off-shoring have been common practice in the various industries for decades. IT outsourcing and off-shoring have become a common business practice for organizations around the globe.

IT auditors can provide an important service by being involved in the acquisition process to highlight risks before the contract is signed. Understanding the software acquisition process, critical contract terms, and supplier service management processes will enable the auditor to be a valuable member of the acquisition team. Auditing can also play an important role by providing an objective review of service management processes. Service expectations can be a very contentious area for an organization due to the trade-offs between service quality and cost. Lastly, the audit of service organizations allows IT auditors to examine whether controls implemented by the user entity are effective to ensure that material errors or fraud in transactions are detected.

Review Questions

1. Why is it important to have a strategy in place? What would be the goal of having such strategy?
2. List the seven basic steps of a software acquisition process.
3. Describe the methods that can be used in gathering system requirements information.
4. What are the advantages and disadvantages for contracted or in-house development?
5. Describe the supplier selection process. What are the basic components of an RFP?
6. Explain what a service-level agreement is. Briefly describe the common types of service-level agreements.
7. When measuring application and infrastructure services, an important measure for both is the number of changes, why?
8. There are many tools available to assist organizations in implementing service management processes. Tools are needed to capture performance, usage metrics from the various platforms, and to consolidate and report on all of this information. Describe common examples of service management tools include.
9. Distinguish between outsourcing and off-shoring.
10. Explain the following relevant terms and concepts when involved in an audit of a service organization.
    a. Service organization.
    b. User entity.
    c. Roles and responsibilities of user auditor.
    d. Roles and responsibilities of service auditor.
    e. Purpose of AICPA's Statements on Standards for Attestation Engagements (SSAE) 16, section AT 801 (PCAOB 324), "Reporting on IT Controls at Service Organizations."
    f. SSAE 16, section AT 801's two types of reports service auditors typically provide.

Exercises

1. Name and summarize control areas that the IT auditor should include in his or her review when examining a software acquisition.
2. As stated in the textbook, outsourcing refers to the transfer of service delivery to a third party, allowing companies to concentrate on core competencies. As the IT Audit Manager, your client asks for advice on outsourcing, specifically whether they should outsource their main financial accounting system. You are well aware of both benefits and risks of outsourcing. Would you advise your client to go ahead and outsource their main financial accounting system? Yes? No? Explain your position.
3. Using an Internet web browser, search for AICPA's Statement on Standards for Attestation Engagements (SSAE) No. 18, and perform the following:
   a. Explain the relevance of SSAE 18 and what does it report on.
   b. Identify advantages of SSAE 18 to auditors.
   c. Contrast SSAE 18 (as appropriate) with SSAE 16.

   Support your reasons and justifications with Audit literature and/or any other valid external source. Include examples as appropriate to evidence your case point. Submit a word file with a cover page, responses to the tasks above, and a reference section at the end. The submitted file should be between 5 and 7 pages long (double line spacing), including cover page and references. Be ready to present your work to the class.

CASE—SOFTWARE ACQUISITION

INSTRUCTIONS: The CFO asked you, IT Audit Manager, to provide guidance to the Payroll Manager (PM) in acquiring software that will automate the process for employees to submit their timesheets. Timesheets are the means by which hourly employees submit their time. Timesheets are approved by PMs, and then processed by payroll department personnel for payment. With the new software, employees will input their time weekly into a computer system. Once employees complete their time sheets, PMs will be able to view and approve them when they log into the system.

TASK: Using the scenario just described, provide answers for the following. You are strongly encouraged to search beyond the chapter (i.e., IT literature and/or any other valid external source) to back up your response. Include examples, as appropriate, to evidence your case point. Submit a word file with a cover page, responses to the task above, and a reference section at the end. The submitted file should be between 6 and 8 pages long (double line spacing), including cover page and references. Be ready to present your work to the class.

   a. What methods would you suggest the PM to gather the requirements for the new system?
   b. How should the PM address and document the system requirements (refer to the outline of the requirements document provided in the chapter).
   c. Which two or three alternative solutions would you recommend the PM to consider.

d. With the assistance of the PM, perform a feasibility analysis using the categories provided in the chapter and one of the alternative solutions described in the previous question.

e. With the assistance of the PM, prepare a risk analysis for the proposed system.

f. Who would you recommend to be on the acceptance testing team?

g. Which tests would you recommend to ensure that system performance requirements are met?

Further Reading

1. Ambrose, C. (2006). *A Sourcing Executive Can Help Optimize Sourcing and Vendor Relationships*, Gartner Research, Gartner, Inc., Stamford, CT.

2. AU-C Section 402. *Audit Considerations Relating to an Entity Using a Service Organization*. www.aicpa.org/Research/Standards/AuditAttest/DownloadableDocuments/AU-C-00402.pdf (accessed September 2017).

3. AU Section 324. *Service Organizations*. https://pcaobus.org/Standards/Auditing/pages/au324.aspx (accessed September 2017).

4. Brown, D. (2013). The SLA conundrum—Executives see green. But everyone else knows it's red inside. www.kpmg-institutes.com/content/dam/kpmg/sharedservicesoutsourcinginstitute/pdf/2012/service-level-agreement-conundrum.pdf

5. Bakalov, R. and Nanji, F. (2007). Offshore application development done right. *Inf. Syst. Control J.*, 5.

6. Benvenuto, N. and Brand, D. (2007). Outsourcing—A risk management perspective. *Inf. Syst. Control J.*, 5.

7. Corporate Executive Board. *Case Studies of Software Purchasing Decisions*, Working Council for Chief Information Officers, February 2013.

8. Deloitte's 2016 Global Outsourcing Survey. Step on it! Outsourcing makes a beeline toward innovation. www2.deloitte.com/us/en/pages/operations/articles/global-outsourcing-survey.html

9. Deloitte's 2014 Global Outsourcing and Insourcing Survey. 2014 and beyond. www2.deloitte.com/content/dam/Deloitte/us/Documents/strategy/us-2014-global-outsourcing-insourcing-surveyreport-123114.pdf

10. Doig, C. 2016. The enterprise software acquisition funnel. *CIO*. www.cio.com/article/3087545/software/the-enterprise-software-acquisition-funnel.html

11. Doig, C. 2016. The payoff from a rigorous software selection. *CIO*. www.cio.com/article/3091810/software/the-payoff-from-a-rigorous-software-selection.html

12. Edmead, M. 2015. Using COBIT 5 to measure the relationship between business and IT. ISACA. www.isaca.org/COBIT/focus/Pages/using-cobit-5-to-measure-the-relationship-between-business-and-it.aspx

13. IT Governance Institute. *Governance of Outsourcing*, IT Governance Domain Practices and Competencies, 2005.

14. Kennedy, C. (July 25, 2017). SSAE 18 vs SSAE 16: Key differences in the new SOC 1 standard. Online Tech. http://resource.onlinetech.com/ssae-18-vs-ssae-16-key-differences-in-the-new-soc-1-standard/

15. KPMG's State of the outsourcing, shared services, and operations industry 2017. HfS Research. www.kpmg-institutes.com/content/dam/kpmg/sharedservicesoutsourcinginstitute/pdf/2017/business-operations-2017-hfs.pdf

16. Kyte, A. (2005). *Vendor Management Is a Critical Business Discipline*, Gartner Research, Gartner, Inc., Stamford, CT.

17. Moreno, H. 2016. How IT service management delivers value to the digital enterprise. *Forbes*. www.forbes.com/sites/forbesinsights/2017/03/16/how-it-service-management-delivers-value-to-the-digital-enterprise/#54ff3bff732e

18. Romney, M.B. and Steinbart, P.J. (2015). *Accounting Information Systems*, 13th Edition, Pearson Education, Upper Saddle River, NJ.
19. Senft, S., Gallegos, F., and Davis, A. (2012). *Information Technology Control and Audit*. CRC Press/ Taylor & Francis, Boca Raton.
20. Singleton, T.W. 2013. How to properly audit a client who uses a service organization—SOC report or no SOC report. ISACA. www.isaca.org/Journal/archives/2013/Volume-1/Pages/How-to-Properly-Audit-a-Client-Who-Uses-a-Service-Organization-SOC-Report-or-No-SOC-Report.aspx
21. SSAE-18—An update to SSAE 16 (Coming 2017). SSAE-16. www.ssae-16.com/ssae-18-an-update-to-ssae-16-coming-2017/ (accessed September 2017).
22. AICPA. Statements on standards for attestation engagements. www.aicpa.org/Research/Standards/AuditAttest/Pages/SSAE.aspx (accessed September 2017).
23. Deloitte. The vendor management program office. Five deadly sins of vendor management. www2.deloitte.com/us/en/pages/operations/articles/vendor-management-program-office-five-deadly-sins-ofvendor-management.html (accessed September 2017).
24. Whittington, O.R. and Pany, K. (2014). *Principles of Auditing & Other Assurance Services*, 20th Edition. McGraw-Hill/Irwin, Boston.
25. Xerox gives EDS $3.2 billion contract. UPI Archives. www.upi.com/Archives/1994/06/14/Xerox-gives-EDS-32-billion-contract/2209771566400/ (accessed September 2017).

Appendix 1: IT Planning Memo

Memo

Date:	[Date]
To:	The Financial Statement Audit File
From:	[IT Auditor Representative], [Office Location]
Subject:	IT Audit Planning

Purpose

The purpose of this memo is to outline the procedures associated with the involvement of the Information Technology Auditors ("IT Auditors") in connection with the financial statement audit ("financial audit") of [company name] (["company abbreviated name" or "the Company"]) for the year [ending or ended] [Month XX, 20XX]. The approach for the IT audit outlined herein serves as a supplement to the financial audit planning memorandum and should be reviewed in conjunction with such working paper.

Planning Discussions

(*The planning meeting between the financial audit team and the IT audit team should be documented in this planning memo. Modify the sections below as applicable.*)

As detailed in the working paper [*working paper reference number*], a discussion with the financial audit Partner, Principal, or Director was held to determine the level of IT audit involvement. (*If an IT auditor has already been involved in the audit, describe previous involvement and/or any relevant planning discussions herein.*) During this planning meeting, risk assessments of areas to be addressed were also discussed along with the nature, extent, and timing of planned tests of controls described further in this planning memo.

IT Audit Team

The IT audit team will consist of the following:

Role	Name
Partner, Principal, or Director	
Manager or Senior Manager	
Senior	
Staff	

Timing

Timing of the IT audit work is scheduled as follows:

1. Planning (starting [*MM/DD/YY*], ending [*MM/DD/YY*])
2. Interim (starting [*MM/DD/YY*], ending [*MM/DD/YY*])
3. Year end (starting [*MM/DD/YY*], ending [*MM/DD/YY*])
4. Sign-off date ([*MM/DD/YY*])

Hours

Hours and costs are based on the estimated time required to complete the IT audit procedures and the level of experience required. Detailed IT audit procedures have been planned with the financial audit team, including discussions regarding the necessary documentation and assistance to be provided by the Company to facilitate the effective and efficient performance of the procedures.

It is estimated that the IT audit procedures will take [##] hours to complete.

The hours incurred are to be charged to: _____ [*Company charge code/number*].

During the course of the IT audit, circumstances encountered that could significantly affect the performance of such audit procedures will be promptly notified to the financial audit team and Company personnel, as appropriate, including any additional hours resulting from such circumstances.

Understand the IT Environment

Meetings with Company personnel will take place in order to gather or update the existing understanding of the IT environment, including significant changes from the prior year. This understanding will be considered as part of the planning process and documented in working paper [*working paper reference number*].

Relevant Applications and Technology Elements

As agreed with the financial audit team, applications are classified as relevant to the audit when they:

- are used to support a critical business process (e.g., revenues, expenditures, payroll, etc.)
- have information generated by the organization (IGO) that is significant for a financial audit test procedure or in the context of any internal controls, such as information used to test a relevant control activity or information used by the Company to perform the control activity
- include application or automated control activities that have been identifying as addressing significant financial audit risks

Relevant applications and their related technology elements have been identified on the following table or documented at [*working paper reference number*].

Relevant Application	Database	Operating System	Network

IT Risks and Controls

IT risks have been identified on the relevant applications based on the understanding obtained from (1) the IT environment, (2) existing application controls, and (3) IGO. Certain control activities will be assessed to determine whether they are adequately designed and operate effectively to address those risks. Refer to working paper [*working paper reference number*] where such controls have been identified and listed.

Relevant Application Controls

In addition to the general control IT areas (information systems operations, information security, and change control management), the IT audit team will test certain relevant application controls. Meetings between the IT audit team and appropriate members of the financial audit team will occur to:

1. understand how application or automated controls work
2. evaluate if they have been adequately designed and implemented
3. assess whether they operate effectively

The relevant application controls to be tested are noted below.

Working Paper Reference #	Relevant Application	Relevant Application Control

Information Generated by the Organization

IGO has been identified and classified as significant for an audit test procedure or in the context of any internal controls. This means that certain information will be used as part of various audit tests of controls and/or organization personnel will use such to perform controls. Given the relevance of this information, the IT audit will include procedures to assess its accuracy and completeness.

Deficiency Evaluation

If deviations or findings result from the IT test procedures performed, they will be assessed to determine their nature and cause, and whether they represent a control deficiency. Evaluation of control deficiencies will be performed in conjunction with the financial audit team. Refer to working paper [*working paper reference number*], where such evaluation will be documented.

Work of Others

(*The work of others may include work from internal auditors, Company personnel (in addition to internal auditors), and third parties. The sample language below focuses on internal audit, and should be tailored if the work of others is utilized.*)

The IT audit team is planning to rely upon the Company's Internal Audit (IA) function to support the IT control procedures. (*This language should be altered if IA will be used in a "direct assistance" capacity versus using IA's own work.*)

If reliance will be placed on certain audit areas performed by IA personnel, the IT audit team will assess and document the competence and objectivity of such IA personnel whose work will be relied upon in order to determine the extent to which such work can be used.

To determine the quality and effectiveness of specific work performed by the internal auditors, the following will be assessed:

- whether the IA work is appropriate to meet the audit objectives
- whether the IA audit work program is adequate and complete
- whether the IA work documentation is acceptable in quality and quantity
- whether results and conclusions are appropriate and consistent with the IA work

Evaluation of Service Organization Controls

(*This section is applicable if there are external service organizations that perform services or general controls relevant for the audit.*)

A service auditor's report will be obtained for the relevant general controls related to the [*relevant application(s)*] application(s) performed by [*name of service organization*]. A review of the report will be performed by the IT audit team to understand the relevant services provided by the service organization. Specifically, the IT audit team will evaluate the service organization controls by:

- assessing the IT controls and related exceptions in the report

■ documenting the IT complementary or locally based user controls specified in the report (*These controls are implemented in the Company and, thus, are not part of the service organization; however, they complement service organization controls. The IT auditor typically document these controls by tying them to the IT audit work performed as part of the IT audit of general controls IT areas.*)

(The table below can be included to summarize information about the relevant service organizations.)

Service Organization	Brief Description of Relevant Service(s) Provided	Service Organization Location	Service Auditor	Report Period	Report Type/ Conclusion

Other Areas of IT Audit Assistance

(This section includes other areas where IT auditors may provide assistance to the financial audit team, including, but not limited to, fraud assistance, tests of business/financial controls, tests of IT entity-level controls, etc.)

Appendix 2: Understanding the IT Environment

Understanding the Organization's Information Technology Environment
[Period under Audit]

The name of the information technology (IT) environment is also the name of the underlying operating system(s) hosting the relevant financial application(s).

IT Environment	

Risks

IT poses specific risks to an organization's internal control, including, for example, unauthorized disclosure of confidential data; unauthorized processing of information; inappropriate manual intervention; system crashes; unauthorized modification of sensitive information; theft or damage to hardware; and loss/theft of information, among others.

Controls

There are two broad groupings of IT controls, both of which are essential to address risks and to ensure the continued and proper operation of information systems. These are as follows:

- *General computer controls or General controls.* They include policies and procedures that relate to applications and support the effective functioning of application controls. General controls cover the IT infrastructure and support services, including all systems and applications. General controls commonly comprise controls over IT areas such as (1) information systems operations, (2) information security, and (3) change control management.
- *Application controls.* These may also be referred to as "automated controls," and apply to the processing controls specific and unique to applications. Application controls are concerned with the accuracy, completeness, validity, and authorization of the data captured, entered, processed, stored, transmitted, and reported.

Relevant Applications

Document the relevant applications associated with the IT environment. Applications to be included in this table are those that impact the generation of financial information (i.e., financial statements).

Application Name(s)	Application Server Name/ Operating System Version	Application Source (e.g., in-house developed, purchased software, hosted by a service organization, etc.)	Related Database Name(s)	Database Server Name/ Operating System Version	Location

Significant Modifications to Applications

Describe significant modifications for the relevant applications listed above, if any, during the period under audit.

Application Name(s)	Description and Date of Significant Modification

Service Organizations

Document information on service organizations related to the IT environment.

Name and Location	Brief Description of Relevant Service(s) Provided	Related Relevant Application to which Services are Performed For

Organization and Personnel

Document whether the organization's approach to information systems and related support activities is centralized or decentralized.

Centralized or Decentralized?	Explain

Document number of staff within the IT department and names and titles of key personnel. Include copy of the IT organization chart, if available.

Number of staff within the IT department	
Names and titles of key IT personnel	

General Computer Controls

Information Systems Operations

The area of information systems operations includes control activities such as data backups and offsite storage, job scheduling, job monitoring and tracking of exceptions, and physical access.

Backups	
1. Describe the existing backup process to protect relevant application information.	
2. Document the name(s) of backup tool(s) used.	
3. Describe the frequency and the type(s) of backup(s) performed.	
4. What is the backups' frequency of rotation to offsite facilities? Briefly describe the offsite facility.	

Automated Job Scheduling Tool	
1. Is there an automated tool used to execute regularly scheduled jobs related to applications, databases, and operating systems, such as scheduled interfaces of data, data purges, table updates, etc.?	
2. Name the automated job scheduling tool used and describe the types of jobs scheduled.	
3. How are changes to the automated job scheduling tool, such as adding, modifying, and deleting jobs and schedules made?	
4. Who can make changes to the automated job scheduling tool, such as adding, modifying, and deleting jobs and schedules?	

Job Processing	
1. Describe the process used to monitor the successful completion of job processing.	
2. Document how such monitoring and review process ensures that exceptions and/or failures identified during job processing are timely resolved.	
3. Name the personnel responsible for the process review and exception tracking mentioned above.	
4. Name the reports that are reviewed and the notification systems and mechanisms in place.	

Physical Security	
1. What methods (e.g., access cards, biometrics, traditional lock and key, security guards, etc.) does the organization employ to restrict physical access to this IT environment, including the data center or computer room, as well as other computing areas where intruders could access information resources? If biometrics authentication is employed, specify whether authentication is through fingerprint, palm veins, face recognition, iris recognition, retina scans, voice verification, etc.	
2. Which users have access to the data center?	
3. Describe the policies and procedures in place for granting access to the data center. Are requests and approvals required and completed before such access is granted?	
4. For an employee that leaves the organization or is transferred to a different department, describe the process in place for changing his/her access to the data center. Consider (1) naming the personnel involved; (2) how are they notified to remove such access to the data center; and (3) how timely access is changed to reflect the employee's new status.	
5. Are user access reviews conducted to support current physical access granted to this IT environment, including the data center that host relevant financial applications, related databases, operating systems, and other repositories for financial information? If so, describe the process and frequency.	

Information Security

The area of information security includes control activities such as security awareness policies and procedures; access requests and user account administration; access terminations; user access reviews; operating system, application, and database security administration (i.e., password parameters); and physical security.

Information Security Policies and Procedures	
1. Are users aware of the information security policies and procedures in place at the organization? If so, how?	
2. Are information security policies and procedures formally written? If so, list the names of the information security policies, procedures, and practices, or add a reference where they can be found.	

Access Administration	
1. For each application listed under the *Relevant Applications* section, document the following: a. Name of personnel responsible for authorizing and modifying user access to information. b. Authorization method for adding and modifying user account access. Is it performed electronically, via a manual form with signature, verbally, etc.?	
2. Describe the process of adding and modifying user account access for the database, operating system, and network infrastructure related to the relevant applications.	
3. Is the process of adding and modifying user account access within relevant applications, databases, operating systems, and networks different for contractors and temporary employees? If so, how?	
4. When employees leave the organization or are transferred within the organization: a. Who notifies IT system administrators? Is there an automated notification mechanism in place (e.g., workflow, etc.) or reliance on Human Resources personnel or the employee's manager to contact the appropriate parties? b. What is the time frame required for notification (e.g., immediately, daily, weekly, monthly, etc.)? c. What is the method of communication to IT system administrators (e.g., via e-mail, phone call, manual forms, etc.)?	
5. Are contractors and temporary employees covered under the termination procedure described above? If not, describe the procedures currently in place.	
6. Is ownership of data explicitly defined? If so, explain how.	

User Access Reviews	
1. Are user access reviews conducted for all relevant applications, related databases, and operating systems, as well as other repositories for financial information? If so, describe the process and frequency.	

Authentication Techniques and Parameters	
1. How do users authenticate or gain access to relevant financial applications? Is it through various layers (i.e., login first to the network, then to the operating system, and finally to the application), or is it through single sign-on? Describe.	

2. For each relevant application, related database, operating system, and network, specify the following system-enforced authentication parameters:

Relevant Application/ Operating System/ Database/ Network (also include the Virtual Private Network (VPN), if applicable)	Enforce Password History (number of passwords remembered)	Minimum Password Age (change interval every 30, 60, 90 days, etc.)	Minimum Password Length (number of characters)	Password Complexity (enabled/ disabled)	Account Lockout (specify number of invalid login attempts)	Other

Note: Typically, a password policy implemented across all systems and applications includes the following recommended minimum logical security authentication parameters:

- Enforce password history of six passwords remembered or greater.
- Minimum password age (or expiration interval) between 30 and 90 days.
- Minimum password length of six characters or greater.
- Password complexity should be enabled to the extent possible for all systems and applications.
- Account lockout between three and five invalid login attempts before accounts lock out.

Access to Information	

1. Does the organization allow external access to or from its computer systems (via external networks)? If so, document:
  a. Who has such access and what is the purpose of such access?
  b. What are the methods used to restrict such access?	

2. Does the organization transmit information across external networks such as the Internet?	
3. Is sensitive or critical data encrypted while being transmitted? If so, how?	
4. Can financial applications or other financial-related information be accessed from remote locations?	
5. Describe methods used to protect the organization's financial applications from unauthorized access from remote locations. If a VPN is used, describe authentication requirements.	

6. Does the organization use a firewall? If so, document the following:
  a. Name of the firewall and its location.
  b. What is the firewall function (one way, two way, or proxy)?
  c. How is the firewall configured, used, and managed?	

7. Can financial applications be accessed wirelessly? If so, describe the methods used to secure such wireless connection.	
8. What is the name of the organization's Internet Website? If applicable, specify whether the site allows users to place orders or pay for goods or services, and where it is hosted (e.g., locally, by a third-party service provider, etc.)	

Change Control Management

The area of change control management includes control activities related to changes implemented in relevant applications, including request approval, prioritization, auditing, and review; implementation of upgrades and patches to operating systems; implementation of applications and databases, including installation of upgrades; and monitoring, security, and change management for the network infrastructure.

Applications

Purchased Applications	
1. Describe the process of selecting and purchasing new applications. Consider the roles and responsibilities of individuals involved in the process.	
2. How are purchased applications customized?	
3. How are new applications or new releases to applications tested and approved prior to implementation from development/test environments into the production environment?	
4. Following implementation, how information is validated for integrity, accuracy, and completeness purposes.	

In-House Developed Applications and Program Changes	
1. Describe the organization's process for developing in-house applications, including the systems development methodologies used and the roles and responsibilities of the individuals involved.	
2. Describe the organization's process for implementing program changes to in-house developed applications. Consider the following: a. How are program changes tested? b. Following testing, how are program changes approved for implementation into the production environment? c. How is the implementation into the production environment performed? Who moves the change into the production environment? d. Following implementation, how is information validated for integrity, accuracy, and completeness purposes.	
3. Do programmers have access to modify code or production data directly in the production environment for the relevant applications? Explain.	
4. Do programmers have access to migrate their own changes from test or development environments into the production environment? If so, why?	
5. Describe the process to maintain version control and to prevent unauthorized changes from being implemented into the production environment.	

Databases

1. Describe the process for acquiring, implementing, and maintaining databases. Consider the roles and responsibilities of individuals involved in this process.	
2. Document the database architecture for the relevant applications (e.g., integrated database used by all relevant applications, multiple separate databases, etc.)?	
3. Name the database management software (e.g., Oracle, IBM DB2, Proprietary, etc.) used by each relevant application.	
4. Does the organization maintain data dictionaries with definitions and representations of the data elements stored in the databases? Describe how the data dictionaries are used. *Definitions and representations of the data may include integrity constraints, stored procedures, general database structure, space allocations, etc.*	

Networks

1. List the relevant applications supported by the network. Include copy of the network diagram or a graphical chart of the network.	
2. Describe the process of implementing configuration changes, upgrades, and/or new network software, including their approval and testing.	

Operating Systems

1. Describe the process for acquiring, implementing, and maintaining operating systems. Consider the roles and responsibilities of individuals involved in this process.	
2. How does the organization assess the impact of implementing new (or modifying existing) operating systems to host the relevant applications?	
3. How are new operating systems or modifications to existing operating systems tested and approved prior to migration into production environments?	
4. Following implementation, how is information validated for integrity, accuracy, and completeness purposes.	

Application Controls

Application controls may also be referred to as "automated controls," and apply to the processing controls specific and unique to applications. Application controls are concerned with the accuracy, completeness, validity, and authorization of the data captured, entered, processed, stored, transmitted, and reported. Describe the application controls currently implemented at the organization. Application controls may include, among others:

- System and/or application configuration controls
- Security-related controls enforcing user access, roles, and segregation of duties
- Automated notification controls to alert users that a transaction or process is awaiting their action
- Automated mathematical calculations to prevent errors
- Input validation checks for data accuracy

Other Information

Electronic Commerce/Emerging Technologies

1. Describe how the organization uses electronic commerce.	
2. Have emerging technologies such as cloud computing, etc. implemented at the organization? If so, describe how? If not, does the organization have plans to implement emerging technologies in the near future?	

Information Generated by the Organization

1. Does the organization use report writer software to create customized reports from relevant application data? If so, name the report writer software. *Customized reports can be generated using database query tools, or by capturing data from a data warehouse.*	
2. Describe the purpose for using the report writer software above, and name the users with access to such report writer software.	
3. Are changes to customized reports in compliance with the change control management process documented within this form? If not, describe current process.	

Future Plans

1. Does the organization plan to upgrade or replace existing relevant applications, databases, network, and/or operating systems in the near future? If so, describe such plans.	

Previous IT Work

1. Has there been any IT-related work performed recently by consultants, internal/external auditors, etc. that may significantly change the understanding documented within this form? If so, describe such work performed (e.g., nature and scope of the work, period covered, results achieved, etc.).	

Appendix 3: Sample IT Audit Programs for General Control IT Areas

IT Audit Program for Information Systems Operations

Batch and Online Processing, Backups, and Physical Access

Audit Control Objective

ISO 1.00 - IT operations support adequate scheduling, execution, monitoring, and continuity of systems, programs, and processes to ensure the complete, accurate, and valid processing and recording of financial transactions.

Audit Control Activities

ISO 1.01 - Batch and/or online processing is defined, timely executed, and monitored for successful completion.

ISO 1.02 - Exceptions identified on batch and/or online processing are timely reviewed and corrected to ensure accurate, complete, and authorized processing of financial information.

Audit Control Objective

ISO 2.00 - The storage of financial information is appropriately managed, accurate, and complete.

Audit Control Activities

ISO 2.01 - Procedures for the restoration and recovery of financial information from backups have been implemented in the event of processing disruption, shutdown, and restart procedures consistent with IT policies and procedures.

ISO 2.02 - Automated backup tools have been implemented to manage retention data plans and schedules.

ISO 2.03 - Backups are properly labeled, stored in an off-site secured environmentally location, and rotated to such facility on a periodic basis.

ISO 2.04 - Tests for the readability of backups are performed on a periodic basis. Results support timely and successful restoration of backed up data.

Audit Control Objective

ISO 3.00 - Physical access is appropriately managed to safeguard relevant components of the IT infrastructure and the integrity of financial information.

Audit Control Activities

ISO 3.01 - A physical access control mechanism is used to restrict and record access to the building and to the computer room (i.e., data center), and authority to change such mechanism is limited to appropriate personnel.

ISO 3.02 - Physical access is authorized, monitored, and restricted to individuals who require such access to perform their job duties. Entry of unauthorized personnel is supervised and logged. The log is maintained and regularly reviewed by IT management.

IT Audit Program for Information Security

Security Administration Function, Security Policies and Procedures, Security Software Tools and Techniques, Unique User Identifiers and Passwords, Security Administrator and Privilege Access, and Information Security Logs

Audit Control Objective

ISEC 1.00 - Security configuration of applications, databases, networks, and operating systems is adequately managed to protect against unauthorized changes to programs and data that may result in incomplete, inaccurate, or invalid processing or recording of financial information.

Audit Control Activities

ISEC 1.01 - The security administration function should be separate from the IT function.

ISEC 1.02 - Formal policies and procedures define the organization's information security objectives and the responsibilities of employees with respect to the protection and disclosure of informational resources. Management monitors compliance with security policies and procedures, and agreement to these are evidenced by the signature of employees.

ISEC 1.03 - Security-related software tools and techniques are in place to restrict and segregate access to sensitive IT functions (e.g., programming, administrative functions, implementation of changes in production environments, etc.) within the systems. Changes related to access are assessed by management for adequate segregation of duties.

ISEC 1.04 - Implementation and configuration of security software tools and techniques are reviewed and approved by management.

ISEC 1.05 - Unique user identifiers have been assigned to users as required on the information security policies and procedures, to distinguish them and to enforce accountability.

ISEC 1.06 - Consistent with information security policies and procedures, local and remote users are required to authenticate to applications, databases, networks, and operating systems via passwords to enhance computer security.

ISEC 1.07 - Passwords must promote acceptable levels of security (consistent with policies and/or best industry practices) by enforcing confidentiality and a strong password format.

ISEC 1.08 - Vendor-supplied passwords embedded on the applications, databases, networks, and operating systems are modified, eliminated, or disabled to avoid security vulnerabilities from being exploited in the systems.

ISEC 1.09 - Administrator, privileged, or super-user account access within the systems are limited to appropriate personnel. Changes to these accounts (e.g., system security parameters, security roles, security configuration over systems, etc.) are logged and reviewed by management.

ISEC 1.10 - Information security logs are configured and activated on applications, databases, networks, and operating systems to record and report security events consistent with information security policies and procedures.

ISEC 1.11 - Reports generated from information security logs (e.g., security violation reports, unauthorized attempts to access information, etc.) are frequently reviewed and acted upon as necessary.

Audit Control Objective

ISEC 2.00 - Adequate security is implemented to protect against unauthorized access and modifications of systems and information, which may result in the processing or recording of incomplete, inaccurate, or invalid financial information.

Audit Control Activities

ISEC 2.01 - Training programs have been established for all personnel within the following areas:

- Organizational security policies
- Disclosure of sensitive data
- Access privileges to IT resources
- Reporting of security incidents
- Naming conventions for user passwords

ISEC 2.02 - System owners authorize user accounts and the nature and extent of their access privileges.

ISEC 2.03 - User account access privileges are periodically reviewed by systems and application owners to determine whether they are reasonable and/or remain appropriate.

ISEC 2.04 - Users who have changed roles or tasks within the organization, or that have been transferred, or terminated are immediately informed to the security department for user account access revision in order to reflect the new and/or revised status.

ISEC 2.05 - Transmission of sensitive information is encrypted consistent with security policies and procedures to protect its confidentiality.

IT Audit Program for Change Control Management

Risk Evaluation, Documentation, Change Request Authorization, Testing, Approval of System Changes, and Implementation into Production Environment

Audit Control Objective

CCM 1.00 - Changes implemented in applications, databases, networks, and operating systems (altogether referred to as "system changes") are assessed for risk, authorized, and thoroughly documented to ensure desired results are adequate.

Audit Control Activities

CCM 1.01 - The business risks and impact of proposed system changes are assessed by management before implementation into production environments. Assessment results are used when designing, staffing, and scheduling implementation of changes to minimize disruptions to operations.

CCM 1.02 - Requests for system changes (e.g., upgrades, fixes, emergency changes, etc.) are documented and approved by management before any change-related work is done.

CCM 1.03 - Documentation related to the change implementation is adequate and complete.

CCM 1.04 - Change documentation includes the date and time at which changes were (or will be) installed.

CCM 1.05 - Documentation related to the change implementation has been released and communicated to system users.

Audit Control Objective

CCM 2.00 - Changes implemented in applications, databases, networks, and operating systems (altogether referred to as "system changes") are appropriately tested. Tests are performed by a group other than the group responsible for the system (e.g., operating systems changes are implemented by someone other than the systems programmer, etc.).

Audit Control Activities

CCM 2.01 - System changes are tested before implementation into the production environment consistent with test plans and cases.

CCM 2.02 - Test plans and cases involving complete and representative test data (instead of production data) are approved by application owners and development management.

CCM 2.03 - A sample of system changes is selected for the period under audit to determine whether documentation supporting the change:

1. is in accordance with installation standards;
2. provides a clear explanation of the change made and the reason for the change; and
3. has been appropriately reviewed and approved by management.

CCM 2.04 - A sample of system changes is selected for the period under audit to determine whether test plans and cases:

1. are in compliance with installation standards;
2. thoroughly tested the implemented change;
3. were reviewed and properly approved; and
4. were tested in a protective environment separate from the production environment.

CCM 2.05 - The names and titles of personnel responsible for implementing system changes are identified. Access to the development or test environments is separate and appropriately restricted to the live or production environment (i.e., adequate separation of duties).

Audit Control Objective

CCM 3.00 - Changes implemented in applications, databases, networks, and operating systems (altogether referred to as "system changes") are appropriately managed to reduce disruptions, unauthorized alterations, and errors which impact the accuracy, completeness, and valid processing and recording of financial information.

Audit Control Activity

CCM 3.01 - Problems and errors encountered during the testing of system changes are identified, corrected, retested, followed up for correction, and documented.

Audit Control Objective

CCM 4.00 - Changes implemented in applications, databases, networks, and operating systems (altogether referred to as "system changes") are formally approved to support accurate, complete, and valid processing and recording of financial information.

Audit Control Activities

CCM 4.01 - Prior to implementation of system changes in live and production environments, documentation of formal acceptance is obtained supporting that testing has been satisfactorily completed, test results were successful and adequate to prevent tampering, and user requirements were met.

CCM 4.02 - Personnel independent from those with access to the development or test environments review changes and deploy them into the live or production environment.

CCM 4.03 - Procedures such as retaining a prior version of the original environment are in place to allow the recovery of such original environment in the event there are problems resulting from the implementation of system changes.

CCM 4.04 - An overall review is performed by management after system changes have been implemented in the live or production environment to determine whether the objectives for implementing system changes were met.

Note: Below are sample IT audit program templates. Do not forget to document at the top of each IT audit program the name of the organization and the period under audit.

IT Area: *INFORMATION SYSTEMS OPERATIONS*

Topics Covered: *Batch and Online Processing, Backups, and Physical Access*

Control Objective	Control Activity	Description of Audit Procedures Performed or Reference to Work Paper(s) Where Procedures Have Been Documented	Effective/Ineffective (if ineffective, refer to the findings work paper)
ISO 1.00 - IT operations support adequate scheduling, execution, monitoring, and continuity of systems, programs, and processes to ensure the complete, accurate, and valid processing and recording of financial transactions.	ISO 1.01 - Batch and/or online processing is defined, timely executed, and monitored for successful completion.		
	ISO 1.02 - Exceptions identified on batch and/or online processing are timely reviewed and corrected to ensure accurate, complete, and authorized processing of financial information.		
ISO 2.00 - The storage of financial information is appropriately managed, accurate, and complete.	ISO 2.01 - Procedures for the restoration and recovery of financial information from backups have been implemented in the event of processing disruption, shutdown, and restart procedures consistent with IT policies and procedures.		
	ISO 2.02 - Automated backup tools have been implemented to manage retention data plans and schedules.		

(Continued)

IT Area: *INFORMATION SYSTEMS OPERATIONS*		
Topics Covered: *Batch and Online Processing, Backups, and Physical Access*		
Control Objective	**Control Activity**	**Description of Audit Procedures Performed or Reference to Work Paper(s) Where Procedures Have Been Documented**	**Effective/Ineffective** (if ineffective, refer to the findings work paper)
	ISO 2.03 - Backups are properly labeled, stored in an off-site secured environmentally location, and rotated to such facility on a periodic basis.		
	ISO 2.04 - Tests for the readability of backups are performed on a periodic basis. Results support timely and successful restoration of backed up data.		
ISO 3.00 - Physical access is appropriately managed to safeguard relevant components of the IT infrastructure and the integrity of financial information.	ISO 3.01 - A physical access control mechanism is used to restrict and record access to the building and to the computer room (i.e., data center), and authority to change such mechanism is limited to appropriate personnel.		
	ISO 3.02 - Physical access is authorized, monitored, and restricted to individuals who require such access to perform their job duties. Entry of unauthorized personnel is supervised and logged. The log is maintained and regularly reviewed by IT management.		

(Continued)

	IT Area: INFORMATION SECURITY		
Topics Covered: Security Administration Function, Security Policies and Procedures, Security Software Tools and Techniques, Unique User Identifiers and Passwords, Security Administrator and Privilege Access, and Information Security Logs			
Control Objective	***Control Activity***	***Description of Audit Procedures Performed or Reference to Work Paper(s) Where Procedures Have Been Documented***	***Effective/Ineffective*** *(if ineffective, refer to the findings work paper)*
ISEC 1.00 - Security configuration of applications, databases, networks, and operating systems is adequately managed to protect against unauthorized changes to programs and data that may result in incomplete, inaccurate, or invalid processing or recording of financial information.	ISEC 1.01 - The security administration function should be separate from the IT function.		
	ISEC 1.02 - Formal policies and procedures define the organization's information security objectives and the responsibilities of employees with respect to the protection and disclosure of informational resources. Management monitors compliance with security policies and procedures, and agreement to these are evidenced by the signature of employees.		

(Continued)

IT Area: *INFORMATION SECURITY*

Topics Covered: *Security Administration Function, Security Policies and Procedures, Security Software Tools and Techniques, Unique User Identifiers and Passwords, Security Administrator and Privilege Access, and Information Security Logs*

Control Objective	Control Activity	Description of Audit Procedures Performed or Reference to Work Paper(s) Where Procedures Have Been Documented	Effective/Ineffective (if ineffective, refer to the findings work paper)
	ISEC 1.03 - Security-related software tools and techniques are in place to restrict and segregate access to sensitive IT functions (e.g., programming, administrative functions, implementation of changes in production environments, etc.) within the systems. Changes related to access are assessed by management for adequate segregation of duties.		
	ISEC 1.04 - Implementation and configuration of security software tools and techniques are reviewed and approved by management.		
	ISEC 1.05 - Unique user identifiers have been assigned to users as required on the information security policies and procedures, to distinguish them and to enforce accountability.		

(Continued)

	IT Area: *INFORMATION SECURITY*		
	Topics Covered: *Security Administration Function, Security Policies and Procedures, Security Software Tools and Techniques, Unique User Identifiers and Passwords, Security Administrator and Privilege Access, and Information Security Logs*		
Control Objective	**Control Activity**	**Description of Audit Procedures Performed or Reference to Work Paper(s) Where Procedures Have Been Documented**	**Effective/Ineffective** *(if ineffective, refer to the findings work paper)*
	ISEC 1.06 - Consistent with information security policies and procedures, local and remote users are required to authenticate to applications, databases, networks, and operating systems via passwords to enhance computer security.		
	ISEC 1.07 - Passwords must promote acceptable levels of security (consistent with policies and/or best industry practices) by enforcing confidentiality and a strong password format.		
	ISEC 1.08 - Vendor-supplied passwords embedded on the applications, databases, networks, and operating systems are modified, eliminated, or disabled to avoid security vulnerabilities from being exploited in the systems.		

(Continued)

IT Area: *INFORMATION SECURITY*

Topics Covered: *Security Administration Function, Security Policies and Procedures, Security Software Tools and Techniques, Unique User Identifiers and Passwords, Security Administrator and Privilege Access, and Information Security Logs*

Control Objective	Control Activity	Description of Audit Procedures Performed or Reference to Work Paper(s) Where Procedures Have Been Documented	Effective/Ineffective (if ineffective, refer to the findings work paper)
	ISEC 1.09 - Administrator, privileged, or super-user account access within the systems are limited to appropriate personnel. Changes to these accounts (e.g., system security parameters, security roles, security configuration over systems, etc.) are logged and reviewed by management.		
	ISEC 1.10 - Information security logs are configured and activated on applications, databases, networks, and operating systems to record and report security events consistent with information security policies and procedures.		
	ISEC 1.11 - Reports generated from information security logs (e.g., security violation reports, unauthorized attempts to access information, etc.) are frequently reviewed and acted upon as necessary.		

IT Area: INFORMATION SECURITY

Topics Covered: *Security Administration Function, Security Policies and Procedures, Security Software Tools and Techniques, Unique User Identifiers and Passwords, Security Administrator and Privilege Access, and Information Security Logs*

Control Objective	Control Activity	Description of Audit Procedures Performed or Reference to Work Paper(s) Where Procedures Have Been Documented	Effective/Ineffective (if ineffective, refer to the findings work paper)
ISEC 2.00 - Adequate security is implemented to protect against unauthorized access and modifications of systems and information, which may result in the processing or recording of incomplete, inaccurate, or invalid financial information.	ISEC 2.01 - Training programs have been established for all personnel within the following areas: • Organizational security policies • Disclosure of sensitive data • Access privileges to IT resources • Reporting of security incidents • Naming conventions for user passwords		
	ISEC 2.02 - System owners authorize user accounts and the nature and extent of their access privileges.		
	ISEC 2.03 - User account access privileges are periodically reviewed by systems and application owners to determine whether they are reasonable and/or remain appropriate.		

(Continued)

		IT Area: *INFORMATION SECURITY*	
Topics Covered: *Security Administration Function, Security Policies and Procedures, Security Software Tools and Techniques, Unique User Identifiers and Passwords, Security Administrator and Privilege Access, and Information Security Logs*			
Control Objective	**Control Activity**	**Description of Audit Procedures Performed or Reference to Work Paper(s) Where Procedures Have Been Documented**	**Effective/Ineffective** *(if ineffective, refer to the findings work paper)*
	ISEC 2.04 - Users who have changed roles or tasks within the organization, or that have been transferred, or terminated are immediately informed to the security department for user account access revision in order to reflect the new and/or revised status.		
	ISEC 2.05 - Transmission of sensitive information is encrypted consistent with security policies and procedures to protect its confidentiality.		

IT Area: *CHANGE CONTROL MANAGEMENT*

Topics Covered: *Risk Evaluation, Documentation, Change Request Authorization, Testing, Approval of System Changes, and Implementation into Production Environment*

Control Objective	Control Activity	Description of Audit Procedures Performed or Reference to Work Paper(s) Where Procedures Have Been Documented	Effective/Ineffective (if ineffective, refer to the findings work paper)
CCM 1.00 - Changes implemented in applications, databases, networks, and operating systems (altogether referred to as "system changes") are assessed for risk, authorized, and thoroughly documented to ensure desired results are adequate.	CCM 1.01 - The business risks and the impact of proposed system changes are assessed by management before implementation into production environments. Assessment results are used when designing, staffing, and scheduling implementation of changes to minimize disruptions to operations.		
	CCM 1.02 - Requests for system changes (e.g., upgrades, fixes, emergency changes, etc.) are documented and approved by management before any change-related work is done.		
	CCM 1.03 - Documentation related to the change implementation is adequate and complete.		
	CCM 1.04 - Change documentation includes the date and time at which changes were (or will be) installed.		
	CCM 1.05 - Documentation related to the change implementation has been released to system users.		

(Continued)

IT Area: *CHANGE CONTROL MANAGEMENT*

Topics Covered: *Risk Evaluation, Documentation, Change Request Authorization, Testing, Approval of System Changes, and Implementation into Production Environment*

Control Objective	Control Activity	Description of Audit Procedures Performed or Reference to Work Paper(s) Where Procedures Have Been Documented	Effective/Ineffective (if ineffective, refer to the findings work paper)
CCM 2.00 - Changes implemented in applications, databases, networks, and operating systems (altogether referred to as "system changes") are appropriately tested. Tests are performed by a group other than the group responsible for the system (e.g., operating systems changes are implemented by someone other than the systems programmer, etc.).	CCM 2.01 - System changes are tested before implementation into the production environment consistent with test plans and cases.		
	CCM 2.02 - Test plans and cases involving complete and representative test data (instead of production data) are approved by application owners and development management.		

IT Area: CHANGE CONTROL MANAGEMENT

Topics Covered: *Risk Evaluation, Documentation, Change Request Authorization, Testing, Approval of System Changes, and Implementation into Production Environment*

Control Objective	Control Activity	Description of Audit Procedures Performed or Reference to Work Paper(s) Where Procedures Have Been Documented	Effective/Ineffective (if ineffective, refer to the findings work paper)
	CCM 2.03 - A sample of system changes is selected for the period under audit to determine whether documentation supporting the change: 1. is in accordance with installation standards; 2. provides a clear explanation of the change made and the reason for the change; and 3. has been appropriately reviewed and approved by management.		
	CCM 2.04 - A sample of system changes is selected for the period under audit to determine whether test plans and cases: 1. are in compliance with installation standards; 2. thoroughly tested the implemented change; 3. were reviewed and properly approved; and 4. were tested in a protective environment separate from the production environment.		

(Continued)

IT Area: *CHANGE CONTROL MANAGEMENT*

Topics Covered: *Risk Evaluation, Documentation, Change Request Authorization, Testing, Approval of System Changes, and Implementation into Production Environment*

Control Objective	Control Activity	Description of Audit Procedures Performed or Reference to Work Paper(s) Where Procedures Have Been Documented	Effective/Ineffective (if ineffective, refer to the findings work paper)
	CCM 2.05 - The names and titles of personnel responsible for implementing system changes are identified. Access to the development or test environments is separate and appropriately restricted to the live or production environment (i.e., adequate separation of duties).		
CCM 3.00 - Changes implemented in applications, databases, networks, and operating systems (altogether referred to as "system changes") are appropriately managed to reduce disruptions, unauthorized alterations, and errors which impact the accuracy, completeness, and valid processing and recording of financial information.	CCM 3.01 - Problems and errors encountered during the testing of system changes are identified, corrected, retested, followed up for correction, and documented.		

(Continued)

IT Area: *CHANGE CONTROL MANAGEMENT*

Topics Covered: *Risk Evaluation, Documentation, Change Request Authorization, Testing, Approval of System Changes, and Implementation into Production Environment*

Control Objective	Control Activity	Description of Audit Procedures Performed or Reference to Work Paper(s) Where Procedures Have Been Documented	Effective/Ineffective (if ineffective, refer to the findings work paper)
CCM 4.00 - Changes implemented in applications, databases, networks, and operating systems (altogether referred to as "system changes") are formally approved to support accurate, complete, and valid processing and recording of financial information.	CCM 4.01 - Prior to implementation of system changes in live and production environments, documentation of formal acceptance is obtained supporting that testing has been satisfactorily completed, test results were successful and adequate to prevent tampering, and user requirements were met.		
	CCM 4.02 - Personnel independent from those with access to the development or test environments review changes and deploy them into the live or production environment.		
	CCM 4.03 - Procedures such as retaining a prior version of the original environment are in place to allow the recovery of such original environment in the event there are problems resulting from the implementation of system changes.		

(Continued)

IT Area: CHANGE CONTROL MANAGEMENT		
Topics Covered: Risk Evaluation, Documentation, Change Request Authorization, Testing, Approval of System Changes, and Implementation into Production Environment		
Control Objective	**Control Activity**	**Description of Audit Procedures Performed or Reference to Work Paper(s) Where Procedures Have Been Documented** / **Effective/Ineffective** *(if ineffective, refer to the findings work paper)*
	CCM 4.04 - An overall review is performed by management after system changes have been implemented in the live or production environment to determine whether the objectives for implementing system changes were met.	

Appendix 4: ACL Best Practice Procedures for Testing Accounting Journal Entries

The audit procedures noted in this appendix create results that may warrant additional investigation. Performing further drill down procedures (e.g., sorting by dollar amount, filtering on account number, stratifying by date, classifying on user, etc.) based on specific audit interest and judgment may still be needed. This would be necessary to identify accounting journal entries exhibiting characteristics of interest to be selected for detail testing.

Typical working paper documentation when testing journal entries includes details of the selected journal entries, results of profiling and other analysis, and descriptions of the test procedures performed. The following are common steps followed by auditors when testing accounting journal entries with ACL:

- Reconcile and format data for ACL-based testing
- Analyze journal entry population
- Identify potential indicators of weak controls over the journal entry posting process

Data Reconciliation and Formatting for ACL-Based Testing

Auditors must reconcile the summary of journal entries provided by the entity (i.e., organization or audit client) to the relevant trial balance (T/B) under audit by general ledger (G/L) account. This is done in order to determine whether the journal entries obtained represent the complete population. There will be situations where data reconciliation is not achieved. When this is the case, the entity needs to provide a new summary of journal entries, or be asked to reconcile between the downloaded journal entries and the T/B. The audit team will then need to perform procedures to ensure the accuracy of the entity's reconciliation.

Once reconciled, the following procedures will assist auditors in formatting and preparing journal entry data for ACL-based analysis and testing:

A. *Balance check to confirm that all debits and credits in the entire journal entry file net to zero. Confirm that all individual journal entries net to zero.* Journal entry detail lines are components of a greater single journal entry. All detail lines of a journal entry, when summarized, should have debits equal to credits.

B. *Reconcile journal entry data to the financial statements on which the audit procedures are being performed.* Reconcile the journal entry data file received to the financial statements (e.g., T/B, etc.) on which audit procedures are being performed to assess the completeness of the journal entry data file used for testing. The journal entry detail file must have, among other fields, account number, account description, beginning balance, as well as debits and credits.

C. *Segregate standard and non-standard journal entries.* Journal entry files, once reconciled to the T/B, should be segregated between standard (automated) and non-standard (manual) journal entries, if possible.

D. *Exclude certain groups of journal entries from further analysis based on the auditor's understanding of the entity's financial reporting process and judgment regarding the risk of material misstatement due to fraud.*

Analysis of the Journal Entry Population

The auditor should analyze the population of journal entries in order to generate summary statistics of such population. This information is useful in gaining an understanding of the composition of the entire journal entry population. Prior to conducting the analyses, it can be helpful to execute high level procedures to obtain an overview of the size and general makeup of the journal detail lines within the journal entry file.

Identification of Potential Indicators of Weak Controls over the Journal Entry Posting Process

Auditors examine journal entry data in order to identify potential deviations in the operating effectiveness of internal controls over the financial closing and reporting process, and the relevant general computer controls. Auditors then flag journal entries of interest for testing either via direct selection or inclusion of subpopulation in sample selection process.

Upon identifying potential indicators of weaknesses in controls over the journal entry posting process, auditors should evaluate such indicators to determine whether they represent deficiencies in the related control activities. Auditors must also document their consideration of the effect of such deficiencies.

The potential indicators below are commonly used to identify deviations on the journal entry data. This list is not all-inclusive, as additional indicators may be used (and tested) based on auditor judgment and entity risk factors. After testing each indicator, the auditor evaluates the results to determine the volume, size, and timing for selections that are to be made in order to gain comfort over the validity of journal entries.

Potential Indicator of Weak Controls	Description of Procedures Performed
A. Suspense accounts	All suspense accounts should be identified and significant suspense accounts or items reviewed by management. Validate that suspense account balances net to zero as expected.

(Continued)

Potential Indicator of Weak Controls	Description of Procedures Performed
B. User authorization	Summarize postings by users during the test period to identify: 1. Unauthorized users, or those users not expected to post journal entries. 2. Inconsistent or unexpected user activity: if a user was terminated, there should be no postings from that user after the termination date. 3. Users who may have exceed their posting authorization level. 4. Users with high dollar and high-/low-count totals. Procedures should be designed to identify journal entries made by individuals who typically do not make entries as a characteristic of interest to consider. A decision is made as to what constitutes "infrequent" based on the professional judgment of the audit team and its understanding of the entity's business processes. 5. Journal entries with blank or missing user IDs. Note: ACL is capable of counting the number of detailed journal entry lines each user entered during the period under review to identify infrequent users.
C. Detail lines with invalid Chart of Account (COA) information	Search relevant characteristics of journal entries for indicators of potential weaknesses in account maintenance. Identify journal entry detail lines which represent activity to accounts not listed in the COA. If a large percentage of accounts on the COA are not used in the financial reporting process, it may be an indicator that the COA is not being maintained regularly.
D. Duplicate detail lines	Identify duplicate journal entry detail lines and duplicate postings to the same account. For example, detail lines which have the same journal entry number and line number represent a potential duplicate entry.
E. Detail lines with an effective date outside the test period	Identify entries in the journal entry file which are not related to activity in the period under review.
F. Detail lines with an invalid postdate	Identify entries which have a postdate which does not make logical sense (e.g., an entry that takes place far in the past or future, etc.).
G. Related party and unusual transactions	Search for related party transactions in the G/L. Management must approve all related party transactions and/or unusual transactions and events, including those that exceed established limits. Board approval is required for specified types of related party and/or unusual transactions, and this approval should be appropriately documented.

(Continued)

Potential Indicator of Weak Controls	Description of Procedures Performed
H. Seldom-used accounts	Review accounts with few number of lines posted in a period. Determination of "few'" is based on professional judgment and understanding of the entity's financial reporting process. Count the number of times each account was used in entries during the period under review.
I. Large journal entry detail lines by account	Identify the largest journal entry detail line for a given account.
J. Entries with key words of audit interest in descriptions	Scan the journal entry description field and identify entries with key words of audit interest. Below are some key words which might be searched. Not all of these key words are applicable to all situations, and there may be other words to consider searching for that are not included in this listing. *adj, adjust, alter, as directed, as requested, bury, capital, ceo, cfo, classif, conceal, confidential, controller, cookie jar, correct, cover, coverup, cover-up, delete, deleted, dummy, early, ebit, ebitda, error, fictitious, fraud, hide, hidden, holdback, immaterial, improper, inappropriate, increase, kitty, manage earnings, manip, misstate, opportunit, plug, recl, reclass, recls, reconcile, reduce, reduct, restate, rev, reversal, reverse, risks, screen, secret, smooth, spread, temp, test, transf, tsfr, txfr*
K. Entries with credits to revenue	Identify journal entries with at least one credit to revenue accounts. Test provides information on the off-setting debit in order to consider whether the debit is unrelated to the credit (e.g., Debit: Fixed Assets and Credit: Revenue, etc.).
L. Entries with credits to expense	Identify journal entries with at least one credit to expense accounts.
M. Entries made on unusual days	Identify entries booked on unusual days. These could be weekend days or holidays, depending on what is considered unusual practice at the entity.
N. Days with posting frequency outside expected range	Review days within the period under review for unusual patterns to identify journal entries for further analysis. This procedure identifies days during the period under review with few or many journal entry postings.
O. Entries posted near the end of the period under review or as post-closing entries that have little or no explanation or description	Identify journal entries which were posted to the accounting system near the end of the period under review, which have little or no description.
P. Detail lines with blank journal entry description	All detail lines of an entry typically have a description. This procedure identifies entries without a sufficient description.

(Continued)

Potential Indicator of Weak Controls	Description of Procedures Performed
Q. Differences between post and effective dates	Examine the time lapse between post and effective dates to identify journal entries for further analysis.
R. Specific dollar amounts	Identify entries with specific dollar amounts, thousands, millions, etc. for further analysis.
S. Frequently used dollar amounts	Identify amounts that are used more often than other amounts.
T. Benford's Law	Identify journal entries with unusual amounts based on Benford's Law analysis. Flag journal entries that may be higher risk for potential future selection and investigation. Benford's Law, also called the first-digits law (can be applied on multiple leading digits), states that in lists of numbers from many real-life sources of data, the leading two digits occur much more often than the others (namely about 30% of the time). Furthermore, the higher the digit, the less likely it is to occur as the leading two digits of a number. This applies to figures related to the natural world or of social significance; be it numbers taken from electricity bills, newspaper articles, street addresses, stock prices, population numbers, or mathematical constants. This test attempts to identify amounts which are skewed due to an attempt to circumvent an unnatural restriction (e.g., a control limiting the size of an entry due to an employee's signing authority, etc.).
U. Entries with recurring ending digits, including round dollar amounts	Identify entries where the amount has recurring ending digits.
V. Managing estimates	Analyze trends in account balances involving management estimates to identify potential bias (a form of management override of controls). Identify entries made near the end of the period to accounts which have balances related to management estimates (e.g., allowances and accruals, etc.).

If any of the aforementioned procedures prompt further analysis and investigation, auditors should use judgment in determining whether the file received from the entity is complete, and that it contains valid information. If there is some concern with the results, a discussion with appropriate entity personnel should take place to determine whether portions of the journal entry data file are to be removed (which also may help in reconciling to the T/B), whether the results identified represent issues to be noted, and/or if a new journal entry file is to be requested.

Appendix 5: IT Risk Assessment Example Using NIST SP 800-30

The risk assessment used in this example is based on the National Institute of Standards and Technology (NIST) SP 800-30 guide, "Guide for Conducting Risk Assessments."[*] The guide is consistent with the policies presented in the Office of Management and Budget Circular A-130, Appendix III, "Security of Federal Automated Information Resources"; the Computer Security Act of 1987; and the Government Information Security Reform Act of October 2000. NIST SP 800-30's risk assessment includes the following nine steps:

1. System Characterization
2. Threat Identification
3. Vulnerability Identification
4. Control Analysis
5. Likelihood Determination
6. Impact Analysis
7. Risk-level Determination
8. Control Recommendations
9. Results Documentation

The steps above are implemented to identify information security risks associated with the selected organization's information system. Step 1 through Step 7 (except Step 4) directly relate to risk assessment. Steps 4, 8, and 9 relate to the identification and implementation of controls that mitigate or reduce risks, as well as documentation of results. Following is an example showing how the NIST risk assessment is implemented in an organization.

Step 1. System Characterization

NIST emphasizes in obtaining a thorough understanding of the IT system, as well as establishing the scope of the risk assessment and the limitations of the IT system being evaluated. The understanding collected should include detailed information to allow identification of potential risks.

---

[*] http://nvlpubs.nist.gov/nistpubs/Legacy/SP/nistspecialpublication800-30r1.pdf.

Overall Description of the Organization, IT Department, and Scope

University XYZ ("the University") is situated on the southern coast of Florida (FL). The University offers associate, baccalaureate, masters, and/or doctoral degrees in arts, humanities, natural and social sciences, as well as in professional areas such as business, education, nursing, law, and medical technology. Currently, there are approximately 15,000 students attending the University.

In regards to IT and related support activities, the University's approach is centralized. That is, the University's computer processing is performed by the IT department, which is the sole provider of technology and telecommunications for the University's departments. Furthermore, the IT department provides data processing and end-user support for the University's systems and applications, including training and documentation of application system controls and procedures. The IT department's organizational structure consists of 30 staff, under the direction of an IT Executive Director.

The scope of this risk assessment is the University's financial application system. The application is called Banner Finance ("Banner") and runs on a Red Hat Enterprise Linux operating system. Refer to Exhibit A5.1.

Collection of Information Relevant for the Risk Assessment

During the risk assessment process, relevant information is gathered via reviews and inspections of documentation, as well as on-site interviews with key management personnel. Key management personnel for purposes of this example include:

- IT Executive Director (ITED)
- Banner Security Administrator (BSA)
- Operations Supervisor (OS)
- Systems Administrator (SA)
- Network Administrator (NA)

When interviewing the ITED and the BSA, it was noted that Banner holds critical and sensitive information about finance, accounting, human resources (HR), and payroll. The BSA further added that users of Banner include finance, accounting, HR, and technical/IT support personnel. Based on review of documentation, the University has several policies and procedures in place related to information systems operations, information security, and change control management.

In regards to the network infrastructure, the NA indicated that the University provides a wide variety of networking resources to all qualified members within the university community. Access to computers, systems, and networks is a privilege which imposes certain responsibilities and obligations, and which is granted subject to university policies, as well as local, state, and federal laws. All users must comply with policies and guidelines, and act responsibly while using network resources.

Physical access to the University's facilities and its data center, according to the ITED and the OS, is restricted through security mechanisms, including (1) biometric devices, (2) security

Exhibit A5.1 Financial Application System in Scope for Risk Assessment Purposes

Application Name	Application Server Name/Operating System Version	Application Server Location
Banner	INB.xyzfl.edu/Red Hat Enterprise Linux	Computer Center, University Headquarters, FL

guards, (3) video surveillance, and (4) visitors' logs. The authority to change the above physical access control mechanisms is limited to the ITED. The OS also stated that the University has implemented various environmental controls in order to prevent damage to computer equipment, and to protect data availability, integrity, and confidentiality. They are as follows: fire suppression equipment (i.e., FM-200 and fire extinguishers), uninterruptible power supplies, alternate power generators, and raised floors.

When asked about logical information security around Banner both, the SA and BSA, agreed on the following:

- Some password settings have been configured although current configuration is not consistent with industry best practices.
- Reviews of user access within Banner are conducted, but not on a periodic basis. Terminated user accounts are removed from Banner, but not in a timely manner. Documentation supporting reviews and removal of user access is not maintained.
- Programmers are restricted to work changes and modifications (i.e., updates and upgrades) to Banner in a test/development environment prior to their implementation in production. However, test results are not reviewed by management (i.e., ITED) nor approved before final implementation in production.

Lastly, Banner information is backed up daily though the OS stated that such daily backup is stored locally as the University has no offsite facility in place for backup storage.

Step 2. Threat Identification

NIST defines a threat as "any circumstance or event with the potential to adversely impact organizational operations and assets, individuals, other organizations, or the Nation through an information system via unauthorized access, destruction, disclosure, or modification of information, and/or denial of service" (p. 8). A threat-source refers to an intentional exploitation of a vulnerability, or an accidental trigger of a vulnerability. Common threat-sources, as indicated by NIST, include the following: natural threats, environmental threats, and human threats.

In order to identify potential threat-sources applicable to the Banner application, the key management personnel listed earlier was interviewed. The following represent the threat-sources identified by management that could potentially exploit the application vulnerabilities:

- Natural threats: hurricanes, earthquakes, and floods
- Environmental threats: system failures, and unexpected shutdowns
- Human threats: unauthorized access by hackers, terminated employees, and insiders (i.e., disgruntled, malicious, negligent, or dishonest employees)

Motivations for human threats, as identified by management, include the following:

- Challenge, ego, and rebellion (hackers)
- Destruction of information, illegal information disclosure, monetary gains, and unauthorized data alteration (terminated employees and hackers)
- Curiosity, ego, intelligence, monetary gains, revenge, unintentional errors, and omissions (insiders)

Exhibit A5.2   Vulnerabilities and Threat-Sources

IT Area	Vulnerability	Threat-Source
IS Operations	1. There is no offsite storage for data backups to provide reasonable assurance of availability in the event of a disaster.	Hurricanes, system failures, and unexpected shutdowns.
Information Security	2. Several of the University's password settings configured on Banner are not consistent with industry best practices.	Unauthorized users (hackers, terminated employees, and insiders).
Information Security	3. Banner application owners do not periodically review user access privileges.	Unauthorized users (hackers, terminated employees, and insiders).
Information Security	4. Terminated user accounts are not removed timely, or not removed at all, from Banner.	Unauthorized users (terminated employees).
Change Control Management	5. Test results for Banner modifications are not approved by management prior to their implementation into the production environment.	Implementation of unauthorized application modifications.

Step 3. Vulnerability Identification

NIST defines vulnerability as a "weakness in an information system, system security procedures, internal controls, or implementation that could be exploited by a threat source" (p. 9). Vulnerabilities around Banner that could be exploited by threat-sources were identified from discussions with management personnel, observations, and inspections of relevant documentation, as recommended by NIST. Documentation reviewed include previous IT risk assessments, as well as IT Audit and Security Reports. Refer to Exhibit A5.2 for a list of five vulnerabilities and threat-source pairs identified per IT area.

Step 4. Control Analysis

This step takes into consideration the controls that are in place, or are planned for implementation by the University in order to reduce or eliminate the probability of a threat exercising a vulnerability. This step is addressed in Steps 8 and 9.

Step 5. Likelihood Determination

NIST states that the following must be considered in order to develop a rating indicating the probability that vulnerabilities may be exercised:

- ■ Motivation and capability of threat-sources
- ■ Nature of the vulnerability
- ■ Existence and effectiveness of current controls

Exhibit A5.3   Likelihood Levels

Likelihood Level	Likelihood Definition	Likelihood-Level Value
Very High	Threat-source is extremely motivated and very capable; controls to prevent the vulnerability from being exercised are non-existent.	1.00
High	Threat-source is highly motivated and sufficiently capable; controls to prevent the vulnerability from being exercised are ineffective.	0.75
Medium	Threat-source is motivated and capable; controls exist that may prevent successful exercise of the vulnerability.	0.50
Low	Threat-source lacks motivation or capability; controls exist to prevent (or at least impede) the vulnerability from being exercised.	0.25
Very Low	There is no threat-source motivation or capability; controls in place prevent the vulnerability from being exercised.	0.10

NIST recommends the following High–Medium–Low definitions to describe the likelihood that vulnerabilities could be exercised by a given threat-source. However, for this example, Very High and Very Low levels have been added to obtain a more granular rating indicating the probability that vulnerabilities may be exercised. Refer to Exhibit A5.3.

Probabilities of Very High = 1.00, High = 0.75, Medium = 0.50, Low = 0.25, and Very Low = 0.10 were assigned for each vulnerability based on management's estimate of their likelihood level.

Step 6. Impact Analysis

According to NIST, impact analysis determines the adverse effect in the IT system resulting from threats that successfully exercise vulnerabilities. The magnitude of the impact cannot be measured in specific units, but can be classified as High, Medium, or Low, as recommended by NIST. In this example, Very High and Very Low magnitudes were also incorporated to obtain a more detailed impact level from threats successfully exercising vulnerabilities. Refer to Exhibit A5.4.

Step 7. Risk-Level Determination

Determination of risk level for a particular vulnerability/threat pair that can be exercised considers:

- The likelihood of a threat-source attempting to exercise a vulnerability;
- The magnitude of the impact from a threat-source successfully exercising the vulnerability; and
- The appropriateness of planned, or existing controls for mitigating or eliminating identified risks.

Exhibit A5.4   Magnitude of Impact Criteria

Magnitude of Impact	Impact Definition	Impact-Level Value
Very High	Results in the loss of extremely priced assets or resources; will violate or harm the University's reputation.	100
High	Results in the loss of highly priced tangible assets or resources; significantly violate or harm the University's reputation.	75
Medium	Results in the loss of costly tangible assets or resources; violate or harm the University's reputation.	50
Low	Results in the loss of some tangible assets or resources; noticeably affect the University's reputation.	25
Very Low	Almost no loss to tangible assets or resources; barely affects the University's reputation.	10

Determination of risk levels (i.e., Risk Rating) is obtained by multiplying the ratings assigned for threat likelihood (i.e., probability) and the threat impact. Determination of these risk levels or ratings is of a subjective nature, as it results solely from management's estimates and opinions (based on knowledge and/or prior experience) when assigning threat likelihood and impact. Exhibit A5.5 shows various degrees or levels of risk, based on NIST, to which an IT system (in this case Banner) might be exposed if a given vulnerability was exercised by a threat. Exhibit A5.5 also suggests necessary actions that management must take for each risk level. For purposes of this example, Very High and Very Low risk levels were incorporated in an effort to obtain granularity of risk ratings.

Exhibit A5.6 illustrates completion of the aforementioned Steps 1–7, including a description of each risk identified, as well as final determination of their levels (i.e., risk rating) for every vulnerability that could be potentially exercised.

Exhibit A5.5   Risk Levels

Risk Level	Risk Rating	Risk Description and Necessary Actions
Very High	100	Corrective measures are mandatory; action plans must be implemented; IT system may not operate.
High	75	Strong need for corrective measures; the IT system may continue to operate, but a corrective action plan must be implemented immediately.
Medium	50	Corrective actions are needed; a plan must be put in place to incorporate necessary actions within a reasonable period of time.
Low	25	Management must determine whether corrective actions are still required, or decide to accept the risk.
Very Low	10	Corrective actions may not be needed, as risks identified may be acceptable.

Exhibit A5.6   Risk Assessment

IT Area	Vulnerability	Threat-Source	Likelihood Determination		Impact		Risk	Risk Rating
			Likelihood Level	Probability	Magnitude of Impact	Impact-Level Value		
IS Operations - Offsite Storage	1. There is no offsite storage for data backups to provide reasonable assurance of availability in the event of a disaster.	Hurricanes, system failures, and unexpected shutdowns.	Medium	0.50	High	75	1. Banner information cannot be recovered in the event of a system failure, impacting the University's ability to report financial information according to established reporting requirements.	37.5
Information Security - Passwords	2. Several of the University's password settings configured on Banner are not consistent with industry best practices.	Unauthorized users (hackers, terminated employees, and insiders).	High	0.75	High	75	2. Security parameters are not appropriately configured, allowing for potential unauthorized user access to the Banner application.	56.25
Information Security - Reviews of User Access	3. Banner application owners do not periodically review user access privileges.	Unauthorized users (hackers, terminated employees, and insiders).	Very High	1.00	High	75	3. Users possess privileges that are not consistent with their job functions, allowing unauthorized or incorrect modifications to Banner's data which could cause management decisions based upon misleading information.	75

(*Continued*)

Exhibit A5.6 (*Continued*)   Risk Assessment

| IT Area | Vulnerability | Threat-Source | Likelihood Determination | | Impact | | Risk | Risk Rating |
			Likelihood Level	Probability	Magnitude of Impact	Impact-Level Value		
Information Security - Terminations	4. Terminated user accounts are not removed timely, or not removed at all, from Banner.	Unauthorized users (terminated employees).	Very High	1.00	High	75	4. Terminated users can gain access to Banner and view or modify its financial information.	75
Change Control Management	5. Test results for Banner modifications are not approved by management, prior to their implementation into the production environment.	Implementation of unauthorized application modifications.	Low	0.25	High	75	5. Banner modifications are not properly authorized. Implementation of such modifications could result in invalid or misleading data.	18.75

As mentioned earlier, Steps 4, 8, and 9 relate to the identification and implementation of controls that mitigate or reduce risks, as well as results documentation. Those steps are addressed in the following sections.

Step 8. Control Recommendations

Risk Mitigation Process

Risk mitigation forms the second half of the risk management methodology. Per NIST, organizations' management must select a reasonable and effective cost approach to implement appropriate IT controls in order to reduce identified risks to acceptable levels. Exhibit A5.7 describes the options available for risk mitigation and risk management strategy.

In conversations with University's management, it was agreed that Risk Planning was the option selected to mitigate the risks identified. Therefore, the mitigation strategy involved preparing a plan which would prioritize, implement, and maintain the necessary controls to address the risks. Management understood that it was appropriate to implement controls to address risks when vulnerabilities can be exercised by threats. That is, management's plan, following NIST, was to incorporate security and protection via implementing controls to either minimize risks or prevent them.

Approach for Control Recommendation/Implementation

In order to select and implement controls to address the risks, management adopted the NIST-recommended approach for control implementation. The approach starts by prioritizing and evaluating recommended controls (RCs) followed by their formal selection and identification of residual risks. The NIST-recommended approach assists in implementing controls that can reduce the risks associated with Banner.

Prioritization and Recommended Controls

Prioritization, according to NIST, refers to the process of establishing significance to the risks identified and the RCs for mitigation. Based on the significance to the risks identified, control

Exhibit A5.7 Risk Mitigation Options

Option	Description
Risk Assumption	Accepts potential risks and continue on with IT operations.
Risk Avoidance	Avoids the risk by eliminating its cause and/or consequence.
Risk Limitation	Limits risks by implementing controls that minimize the impact of a vulnerability exercised by a threat.
Risk Planning	Manages risks through a risk mitigation plan that prioritizes, implements, and maintains controls.
Research and Acknowledgment	Lower risks of loss by acknowledging vulnerabilities and researching controls to correct them.
Risk Transfer	Transfers the risk to compensate for potential losses.

recommendations are established. Management listed all potential RCs that could reduce the risks identified, and assigned priority to those controls using rankings ranging from Very High to Very Low levels (refer to Exhibit A5.8). Management also acknowledged that implementing all RCs may not be the most appropriate and feasible option. Further analyses related to feasibility, effectiveness, and cost–benefit were performed in the following sections for each RC in order to determine the most appropriate ones for reducing risks.

Feasibility and Effectiveness Evaluations of Recommended Controls

Management acknowledged that the control recommendation process involves selecting a combination of controls from technical, management, and operational categories that could potentially reduce risks around Banner. The following are descriptions of each category of control based on NIST.

- *Management Security Controls.* These controls manage and reduce the risk of loss while protecting the organization's mission. Management security controls take the form of policies, guidelines, and standards to fulfill the organization's goals and missions.
- *Technical Security Controls.* These controls relate to the configuration of parameters within applications, systems, databases, and networks to protect against security threats over critical and sensitive information, as well as IT system functions.

Exhibit A5.8 Recommended Controls and Assigned Priority

IT Area	Risk	Risk Level	Recommended Control (RC)	Action Priority
IS Operations - Offsite Storage	1. Banner information cannot be recovered in the event of a system failure, impacting the University's ability to report financial information according to established reporting requirements.	Medium	RC1. Backups of Banner financial data are archived off-site to minimize risk that data is lost.	Medium
Information Security - Passwords	2. Security parameters are not appropriately configured, allowing for potential unauthorized user access to the Banner application.	High	RC2. The identity of users is authenticated to Banner through passwords consistent with industry best practices minimum security values. Passwords must incorporate configuration for minimum length, periodic change, password history, lockout threshold, and complexity.	High

(Continued)

Exhibit A5.8 (*Continued*) Recommended Controls and Assigned Priority

IT Area	Risk	Risk Level	Recommended Control (RC)	Action Priority
Information Security - Reviews of User Access	3. Users possess privileges that are not consistent with their job functions, allowing unauthorized or incorrect modifications to Banner's data which could cause management decisions based upon misleading information.	Very High	RC3. Banner owners authorize the nature and extent of user access privileges. RC4. The ability to make modifications to Banner security parameters, security roles, or security configuration is limited to appropriate personnel. RC5. User access privileges within Banner are periodically reviewed by application owners to verify that access privileges remain appropriate and consistent with job requirements.	Very High
Information Security - Terminations	4. Terminated users can gain access to Banner and view or modify its financial information.	Very High	RC6. The security administrator is notified of employees who have been terminated. Access privileges of such employees are immediately changed to reflect their new status.	Very High
Change Control Management	5. Banner modifications are not properly authorized. Implementation of such modifications could result in invalid or misleading data.	Low	RC7. Modifications to Banner are tested and approved by management prior to their implementation in production in accordance with test plans and results. RC8. User and other requests for modifications to Banner, including upgrades, fixes, and emergency changes, are documented and approved by management to verify that all changes to the production environment were documented, tested, and authorized.	Low

■ *Operational Security Controls.* These controls make certain that security procedures governing the use of the organization's IT assets (i.e., Banner) are adequately implemented consistent with the organization's goals and mission. Operational security controls address operational deficiencies that could result from potentially exercised vulnerabilities.

Category determination and discussions of each RC per IT area were performed with the assistance of management and are documented in Exhibit A5.9.

Cost–Benefit Analysis for Recommended Controls

Cost–benefit analysis should be conducted following identification of RCs and evaluation of their feasibility and effectiveness. The objective of the cost–benefit analysis is to support that the cost of implementing the control is justified by a reduction in the level of risk. That is, cost–benefit analyses should ensure that controls are implemented in a cost-effective way.

Management understood appropriate that the benefits of the selected controls must be evaluated in terms of their impact in risk reduction. Along the same line, the effect of not implementing controls needs to be assessed to support whether Banner can continue operating effectively without implementing the controls. Management must determine the minimum level of acceptability for the risks identified, as well as, the impact of each selected control to determine their effect on Banner. Assessments of the potential controls in relation to the risks can be done following the rules below, as suggested by NIST:

■ Rule 1. If control reduces risks more than needed, consider an alternate, less expensive control.
■ Rule 2. If the cost of the control is higher than the risk reduction provided, consider identifying additional controls.
■ Rule 3. If the control does not provide a significant risk reduction, consider identifying additional controls.
■ Rule 4. If the control does provide a significant risk reduction and it is cost-effective, implement the control.

Below is an overall summary of the cost–benefit analysis procedures management performed for each RC:

■ Assessed the impact of implementing versus not implementing the RC.
■ Estimated the cost of implementation taking into account the following factors, as applicable:
  – Additional purchases of hardware and software required.
  – Reduced operational effectiveness resulting from no implementation.
  – Reduced system performance, functionality, or security from no implementation.
  – Cost of implementing new/revised policies, procedures, standards, etc.
  – Cost of hiring (or training) personnel to implement RCs.

■ Evaluated potential benefits over costs to support implementation of the new control, and communicate such to the University's senior management and Board of Directors.

Once the cost–benefit analysis was performed, controls were formally selected (refer to Exhibit A5.10). However, because risks cannot be eliminated completely, but reduced to acceptable levels,

Exhibit A5.9   Recommended Control: Category Determination and Discussion

IT Area: Recommended Control	Control Category and Discussion
IS Operations - Offsite Storage: RC1. Backups of Banner financial data are archived off-site to minimize risk that data is lost.	RC1 is part of the Management Security Controls and the Operational Security Controls categories. RC1 has the capability of supporting the continuity of operations and business resumption during emergencies or disasters. Equally important, RC1 ensures that Banner's backed up data gets rotated to an off-site location in order to minimize the risk that significant data could be lost during a particular event (e.g., hardware failure, computing environment disaster, etc.). The recovery of lost data could require significant effort or be virtually impossible in certain situations.
Information Security - Passwords: RC2. The identity of users is authenticated to Banner through passwords consistent with industry best practices minimum security values. Passwords must incorporate configuration for minimum length, periodic change, password history, lockout threshold, and complexity.	RC2 belongs to the Technical Security Controls category. Strong authentication controls through configuration of logical security settings (i.e., passwords) verify the identity of the employee to ensure it is valid and genuine. More important, strong password controls promote adequate security and protect Banner's financial data from threats, such as, deliberate attacks by malicious persons or disgruntled employees trying to gain unauthorized access in order to compromise system and data integrity, availability, or confidentiality. Strong passwords may also protect Banner's financial information from unintentional acts, such as negligence and errors, to circumvent system security.
Information Security - Reviews of User Access: RC3. Banner owners authorize the nature and extent of user access privileges. RC4. The ability to make modifications to Banner security parameters, security roles, or security configuration is limited to appropriate personnel. RC5. User access privileges within Banner are periodically reviewed by application owners to verify that access privileges remain appropriate and consistent with job requirements.	RC3, RC4, and RC5 pertain to the Management Security Controls as well as the Technical Security Controls categories. They make certain that access to users is granted following the principle of "least privilege" (also known as the principle of minimal privilege), which basically requires that users must be able to access only the information and resources necessary to carry on their daily tasks. In other words, access granted to users must be legitimate and consistent with job responsibilities. Failure to review user access levels on a periodic basis may not allow the University to detect and correct unauthorized access in a timely manner.

(Continued)

Exhibit A5.9 (*Continued*) Recommended Control: Category Determination and Discussion

IT Area: Recommended Control	Control Category and Discussion
Information Security - Terminations: RC6. The security administrator is notified of employees who have been terminated. Access privileges of such employees are immediately changed to reflect their new status.	RC6 belongs to the Technical Security Controls category. Notification of employee resignations or terminations should be communicated, in a timely basis, to IT personnel and must include specific systems and/or applications so that access may be removed accurately. Failure to notify IT personnel may result in terminated employees having access rights capable of executing different types of financial transactions, exposing the University to malicious attacks, damages, intrusions, and/or misappropriation of assets. Furthermore, having active accounts of terminated employees within Banner for a period longer than necessary allows opportunities for manipulation of confidential and/or sensitive information, security breaches, etc. This may result in former employees with potential access rights for executing unlimited types of transactions (e.g., fraud accounting, disbursements, etc.).
Change Control Management: RC7. Modifications to Banner are tested and approved by management prior to their implementation in production in accordance with test plans and results. RC8. User and other requests for modifications to Banner, including upgrades, fixes, and emergency changes, are documented and approved by management to verify that all changes to the production environment were documented, tested, and authorized.	RC7 and RC8 are part of the Management Security Controls category. Change control is critical to have highly reliable systems that meet the defined service levels of the University and should be formalized through adequate documentation. It is necessary that each change be controlled throughout its life cycle and integrated into the production environment in a systematic and controlled manner. The primary objective is to maintain the integrity and reliability of the production environment, while making changes for the user community. Failure to enforce appropriate change controls can result in operational disruptions, degraded system performance, or compromised security. Furthermore, lack of documentation may result in the inability to track where changes have been made, thus, delaying the correction of problems. Proper documentation supports test results, including resolution of problems encountered, as well as approvals to implement changes into live environments.

Exhibit A5.10 Selection of Controls, Results, and Residual Risks

IT Area	Risk	Selected Control	Results/Residual Risk
IS Operations - Offsite Storage	1. Banner information cannot be recovered in the event of a system failure, impacting the University's ability to report financial information according to established reporting requirements.	RC1. Backups of Banner financial data are archived off-site to minimize risk that data is lost.	Risk reduced to acceptable levels/No residual risk.
Information Security - Passwords	2. Security parameters are not appropriately configured, allowing for potential unauthorized user access to the Banner application.	RC2. The identity of users is authenticated to Banner through passwords consistent with industry best practices minimum security values. Passwords must incorporate configuration for minimum length, periodic change, password history, lockout threshold, and complexity.	Management configured three of five security parameters (i.e., minimum length, password history, and lockout threshold). Periodic change and password complexity were not configured/Residual risks remained.
Information Security - Reviews of User Access	3. Users possess privileges that are not consistent with their job functions, allowing unauthorized or incorrect modifications to Banner's data which could cause management decisions based upon misleading information.	RC5. User access privileges within Banner are periodically reviewed by application owners to verify that access privileges remain appropriate and consistent with job requirements.	Management stated that an annual user access review within Banner would be implemented/Residual risks remained.
Information Security - Terminations	4. Terminated users can gain access to Banner and view or modify its financial information.	RC6. The security administrator is notified of employees who have been terminated. Access privileges of such employees are immediately changed to reflect their new status.	Risk reduced to acceptable levels/No residual risk.
Change Control Management	5. Banner modifications are not properly authorized. Implementation of such modifications could result in invalid or misleading data.	RC7. Modifications to Banner are tested and approved by management prior to their implementation in production in accordance with test plans and results.	Risk reduced to acceptable levels/No residual risk.

there were some risks that still remained after formal selection and implementation of the controls. These remaining risks are called "residual risks" and are discussed next.

Residual Risk Overview

NIST states that no IT system, including its systems and applications, is risk free. Although implemented controls can mitigate or reduce risks, they simply cannot eliminate all risks. Any risk remaining after implementation of new or enhanced controls is referred to as a residual risk. NIST also states that reductions in risks resulting from new or enhanced controls can be analyzed by organizations in terms of the reduced threat likelihood or impact, as these two parameters define the mitigated level of risk to the organizational. NIST further suggests that implementation of new or enhanced controls can mitigate risks by:

- eliminating vulnerabilities through minimizing possible threat-source/vulnerability pairs;
- adding targeted controls to reduce capacities and motivations of threat-sources; and by
- reducing the magnitude of the adverse impact through limiting the extent of vulnerabilities.

Exhibit A5.11 illustrates the relationship between control implementation and residual risk.

Step 9 describes the results of this risk assessment, including the residual risks remaining in Banner after mitigation has been applied, as well as a plan for managing such residual risks.

Step 9. Results Documentation

As indicated above, implemented controls may lower risks, but not necessarily eliminate them. Residual risks may remain after all other known risks have been countered, factored, or reduced, exposing the organization to loss.

Exhibit A5.10 shows the effect from implementing controls over the risks identified, as indicated by management. The following summarizes such effect once the selected controls were implemented:

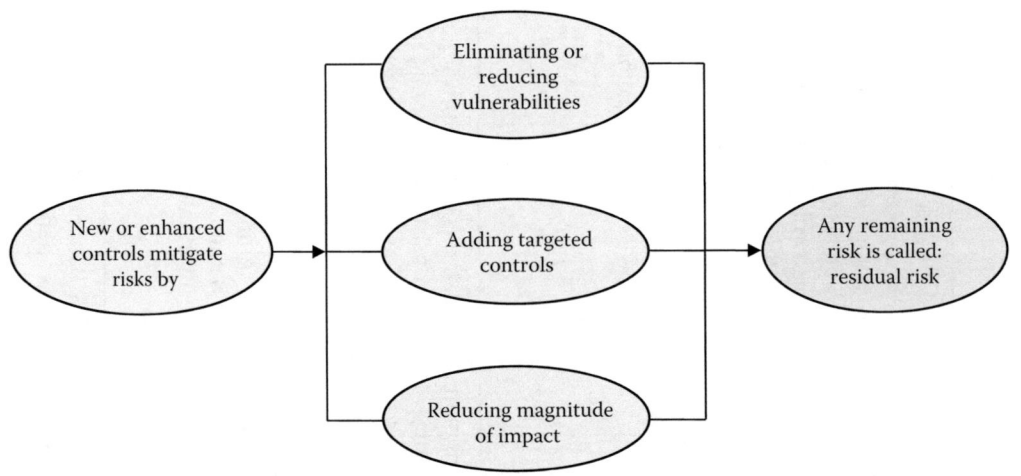

Exhibit A5.11   Implementation of new or enhanced control and residual risk.

- Risk 1, Risk 4, and Risk 5 related to recovery of Banner data, access of terminated user accounts, and unauthorized implementation of Banner changes, respectively, were reduced to acceptable levels with no remaining residual risks. No further procedures were deemed necessary.
- Risk 2, related to configuration of inappropriate security parameters (i.e., passwords), as well as Risk 3 (i.e., inconsistencies in user access) was not reduced to acceptable levels resulting in some residual risks remaining, which are discussed below.

For Risk 2, management configured three out of five password parameters (i.e., minimum length, password history, and lockout threshold). However, periodic password change and complexity settings were not configured. Management acknowledges that periodic password change (or password expiration) and complexity may be a source of frustration to users, who are often required to create and remember new passwords every few months for various accounts. Therefore, users tend to choose weak passwords and use the same few passwords for many accounts. Management further stated that the costs for configuring these two password settings were not justifiable. As a result, a residual risk remained still exposing Banner financial data to threats like malicious attacks, damages, intrusions, and/or manipulation.

After several discussions regarding the residual risks, management expressed interest over improving current Banner password settings in the near future. In the meantime, the following constitutes management's plan and current safeguards to manage and/or mitigate the residual risks related to Risk 2:

- Current password settings configured at the local area network (LAN) are consistent with industry best practices and address settings such as minimum length, password history, lockout threshold, password change, and complexity. This current control helps managing and mitigating the residual risk in Banner, as users need to authenticate to the LAN before accessing the Banner application. The LAN serves as the first level of authentication for Banner users, and those settings are effectively configured.
- Information security tools over Banner are administrated and implemented to restrict access, as well as, record and report security events (e.g., security violation reports, unauthorized attempts to access information resources, etc.).
- Banner application owners authorize access for new users and/or users that have changed roles and responsibilities.
- Banner users are required to have a unique user identifier in order to distinguish one user from another and to establish accountability.
- Banner terminals and work stations are protected by time-out facilities, which are activated after an appropriate, predetermined period of inactivity has elapsed.

In regards to Risk 3, management stated that only one user access review would continue to be performed during the year as the cost for performing additional/periodic reviews of user access is not justifiable. As a result, the following residual risks remained as acknowledged by management:

1. Lack of periodic reviews of user access levels may result in employees having access rights to record and authorize different types of transactions (e.g., accounting transactions, journal entries, etc.), thus, exposing the University to a risk of fraud, manipulation of information, or misappropriation and errors.
2. Failure to review user access levels on a periodic basis may not allow the University to detect and correct unauthorized access in a timely manner.

In order to come up with a plan to identify safeguards to manage and/or mitigate such residual risks, management pointed to the following procedures which were currently in place:

- Banner application owners authorize access for new users and/or users that have changed roles and responsibilities.
- The ability to make modifications to Banner security parameters, security roles, or security configuration is limited to appropriate personnel.
- Information security tools over Banner are administrated and implemented to restrict access, as well as record and report security events (e.g., security violation reports, unauthorized attempts to access information resources, etc.).
- There are effective procedures in place to ensure that production systems are available for the execution of processing and that financial data can be recovered should any disruptions in processing occur.
- There are effective change control procedures to ensure that Banner modifications are tested in separate test/development environment before implementation into the production environment.

Management further expressed their intention to segregate Banner users in groups. A schedule will be prepared to conduct additional, periodic reviews of access for several groups of users during the year. Access of all these groups of users must be reviewed and approved during a period of two years. Documentation supporting those reviews will also be maintained as evidence.

The aforementioned control activities are helpful in managing and mitigating the residual risks resulting from password settings, as well as the lack of periodic user access reviews. These additional safeguards compensate the exceptions previously noted and can mitigate the risks of unauthorized users gaining access to Banner financial data.

Overall, results of this risk assessment exercise were satisfactory, meaning that IT controls are functioning effectively. That is, controls in place not only mitigate risks, but cover effectiveness and efficiency of operations, reliability and completeness of accounting records, and compliance with applicable laws and regulations, among others. Management expressed great satisfaction with the results of both, risk assessment and mitigation exercises, and further indicated that the results obtained around the Banner financial application were consistent with previous risk assessments and evaluations. Particularly, the results of the risk exercises performed motivated the University's IT management to continue providing users safe and reliable financial information by constantly protecting the information's confidentiality, integrity, and availability through effective IT controls.

Appendix 6: Sample Change Control Management Policy

	Change Control Management Policy Policy Number: XYZ-WI-CCM-001 Version: 1.0 Department: Information Technology		
Name	Position	Signature	Date
Prepared by:	IT Manager and/or Administrator		XX/XX/20XX
Reviewed by:	Director of IT Infrastructure		XX/XX/20XX
Reviewed by:	Systems Development Manager		XX/XX/20XX
Approved by:	Vice President of IT		XX/XX/20XX

Mission

Promote an IT environment and infrastructure that is reliable and available.

Purpose

The purpose of the Change Control Management policy is to:

- Provide a formal methodology in which changes relating to the Organization's production environment are documented.
- Require that changes are properly requested, worked, and approved prior to their installation or implementation in the production environment.
- Ensure that all parties affected by the change are timely notified.
- Ensure that installation or implementation of changes is scheduled and coordinated accordingly within the organization.

Scope

This policy covers the change control management process, including how changes (see *Change* in "Definitions" section) relating to operating systems, applications, databases, networks, hardware, and infrastructure (altogether referred to as "systems") should be implemented within the IT production environment, or the environment that the Organization relies on in order to conduct normal business operations and/or safeguard the IT environment.

Definitions

Change Control Management. Process in place to promote quality service and to improve daily operations of the organization. The process ensures system changes are implemented in an effective, efficient, and consistent manner in order to reduce any negative impact resulting from the implementation of such changes.

Change Request Form or CRF. Formal documentation and notification of a change (and its particularities) submitted by the individual requesting the change (i.e., Change Requester). The CRF serves as the written communication and agreement between the requester and the IT application/development staff.

Change. Modification needed to transform or alter the operating environment or procedures of the organization. A change could potentially impact the reliability and availability of the IT infrastructure, computing environment, and data used in conducting normal internal and external business operations. Changes can be classified as normal (i.e., impact and risk of the change is not significant to the organization); significant (i.e., high risk and high impact to the organizations users); or as emergency changes (i.e., changes needed to correct errors or unexpected issues within any system of the IT infrastructure that would otherwise result in an adverse impact to business operations). Refer to the "Types of Changes" section.

Change Control Board or CCB. Committee created to assess change requests for business need (e.g., cost versus benefit, priority, potential impact, etc.), and make decisions regarding whether proposed changes should be implemented. Decisions may take the form of recommendations for implementation, further analysis, deferment, or cancellation of the change. It is common for the CCB to meet twice a month, or as needed (as is the case for emergency changes), become part in the scheduling of all changes, and be available for consultation should emergency changes are required. Upon successful evaluation, the CCB approves the CRF, and assigns a Change Owner to administer the implementation and closure of the change.

Types of Changes

Below is a list of circumstances that may trigger common system changes within organizations.

- Upgrades to hardware (e.g., periodic maintenance, users specific needs, hardware changes, additions, deletions, reconfigurations, relocations, preventive, emergency maintenance, installation of servers, etc.)
- Upgrades to software (e.g., periodic maintenance, users specific needs, keeping a system up during a normal shutdown period, temporary or permanent system fixes, new releases, new versions, database table changes, modifications to libraries, changes to operating/scheduling jobs, fixing bugs, correcting errors, etc.)

- Purchase of new hardware and/or software systems
- Implementation of information systems (e.g., testing and implementation of new applications, new operating systems, new versions, new releases, data conversions and/or migrations, etc.)
- Network systems (e.g., additions, modifications, and/or deletions of network lines, routers, network access, servers, user access, etc.)
- Environmental changes (e.g., responding to sudden interruptions or failures in power, uninterruptible power supply systems, generators, air conditioning, electrical work, facility maintenance, security systems, fire control systems, etc.)

Responsibilities

Organization users are responsible to use the CRF when addressing system changes. The IT Department's responsibility is to guide the change process (ensuring compliance with user requirements) while maintaining close communication with requesters, IT personnel, and management during the modification, testing, and implementation process. Following are descriptions of the responsibilities for each role.

Change Requester

User responsible to initiate and submit the request for a change. The request is sent to the Change Register using the required form (i.e., CRF).

Change Register

Receives the submitted CRF from the Change Requester. Upon receipt, assesses the change request to ensure accuracy and completeness of all necessary information related to the change. The Change Register's responsibilities include:

- Classifying the change request according to the change type (refer to "Types of Changes" section).
- Classifying the request based upon the business requirements and established policies as either normal, significant, or emergency (refer to *Change* in "Definitions" section). If the change is classified as an emergency change, it is referred to the Change Control (or Project) Manager. Depending on how critical the change is, the Change Control (or Project) Manager will conduct meetings, as appropriate, to discuss the emergency change and to obtain the required authorizations for immediate implementation.
- Assessing the change to determine operational impact, risks, and priority. The Change Register will identify the change's impact on infrastructure, systems, and applications, verifying that all equipments (e.g., software, hardware, etc.) required to work with the change are available, as well as perform the necessary assessments or risks (i.e., risk benefits and risk analysis) related to the change. The Change Register will also determine priority based on the results of the impact and the urgency of the change, as well as the assessment of risk. Priority will account for the number of affected users, IT resources required, systems involved, hours required, etc.
- Upon determining that all change request information is accurate and complete, the Change Register submits the CRF to the Change Control Reviewer for review and approval.

Change Control Reviewer

Receives the completed CRF for review from the Change Register. Responsibilities include:

- Ensuring that all documentation is complete and accurate, that the CRF includes a detailed description of the change, the business justification, and impact if not implemented, and that all applicable parties are aware of any possible impact. The Change Control Reviewer also reviews and corroborates that documentation related to change type, classification, operational impact, risk, and priority is complete and accurate.
- If the CRF is incomplete or needs additional information/justification, it will be forwarded back to both, Change Requester and Change Register, indicating the missing or inaccurate information. Once the CRF has been reviewed and modified as necessary, the Change Register re-submits it to the Change Control Reviewer.
- If the CRF is accurate and complete, the Change Control Reviewer forwards it to the Change Control (or Project) Manager for approval.

Change Control (or Project) Manager

The Change Control (or Project) Manager oversees the entire change control management process, and is responsible for its continual improvement. Other responsibilities include:

- Assessing the CRF, and determining potential impact to the department requesting the change, as well as to the entire organization's infrastructure.
- Notifying all affected parties about the upcoming change, and about potential conflicts that will need resolution.
- Coordinating, scheduling, and running meetings with the CCB related to change control management, specifically to evaluate, approve, authorize and/or reject the requested change(s).
- Grouping all outstanding CRFs awaiting evaluation or further consideration or action.
- Once the request is approved, the Change Control (or Project) Manager coordinates for support, if needed, to resolve problems identified or to answer any questions during or immediately after installation/implementation.
- With assistance of CCB members, the Change Control (or Project) Manager reviews all implemented changes to ensure compliance with objectives, and to identify actions and opportunities to correct problems and improve quality service, respectively. The Change Control (or Project) Manager identifies and refers back any changes that have failed or have been backed out.

Change Owner

The Change Owner is responsible for the effective planning, administration, execution, and implementation of all approved changes. He or she prepares and forwards the implementation plan to the Implementation Team for building, testing, and validation of the change. The plan includes the scheduling of resources and the assignment of required tasks. Other tasks include:

- Receiving approved CRFs from the CCB.
- Communicating the final implementation date to the Change Control (or Project) Manager.

- Updating CRFs, if/when needed.
- Instructing the Implementation Team to proceed with the change implementation.
- Identifying if any of the changes are classified as emergency changes. Note: Emergency changes are subject to the same test or quality assurance activities as normal changes, except that controls, supervisory tasks, and rollback plans/procedures must be in place at all times to protect the production environment from unauthorized updates.
- Gathering approvals prior to implementation of the change into the production environments. These final approvals relate to:
  - Completion of all testing, including documentation of effective test results
  - Implementation date
  - Quality assurance
  - Completion of all change-related documentation (e.g., user/training materials, configuration of information security, backup and restoration procedures, etc.).
- Communicating identified problems to the Change Control (or Project) Manager and together determine if implementation of the approved change should continue or not. If implementation of the change must not continue, then the change must be rolled back. If the change implementation is successful, the Change Owner updates the CRF with results.
- Conducting reviews of the change after implementation (i.e., post-implementation reviews). These reviews address, among others, if:
  - the change met or not its original objectives;
  - implementation of the change was performed as scheduled and within budget;
  - all affected parties were properly informed on progress, implementation date, and results of the implementation; and
  - the change implementation was or not rolled back (if so, documentation, such as, CRFs must be appropriately revised and updated).
- Reporting to, and discussing overall results from implementation of the change with both, Change Control (Project) Manager and the CCB.
- Closing the CRF.

Revision

Version	Description of Revision Performed	Performed By	Date
1.00			XX/XX/20XX

Appendix 7: Sample Information Systems Operations Policy

	Information Systems Operations Policy Policy Number: XYZ-WI-ISO-001 Version: 1.0 Department: Information Technology		
Name	Position	Signature	Date
Prepared by:	IT Operations Supervisor and/or Administrator		XX/XX/20XX
Reviewed by:	Director of IT Infrastructure		XX/XX/20XX
Reviewed by:	Information Systems Operations Manager		XX/XX/20XX
Approved by:	Vice President of IT		XX/XX/20XX

Mission

Promote an IT environment and infrastructure that is reliable and available.

Purpose

The purpose of the Information Systems Operations policy is to:

- Provide a formal methodology in which operations relating to the organization's information systems (IS) are documented.
- Ensure the effective operation of computer systems and programmed procedures.
- Implement, execute, and enforce controls and procedures to protect IS operations.

◼ Promote awareness among all organization's users that IS equipment should be protected at all times against unauthorized use, theft, physical damage, or environmental damage. IS equipment consists of desktop, workstations, servers, laptops, and notebooks computers; networks; mobile devices; external drives; USB drives; and backup tapes; among others.

Scope

This policy covers IS processes, procedures, and controls to be implemented to conduct normal business operations while safeguarding the IT environment.

Responsibilities

1. Users:
   a. Organization users are fully responsible for reading, understanding, and complying with all policies and procedures described within this document.
2. Supervisors/Administrators:
   a. Designated as application owners (custodians) of organization's systems, including applications, databases, operating systems, and networks.
   b. Ensure users strictly adhere to and comply with the policies and procedures described within this document.
   c. In connection with human resources personnel:
      i. authorize user access and assign custody of information.
      ii. review user access on a periodic basis; modify such access as needed.
      iii. ascertain that training is provided to all users.
      iv. ascertain that policies and procedures are distributed and communicated to all users.
   d. Design and implement physical access controls and mechanisms to safeguard IS specific areas and equipment.
   e. Approve backup schedules, frequencies, and rotation.
   f. Approve backup results and correction to backup exceptions in a timely manner.
   g. Ensure policies and procedures related to IS operations are revised and updated accordingly.
3. Information Systems Operations personnel:
   a. Comply with and strictly adhere to all policies and procedures described within this document.
   b. Receive from users (and document) notification of identified failures and events related to IS operations.
   c. Design and implement backup schedules, frequencies, and rotation.
   d. Execute and monitor backup procedures.
   e. Communicate exceptions noted, if any, to supervisors and administrators for timely correction.
   f. Maintain updated backups for all system and application data.
   g. Keep backups safe in adequately controlled environments or locations.
   h. Develop and document business continuity plan (BCP) and disaster recovery plan (DRP). In connection with security personnel, perform tests for both plans.

4. Human Resources Department:
   a. In connection with operations and security personnel, grant, remove, or modify physical access, as appropriate, to organization's sensitive facilities (i.e., data centers, computer rooms, network equipment, and other computer-related or sensitive building/physical areas that host relevant financial applications, databases, operating systems, and other repositories for financial information).
5. Management:
   a. Review and approve policies, procedures, and responsibilities stated in this document.
   b. On an annual basis, review, revise, and approve BCP and DRP. Review and approve test results.
   c. Promote sound policies and procedures that foster business practices that are trustworthy, reliable, effective, efficient, consistent, ethical, and compliant with all applicable laws and regulations.

Policy: Information System Operations Guidelines

Overall guidelines, strategies, and/or best practices should be implemented to govern sound and safe IS operations at the organization. As required by this policy, the organization should inform users that:

■ IT Department is responsible to provide secure, accurate, and complete IS services to the organization.
■ IT Department maintains guidelines, policies, and procedures necessary to adequately control IS operations at the organization.
■ Guidelines, policies, and procedures should be clearly defined, formally approved, implemented, and revised at least once a year (or when appropriate) to reflect changes within the processing environment.
■ Updated guidelines, policies, and procedures should be communicated and distributed to all organization users.
■ IS operation activities must be planned and scheduled accordingly (e.g., daily, weekly, monthly, etc.) consistent with policies and procedures.
■ IS supervisors or administrators must ensure the availability of all necessary supplies, hardware, software, as well as other related equipment and documentation for users to execute their tasks and duties.
■ Maintenance for all software, hardware, and related IS equipment must be properly coordinated, scheduled, and executed to ensure accurate and complete function and operation.
■ Food, liquids, and the like are prohibited at the IT Department, data center, and/or nearby any IS equipment.

Policy: Safeguarding of Information Systems

IS equipment and related documentation should be protected at all times against unauthorized use, theft, physical damage, or environmental damage. As required by this policy, the organization should inform users that:

■ IS equipment must not be left unprotected.
■ Users must keep user ids and passwords safe and protected. The lack of doing so may result in a serious security breach to the organization's sensitive and confidential data.

- IS equipment must not be used or placed in an unsafe physical area exposing such equipment to potential physical damage.
- Unless explicitly authorized, users cannot add, delete, or modify application programs.
- Unless explicitly authorized, users cannot remove peripheral devices or change/delete configurations settings (e.g., sound settings, screen appearance, etc.) in any organization's computer, workstation, server, mainframe system, etc.
- To protect organization's sensitive facilities, the following conditions or controls should be implemented:
  - Alternate power supplies to prevent system downtime and crashes in case the main power supply fails.
  - Alternate air conditioning systems in case the main air conditioning system fails.
  - The data center room should be placed in a safe place with a raised floor constructed above the building's original concrete slab floor, leaving an open space between the two for wiring or cooling infrastructure.
  - Fire suppression equipment systems (e.g., fire sprinkler system, gaseous fire suppression, condensed aerosol fire suppression, etc.) that do not require human intervention should be installed to control and extinguish fires.
  - Cabinets storing files, documents, backup media, software, and/or any other sensitive information or equipment should be secured and fire-resistant.

Policy: Designated Application Owners

Application systems, including IS equipment, must have at least one designated owner. As required by this policy, the organization should inform users that:

- The designated owner, authorized by supervisors or administrators, is responsible for the business results of his/her application system, business use, and all related information.
- Changes or modifications to the application system are not to be made without the consent of the designated owner.
- Non-designated users should not perform modifications or changes to application systems unless explicitly authorized by (and with the consent of) the designated owner.
- Owners designated for application systems and their information have the authority and responsibility to:
  - Classify information and its value.
  - Authorize access to information to delegates (custodians).
  - Identify required controls and communicate such controls to custodians and/or users of the information.
  - Determine (and enforce) statutory requirements for data privacy and retention.
- Designated application owners are responsible to protect the privacy and confidentiality of the information owned.
- Appropriate forms must be documented, completed, and authorized. Forms must include names of designated application owner and application system, as well as a description of related information owned and/or produced. Both, designated application owner and appropriate supervisor/administrator must sign the form.

Policy: Application and System Users

The term user is defined as every employee who utilizes the organization's application systems and IS equipment. Users are authorized by designated application owners to read, create, edit, update, or delete information, as appropriate. Users are required for the safeguarding of the information and the application systems utilized in their daily duties and tasks. As required by this policy, the organization should require users to:

- Be held accountable for the proper usage and protection of all organization's application systems and information.
- Utilize the information only for the purpose intended by the designated application owner.
- Comply with all policies, procedures, and controls implemented by the designated application owner.
- Ensure that classified, private, or sensitive information is not disclosed to anyone without permission of the designated application owner.
- Make certain that individual passwords are not disclosed to, or used by, unauthorized users.

Policy: Physical Security and Access to Organization's Sensitive Facilities

Specific rules, procedures, and controls have been designed and implemented for granting, monitoring, modifying, and removing physical access to the organization's sensitive facilities. As required by this policy, the organization should inform users that:

- Physical access granted to all organization's sensitive facilities should be previously authorized, documented, managed, and monitored.
- Physical access to organization's sensitive facilities should be restricted to users who require such access to perform their job duties. Such restricted access helps prevent malicious or accidental destruction of IS equipment and organization's sensitive facilities.
- Access granted to external support personnel (e.g., vendors, contractors, suppliers, technical personnel, building maintenance personnel, etc.) to the organization's sensitive facilities must be properly authorized and documented. Such access granted must be consistent with job purpose and responsibilities.
- Granting access to organization's sensitive facilities must follow proper approval from supervisor or administrator.
- A physical access control mechanism (e.g., access cards, biometrics, traditional lock and key, security guards, etc.) is used to restrict and record access to the building and to organization's sensitive facilities, and authority to change such mechanism is limited to appropriate personnel.
- Physical access to organization's sensitive facilities is granted through ID cards or biometrics authentication.
- Biometrics authentication is employed through fingerprint recognition.
- ID cards must not be transferred or used by unauthorized users in order to bypass physical security mechanisms and controls.

- ID cards no longer used/required are to be returned immediately to the supervisor or administrator.
- Lost or stolen ID cards must be notified immediately to the supervisor or administrator. Charges will be made for all lost, stolen, or not returned ID cards.
- Entry of unauthorized personnel to organization's sensitive facilities is supervised and logged, and such log is maintained and regularly reviewed by IT supervisors or administrators.
- Visitors entering the organization's sensitive facilities will be required to sign in and out to evidence their visit, as well as document the purpose of such visit.
- Visitors must be escorted by authorized personnel at all times.
- Access records and visitor logs for all organization's sensitive facilities must be reviewed on a periodic basis. All unusual activity must be monitored and investigated accordingly. Any unusual access identified should be notified and immediately removed.
- User access reviews occur frequently to support current physical access granted to organization's sensitive facilities.

Policy: Communication of Downtime and System Failures

A notification method is instituted to reduce systems' downtime, outages, communication failures, power failures, crashes, and interruptions (altogether referred to as failures). As required by this policy, the organization should inform users that:

- IS failures identified are immediately communicated to IS operations personnel.
- Failures are logged and documented by IS operations personnel, and such documentation must describe:
  - conditions that existed at the time of the failure
  - any unusual activity noted prior to the failure, and
  - potential solutions received from managers, supervisors, administrators, vendors, consultants, etc.
- Following appropriate approval, corrections and solutions are coordinated and put in place.
- Authorized correction and solutions are documented and made available to appropriate personnel. Correction of failures is documented, closed, and available for future reference.

Policy: Backup Procedures

Backup policies and procedures are implemented to safeguard the organization's data, information, applications systems, and software. As required by this policy, the organization should inform users that:

- Backups are planned, consistently scheduled, and performed using effective and efficient backup software, tools, and/or technologies.
- Copies of backed up data, information, application systems, and software are kept for continuity planning purposes.
- In case of malfunction, interruptions, or failures, one alternate backup unit must be available, at a minimum, to replace the main backup unit.
- Backup media must be properly identified and labeled.
- Backup media must be securely stored and restricted to authorized personnel. Physical access controls must be observed and complied with related to the storage of backups.

- Prior to performing backups, information must be classified according to purpose. For critical and/or sensitive information, a minimum of two backup copies are generated (one copy is delivered to offsite storage facilities; the other copy is maintained internally). Information that is classified as non-critical or non-sensitive is evaluated, backed up, and stored offsite if/when needed. Typically, maintaining one copy of such type of information is sufficient.
- Backups of critical and/or sensitive information are also protected through encryption methods.
- Full system backups are to be executed prior to any significant modification being made to relevant information or application system.
- Frequency of backups (i.e., daily, weekly, monthly, quarterly, semi-annually, and yearly) is established based on how sensitive and critical the information is.
- Retention of backups, which based on frequency, is established as follows:
  - Daily—7 days
  - Weekly—30 days
  - Monthly—90 days
  - Quarterly—365 days
  - Semi-annually—5 years
  - Annually—10 years
- The backup execution process is monitored, logged, and includes description of the information or application system being backed up.
- Backup restoration tests are validated and executed. Results are monitored and approved.

Policy: Business Continuity Plan

An established BCP describes the processes, steps, and procedures to be carried out in the event of an emergency (e.g., natural disaster, failures, unplanned interruptions to normal business operations, etc.) to achieve a timely recovery and availability of all essential business processes, including the IS. As required by this policy, the organization should inform users that:

- Management approves development of the BCP, and oversees its implementation.
- The BCP has been prepared following results from risk, impact, and cost analyses and assessments related to losses of information and IS services.
- Business priorities and critical needs have also been considered and incorporated into the BCP.
- Personnel responsibilities have been assigned, as appropriate. Strategies and procedures for recovery have been documented.
- The BCP is updated and tested on a regular basis by IS operations personnel.
- Test criteria, conditions, and frequency have been established and added to the BCP.
- Test results are evaluated and gaps, weaknesses, or deficiencies identified are addressed.
- Tests and test results are shared with management for review and approval.

Policy: Disaster Recovery Plan and Tests

A DRP is a survival tool to help the organization respond to threats and recover in the wake of an event that disrupts normal business operations. Testing of the DRP should be conducted on a periodic basis to validate assumptions, and to update the plan based on the constantly changing environment. Testing also provides the opportunity to practice recovery procedures and identify missing elements. Tests conducted must be available to assess technical and non-technical procedural aspects of the organization's DRP. Tests also reduce the opportunity for miscommunication when the plan is implemented during a real disaster. Further, testing offers management an opportunity to spot weaknesses and improve procedures. As required by this policy, the organization should inform users that:

- IT security and operations personnel are responsible for developing, documenting, and testing the DRP.
- Disasters considered in the DRP include natural disasters (e.g., earthquakes, tsunamis, hurricanes, tornados, flood, fires, etc.) or unnatural (e.g., cyberattacks, disruption of service, fraud, terrorism, market collapse, etc.), both of which create economic chaos and severe business interruptions.
- The DRP is supported and approved by management, and should address components such as:
  - Objectives and mission statement
  - Critical support services and resources required to deliver products or services
  - List of all critical hardware and software applications
  - Minimum service level requirements for information and telecommunications
  - Key personnel and minimum staff required to operate during the disaster
  - Employee relocation plans to alternate work sites
  - Alternate processing facilities and steps to transfer necessary resources to such facilities
  - Program and data backups stored off-site
  - Full and incremental program and data backups
  - Disaster recovery chairperson and committee(s) appointed
  - Contact lists, emergency telephone numbers, and alert systems with steps to follow
  - Insurance coverage
  - Tests and drills
  - Up-to-date system and operation documentation
- A test plan, procedures, and appropriate test scenarios should be developed and documented consistent with DRP purpose, goals, and/or objectives.
- A test plan, procedures, and appropriate test scenarios should be approved by management.
- Tests should be executed at least once a year.
- Tests should be coordinated and agreed with alternate off-site location(s).
- Tests should include full restoration of backup media and validation that data remains complete and accurate.
- Test results should be documented and any gaps, weaknesses, or deficiencies resulting from those tests should be addressed and corrected on a timely basis.

Revision

Version	Description of Revision Performed	Performed By	Date
1.00			XX/XX/20XX

Appendix 8: Auditing End-User Computing Groups

Audit of End-User Computing Groups

Once it is determined that an audit of an end-user computing (EUC) group is required, the IT auditor, in conjunction with management, defines objectives, methodology, and scope and content, before preparing the audit plan. The audit objectives may cover specific applications, end-user support, financial issues, and/or strategic information to be reported to the management.

Audit Objectives

PC applications have grown from individuals creating personal productivity tools into critical applications that are used by the entire organization. Management may not fully realize the importance of EUC groups to the organization to dedicate the necessary resources for a complete and thorough applications audit. However, it is essential to have management's support to overcome any obstacles put forth by the EUC groups. End users tend to think of their PCs as personal property, and they may be resentful of an intrusion by auditors. However, end user's cooperation can be gained, in part, by explaining the objectives of the audit. A common audit objective would be to evaluate the effectiveness of application and general controls over EUCs (i.e., critical Excel spreadsheets, access databases, and other data analysis and reporting tools). Other objectives of an EUC audit may require assessment and testing to ensure that:

- The organization's EUC applications are appropriately managed and stored, consistent with internal policies, procedures, standards, and guidance, to ensure their hosted data is or remains complete, accurate, and valid.
- Security is appropriately configured, implemented, and administered on all EUC applications to safeguard against unauthorized access. Unauthorized modifications can cause information or its processing to be incomplete, inaccurate, or invalid.
- Sensitive, private, and/or confidential information hosted in the organization's EUC applications are adequately protected to prevent authorized access, data breaches, and to make certain that such type of information is not compromised or disclosed without authorization.

Audit Methodology

The method used to conduct the EUC group audit depends on the environment being reviewed and the agreed-upon audit objectives. An inventory of end-user applications can be used to gain a general understanding of the EUC group. The auditor should discuss this inventory with management to determine what type of audit should be performed. For example, a more formal audit can be used if a specific application is being evaluated for reliance on financial information, whereas a statistical audit that collects sample data from transactions or supporting logs can confirm end-user practices. Auditors could also perform a quick, informal assessment by interviewing the IT staff about their impressions of the EUC group.

Audit Scope and Content

Defining the EUC group for a particular environment will determine the audit scope and content of the audit. The scope limits the coverage of the audit to a particular department, function, or application. The content defines what aspects of a particular area are covered. Depending on the audit objective, the content covers general controls, application controls, hardware and software acquisition, systems development controls, change controls problem management, or disaster recovery.

Audit Plan

The audit plan details the objectives, methodologies, scope and content, as well as the procedures to fulfill those objectives. Like any audit, an audit of an EUC group begins with a preliminary survey or analysis of the control environment by reviewing existing policies and procedures. During the audit, these policies and procedures should be assessed for compliance and operational efficiency. The preliminary survey or analysis should identify the organization's position and strategy for the EUC group and the responsibilities for managing and controlling it. The following audit procedures are performed to gather the necessary evidence on which to base audit findings, conclusions, and recommendations.

- *Evidence gathering.* A review of any documentation that the end-users group uses.
- *Inquiry.* Conducting interviews with end users and any IT support technicians.
- *Observation.* A walk-through to become familiar with department procedures and physical assess controls.
- *Inventory.* A physical examination of any inventoried goods or products on hand in the EUC group.
- *Confirmation.* A review of the end users' satisfaction surveys that were handed out and completed during the preliminary audit planning stages.
- *Analytical procedures.* A review of data gathered from statistical or financial information contained in spreadsheets or other data files.
- *Mechanical accuracy.* A review of the information contained in any databases used by the EUC group through testing procedures.

After the evidence is gathered, the auditor should assess control strengths and weaknesses, considering the interrelationships between compensating and overlapping controls. These controls should be tested for compliance and to ensure that they are applied in accordance with management's policies and procedures. For example, management policy may state that end users should change their passwords periodically to protect information resources. To test for compliance, the auditor identifies the controls that force password changes. Substantive tests determine the adequacy of these controls to prevent fraudulent activity. For example, software piracy puts the company at risk for fines and the potential loss of goodwill. Reviewing the directories on the LAN and PC drives for unlicensed software would assess the effectiveness of controls to ensure that only property-licensed software is installed.

The audit plan should address the review of policies and procedures, ensure that there is enough and available documentation supporting EUC processes and procedures, define audit tests to be performed, and describe the content of the final audit report. These are described next.

EUC Policies and Procedures

IT should have policies or guidelines that cover EUC groups. These should be designed to protect company data. IT should also have standards to ensure that end users are not using hardware or software that is not supported by them. There should be an EUC policy that encompasses and is applicable to all EUC groups. If only departmental policies exist, each policy should be similar to ensure continuity between departmental policies. A companywide policy related to EUC groups should cover:

- Assignment of ownership of data
- User accountability
- Backup procedures
- Physical access controls to PCs
- Appropriate documentation of all EUC groups' applications and adequate documentation changes and modifications
- Segregation of duties

EUC Documentation

Because many end users develop their own applications, often there is little or no documentation apart from the end user's own notes. Another audit concern is that several end users may be developing the same type of application independently of each other, which is an inefficient use of computer resources. For example, if an end user in accounting is developing an application that is already in use in payroll, there should be some type of documentation or reference for end users to consult to prevent duplication of effort.

Another problem posed by inadequate documentation is illustrated by this example. An end user has developed several applications that have become crucial to the operation of the organization. This individual has left the company without leaving any documentation on those applications, and other end users must use this application and make modifications to it.

End users must assume responsibility for the maintenance of documentation for their systems and applications, and ensure that it is complete, current, and accurate. The IT auditor can perform an effective management advisory role by highlighting and emphasizing the importance of EUC documentation.

EUC Audit Testing

The auditor must address many considerations that cover the nature, timing, and extent of testing. The auditor must devise an auditing testing plan and a testing methodology to determine whether the previously identified controls are effective. The auditor also tests whether the end-user applications are producing valid and accurate information. For microcomputers, several manual and automated methods are available to test for erroneous data. An initial step is to browse the directories of the PCs in which the end-user-developed application resides. Any irregularities in files should be investigated.

Depending on the nature of the audit, computer-assisted techniques could also be used to audit the application. The auditor should further conduct several tests with both valid and invalid data to test the ability and extent of error detection, correction, and prevention within the application. In addition, the auditor should look for controls such as input balancing and record or hash totals to ensure that the end user reconciles any differences between input and output. The intensity and extent of the testing should be related to the sensitivity and importance of the application. The auditor should be wary of too much testing and limit his or her tests to controls that cover all the key risk exposures and possible error types. The key audit concern is that the testing should reveal any type of exposure of sensitive data and that the information produced by the application is valid, intact, and correct.

EUC Audit Report

The audit report should inform management about the results of the review of the EUC group. It can also suggest support for resources to enhance end-user controls. In addition, the audit report should recommend policies and procedures that could strengthen end-user controls. Finally, the audit report should convince end users and management of the need for controlling EUC by identifying the importance of the information and assets stored on the PCs and LANs and by pointing out the risks to those assets.

The auditor's report should also recommend management of the types of controls that are needed to increase efficiency and decrease risk and exposure. These recommended controls should be defined in a cost versus benefit manner and should be expressed in terms that management will understand: How much it will cost the company if these types of controls are not in place? or how much the company can save if such controls are in place? After these recommendations have been made, approved, and implemented, the auditor should reevaluate the controls to ensure that they have been implemented and are effective.

Appendix 9: Recommended Control Areas for Auditing Software Acquisitions

To offset risks associated with acquiring software, recommended control areas, such as the ones below, should be examined by IT auditors.

1. Alignment with the company's business and IT strategy.
2. Definition of the information requirements.
3. Feasibility studies (e.g., cost, benefits, etc.).
4. Identification of functionality, operational, and acceptance requirements.
5. Conformity with existing information and system architectures.
6. Adherence to security and control requirements.
7. Knowledge of available solutions.
8. Understanding of the related acquisition and implementation methodologies.
9. Involvement and buy-in from the user.
10. Supplier requirements and viability.

Alignment with the Company's Business and IT Strategy

Any system development project, whether the system is developed by the organization or purchased elsewhere, should support the organization's business and IT strategy. The business requirements associated with the solution being sought should link to goals and objectives identified in the company's business and IT strategy. For further information about IT alignment with the business, see Chapter 5.

Definition of the Information Requirements

System and information requirements should be evaluated to determine if they are current and complete. Owing to the fast pace of business, requirements can change quickly. Consequently, requirements that are gathered too far in advance of the actual purchase may not capture any changes in business requirements or newly available technical features.

The biggest challenge with defining system requirements is getting them to be complete. The requirements for a system can never be 100% complete. Conversely, revising requirements throughout the system acquisition process can result in change in scope, expectations, cost, and consequently, success.

Feasibility Studies (e.g., Cost, Benefits, etc.)

Feasibility studies should be reviewed to ensure that the selected solution not only meets the requirements but also is compared and contrasted with the feasibility of the other solutions. Related controls and risks are illustrated in the following paragraphs using economic and technical feasibility as examples.

Economic feasibility should be reviewed and approved by an involved and knowledgeable sponsor before the final decision to ensure that the "make versus buy" question is effectively evaluated. Management should formally sign off on the cost–benefit analysis. In one circumstance, a government agency purchased a software solution without traceable documentation of the alternatives reviewed and the related cost–benefits associated with each. Consequently, regulators scrutinized the competency of senior management and the fairness of the selection process.

There are multiple examples of companies that prepare misleading cost–benefit studies that are based on immeasurable benefits and incomplete costs. Benefits are often presented in terms of functions that are not measured in the current environment. As a result, it is difficult to prove benefit in the new environment. Indirect or in-kind costs are often excluded from cost estimates. Examples include staff costs associated with reassigning staff from their regular duties to the implementation project. These costs can include fees for temporary staff, loss of revenue associated with reduced service due to reduced staff, or changes in employee compensation resulting from increased job responsibilities or expected skills.

Technical feasibility should be reviewed and approved by an involved and knowledgeable sponsor before the final decision to ensure the organization's ability to implement and support the selected solution. In one example, a company's CFO purchased a financial ERP package without consulting the company's technology division in advance. The technology division was placed in the position of scrambling to incorporate the package into the existing architecture following the purchase. Consequently, changes were required to the design of the technical infrastructure resulting in unplanned hardware and software purchases.

Identification of Functionality, Operational, and Acceptance Requirements

Requirements need to extend beyond end-user expectations. They should include the internal functionality of the system with consideration for operational and acceptance requirements. Examples of functionality that can be missed include printing requirements or the business-specific algorithms for calculations. In one example, a company unexpectedly had to change its check printing process due to the implementation of a software package. Because the checks were now completely laser printed, the printed check had to include the account information in optical characters as well as the associated bar code. The banking information on the new check had to be reviewed and approved by the bank. Additionally, the font size had to be adjusted appropriately for the post office to systematically sort the mail. Finally, the checks were jamming in the mail sorter at the post office. Because the check was now a self-enclosed mailer without a separate envelope, the paper used to print and mail the check became stressed from being passed through the folding machine, postage meter, and the mail sorter. The company subsequently purchased a different folding machine and implemented other alternatives to apply the postage.

Acceptance criteria should be specific with detailed measures. Acceptance plans should include inspections, functional tests, and workload trials. Without acceptance guidelines, the selected solution may not comply with user requirements, performance expectations, or the terms of the contract. There are many cases where inadequate acceptance guidelines have resulted in business interruptions from inadequate system performance or nonworking functions.

Conformity with Existing Information and System Architectures

This control area is directly correlated with the evaluation of technical feasibility and the business's information elements. As mentioned earlier, conformity with the existing system architecture is critical. In addition, the feasibility analysis defines the constraints or limitations for the system. Feasibility should be assessed in the following categories: economic, technical, operational, schedule, legal or contractual, and political. Feasibility can include those things that are tangible, intangible, one-time, or recurring.

In one case, a company selected a software package that did not accommodate the company's method for recording the commission for its sales agents in the general ledger. Consequently, the company chose to modify the software's structure for the chart-of-accounts structure, which resulted in changes and maintenance to the basic data structure of the product, as well as all the associated code and screens.

Adherence to Security and Control Requirements

A complete understanding of the company's security and control requirements is needed to ensure that the selected solution is appropriate. Company security policies and applicable regulations need to be reviewed during the selection process to ensure that security and control requirements are considered in the selection process. The company security officer should be involved in defining the security and control requirements as well as participate in the selection process.

System acquisitions and implementations become more difficult when these requirements are not well understood or documented. The results will be missed security functionality or poorly implemented security. In instances where the security policy or requirements are not well documented, it would be wise to have them documented and approved by senior management before the selection process to ensure that security and control requirements are met.

In situations where there are gaps in security between requirements and the evaluated solutions, cost and benefits of controls should be evaluated to ensure that the costs do not exceed the benefits. This provides an opportunity for risks and controls to be revised and updated to reflect changes in business and technology. The security officer and management should participate in and approve any changes to security requirements or selected controls.

Knowledge of Available Solutions

Often, system development and acquisition efforts become more focused on a specific solution due to the knowledge or experience of the participants. By focusing on a specific end result, other alternatives are not considered. By not considering other alternatives, the selected solution may increase cost, scope, or the timeline for the project because they did not meet basic requirements, such as incompatibility with the current company infrastructure or business practices. Specifically, it takes additional time and resources to integrate the selected solution into company technical infrastructure or business practices.

Understanding of the Related Acquisition and Implementation Methodologies

Acquisition methods of an organization can be very specific or general based on a variety of factors like government regulations. As an example, government acquisition guidelines require that equal opportunity be provided to all potential providers. Consequently, there are specific requirements for advertising RFPs, evaluating bidders, as well as awarding contracts. If these guidelines are not followed, agreements may not be considered valid.

Selected implementation methods may be inadequately understood, and this may introduce risk to the deadlines, scope, and costs of the project. As an example, a company selected an implementation partner to assist it in converting its legacy billing system into a state-of-the-art system using recent programming technologies. The company's experience and culture was based on traditional mainframe technologies and the traditional waterfall system development methodologies. The staff was ill equipped to understand (and participate in) the recent approaches employed by the selected implementation partners. The company actually experienced failed contracts with several implementation partners and there were subsequent lawsuits filed by the company and implementation partners.

Involvement and Buy-In from the User

User involvement and buy-in is critical. Without user involvement, requirements will be missed and they will not support new systems. There is an increased awareness of the criticality of user support and buy-in. As a means for increasing user support, many projects are now including communication and business change management as part of their project plans.

Change management, in this context, includes people, organizations, and culture. A culture that shares values and is open to change contributes to success. To facilitate the change process, users should be involved in the design and implementation of the business process and the system. Training and professional development supports this as well.

System-implementation success relies on effective communication. Expectations need to be communicated. Communication, education, and expectations need to be managed throughout the organization. Input from users should also be managed to ensure that requirements, comments, and approvals are obtained. Communication includes the formal promotion of the project team as well as the project progress to the organization. Employees should also know the scope, objectives, activities and updates, and the expectation for change in advance.

Supplier Requirements and Viability

The acquisition process should ensure that the selected supplier meets the supplier requirements of the organization as outlined in the proposal. As mentioned in Chapter 13, these requirements include:

■ Stability of the supplier company
■ Volatility of system upgrades
■ Existing customer base
■ Supplier's ability to provide support
■ Required software in support of the supplier system
■ Required modifications of the base software

To determine the viability of the supplier, elements such as financial condition, risk of acquisition, likelihood of exiting the market, and reputation for responsiveness during problems should also be evaluated.

Appendix 10: Glossary

Accounting: Process of identifying, measuring, and communicating economic information about an entity for decisions and informed judgments.

Account Receivable: Refer to *Receivable*.

Accuracy: The quality or state of being correct or precise.

Actuarial Services: Method by which corporations define, evaluate, and plan for the financial impact of risk. Actuaries use mathematical and statistical models to assess risk in the insurance and finance industries. Actuarial science is used to evaluate and predict future payouts for insurance and other financial industries such as the pension industry.

Advisory: Being in a capacity to provide advice or opinions. Advisory committees or similar groups also have the authority to render a decision or judgment on an issue, in addition to providing opinions. For example, an advisory committee to a board of directors may have the ability to decide whether or not certain restrictions or regulations are adequately being met.

American Institute of Certified Public Accountants (AICPA): Represents the CPA profession nationally regarding rule-making and standard-setting, and serves as an advocate before legislative bodies, public interest groups, and other professional organizations. The AICPA develops standards for audits of private companies and other services by CPAs; provides educational guidance materials to its members; develops and grades the Uniform CPA Examination; and monitors and enforces compliance with the profession's technical and ethical standards; among other responsibilities.

Assessment: Evaluation or estimation of the nature, quality, or ability of someone or something.

Assets: Properties and other things of value owned and controlled by an economic unit or business entity. Examples include cash, equipment, copyright, building, and land.

Assurance: Positive declaration intended to give confidence; a promise.

Attest or Attestation: To bear witness to; certify; declare to be correct, true, or genuine; declare the truth of, in words or writing, especially affirm in an official capacity.

Audit: Examination and evaluation of the financial statements of an organization. This is done objectively, to make sure that the records are an accurate representation of the transactions. It can be done internally by employees of the organization, or externally by an outside auditing firm.

Audit Committee: An operating committee of a company's board of directors that is in charge of overseeing financial reporting and disclosure. All U.S. publicly traded companies must maintain a qualified audit committee in order to be listed on a stock exchange.

Audit Engagement: Audit performed by the auditor. More specifically, it refers only to the initial stage of an audit during which the auditor notifies the client he has accepted the audit work and clarifies his understanding of the audit's purpose and scope.

Audit Finding: Refer to *Finding*.

Auditee: Person or entity that is being audited.

Auditor: Person or firm who conducts an audit. Refer to *Audit*.

Audit Process: An audit process starts with a risk assessment, followed by an audit plan and a detail of the procedures to be performed. The process ends with documentation and communication of the audit results.

Audit Report: Formal communication issued by the auditors describing the overall results of the audit. The audit report should include (at a minimum) the audit scope, objectives, a description of the audit subject, a narrative of the audit work activity performed, and conclusions. To be effective, audit reports must be timely, credible, readable, and have a constructive tone. No one element is more important than the others. They all work together to provide a professional report that is accepted and acted upon.

Audit Scope: Refer to *Scope*.

Audit Trail: Paper or electronic trail that gives a step-by-step record or documented history of a transaction. It allows the trace of data to its source. In accounting terms, audit trails assist in tracing financial data from the general ledger to the source document (e.g., invoice, bank statement, voucher receipt, etc.).

Batch Processing: Type of processing of data which requires minimum human interaction. It is a non-real time processing of data, instructions, or materials. Batch process jobs can run without any end-user interaction or can be scheduled to start up on their own as resources permit. Batch processing is used for transmitting large files or where a fast response time is not critical. The files to be transmitted are gathered over a period and then send together as a batch.

Benchmarking: Process of measuring the quality and performance of an organization's policies, products, programs, strategies, services, or processes against those of peer organizations with similar standard measurements. Benchmarking helps organizations in determining how other organizations achieve their high-performance levels, and use this information to improve their own performance.

Blackout Period: A period of at least three consecutive business days but not more than 60 days during which the majority of employees at a particular company are not allowed to make alterations to their retirement or investment plans. A blackout period usually occurs when major changes are being made to a plan. For example, the process of replacing a plan's fund manager may require a blackout period to allow for necessary restructuring to take place.

Blockchain Technology: Digital ledger of economic transactions that is fully public and continually updated by countless users. It is a list of continuous records in blocks. A blockchain database contains two types of records: transactions and blocks. Blocks hold batches of transactions. The blocks are time-stamped and linked to a previous block. Transactions cannot be altered retroactively.

Board of Directors: A group of individuals that are elected as, or elected to act as, representatives of the stockholders to establish corporate management related policies and to make decisions on major company issues.

Bookkeeping: Process of recording day-to-day financial transactions of a business, including purchases, sales, receipts, and payments by an individual person or an organization/ corporation. Part of the process of accounting in business.

Boot Sector: A dedicated region or sector within a hard drive or any other data storage device that contains code instructions needed to "kick off" a computer system.

Business Continuity: A Business Continuity Plan (BCP) is a formal document that describes the organization strategy, process, and procedures that should be implemented in the event of a disaster. The BCP should recognize threats and risks facing the organization, and should document specific procedures to prevent, and recover from, those threats and risks in order to protect organizational assets and maintain functional operations.

Chief Audit Executive (CAE): Senior-level independent executive responsible for the company's internal audit. CAE's have comprehensive knowledge of the business, and are concerned mainly with the company's internal control structures, including effectiveness and efficiency of operations, reliability of financial reporting, and compliance with relevant laws and regulations.

Chief Executive Officer (CEO): Most senior-level executive, officer, leader, and/or administrator in a company. CEOs are responsible for managing and administering the overall operations and resources of a company, including making major and significant decisions. CEO's tasks focus on maximizing the value (e.g., market share, share price, sale revenues, etc.) of the company. CEOs report to the company's board of directors.

Chief Financial Officer (CFO): Senior-level executive responsible for managing, tracking, and monitoring financial and accounting activities of a company. CFOs analyze the company's financial strengths and weaknesses, assess financial risks, and propose corrective actions. CFOs ensure that the company's financial reporting is performed timely and completely. CFOs typically report to (and assist) CEOs with budgeting, forecasting, and cost-benefit analyses, among others.

Chief Information Officer (CIO): Senior-level executive responsible for the administration, design, and implementation of IT and computer systems in place to benefit and support the organization's goals and objectives. CIOs typically report to the CEOs, COOs, or CFOs.

Chief Information Security Officer (CISO): Senior-level executive responsible for developing and implementing an information security program to ensure information assets and technologies are adequately protected. Such security program must establish strategies, policies, and procedures designed to protect enterprise communications, systems, and assets from both internal and external threats. The CISO may also work alongside the CIO to procure cybersecurity products and services and to manage disaster recovery and business continuity plans. The CISO is also usually responsible for information-related compliance.

Chief Operating Officer (COO): Senior-level executive responsible for the daily management, administration, and operation of the company. That is, COOs are focused on executing the company's business plans according to its business model. COOs are also referred to as Executive Vice President of Operations. COOs report to the CEO, and are considered second in command within the company.

Chief Risk Officer (CRO): Senior-level executive responsible for identifying, analyzing, and mitigating internal and external events that could threaten a company. The CRO works to ensure that the company is compliant with government regulations and reviews factors that could negatively affect investments or a company's business units. The CRO typically reports directly the CEO.

Chief Technology Officer (CTO): Senior-level executive responsible for the management of scientific, research and development, and technological needs within an organization. The CTO typically reports directly the CEO.

Client Devices: A client device refers to a program, desktop computer, workstation, or a user that is capable of obtaining information and applications from a server (referred to as a client/server relationship). An example would be a Web browser requesting information (in the form of a Web page, etc.) from servers all over the Web. The browser itself acts as the client, while the computer handling the request and sending back the requested information is the server.

Cloud (or Cloud Computing): Cloud, in simple terms, refers to the Internet. Cloud computing allows data and programs to be stored, and accessed from, the Internet instead of the local computer's hard drive. This type of Internet-based computing provides shared computer processing resources and data to computers and other devices on demand.

Compliance: Ensure that the organization's financial matters are handled in accordance with federal laws and regulations. Processes must be implemented for recording, verifying, and reporting the value of a company's assets, liabilities, debts, and expenses to ensure compliance.

Computer Forensics: Practice of collecting, analyzing, and reporting on digital data in a way that is legally admissible. It can be used in the detection and prevention of crime and in any dispute where evidence is stored digitally.

Computer Virus: Refer to "virus" definition.

Configuration Items (CIs): Fundamental structural unit of a configuration management system. Examples of CIs include individual requirements documents, software, models, and plans.

Consistency: Conformity in the application of something, typically that which is necessary for the sake of logic, accuracy, or fairness.

Control or Control Procedure: Policies and procedures established to provide reasonable assurance of the success of management control.

Control Process: Direction for organizational control that derives from the goals and strategic plans of the organization.

Credit Risk: The risk to the lender that a borrower may fail in making the required loan payments. Risk to the lender includes lost principal and interest, disruption to cash flows, and increased collection costs.

Cryptography: Method of storing and transmitting data in a particular form so that only those for whom it is intended can read and process it.

Cyber Campaigns: Series of related cyber operations performed in a given network environment by either a single actor, or as a combined effort of multiple actors. Such cyber operations typically include planning and coordinating cyberattacks aimed toward a single, specific, strategic objective or result.

Cybersecurity: Technologies, processes, and practices designed to protect networks, computers, programs and data from attack, damage, or unauthorized access.

Cyberterrorism: Politically motivated use of computers and information technology to cause severe disruption or widespread fear in society.

Cyberwarfare: Internet-based conflict involving politically motivated attacks on information and information systems. Cyberwarfare attacks can disable official websites and networks, disrupt or disable essential services, steal or alter classified data, and cripple financial systems, among others.

Data Cartridge: Refer to *Tape Cartridge*.

Data Repositories: Refer to a destination designated for data storage. Specifically, a data repository refers to a particular kind of setup within an overall IT structure, such as a group of databases, where an enterprise or organization has chosen to keep various kinds of data. Data repositories keep certain population of data isolated so that it can be mined for greater insight or business intelligence or to be used for a specific reporting need.

Deficiency (or Control Deficiency): Exists when the design or operation of a control does not allow management or employees, in the normal course of performing their assigned functions, to prevent or detect misstatements on a timely basis.

Denial of Service (or Denial of Service Attack or DoS Attack): Attack designed to disable a network by flooding it with useless traffic.

Differential Backup: Considered a partial backup. It is a copy of all changes made since the last *full* data backup. A differential backup captures only the data that has changed (differences) since the last *full* backup, therefore, it is completed faster, and require less media to store the backup.

Disk Packs: Core component of a hard disk drive. Storage device for a computer that consists of a stack of magnetic disks that can be handled and stored as a unit.

Due Care (or Due Professional Care): Imposes a responsibility upon each professional within an independent auditor's organization to observe the standards of field work and reporting.

Encryption: Effective process to achieve data security by converting or translating data into a secret code, especially to prevent unauthorized access. To read (or decrypt) an encrypted file, access to a secret key or password is needed. Unencrypted data is called plain text while encrypted data is referred to as cipher text.

Entity-level Risk: Refers to a risk that can affect multiple cycles and financial statements areas of an entity.

Examination: A detailed inspection or investigation.

Extend (or Extensions): Extending or performing extensions on an invoice transaction, for instance, involves verifying that the number of units of each invoice item times its unit cost reconciles with the total dollar amount for each item. To illustrate, if 20 units of invoice item A have a per unit cost of $10, the total cost for invoice item A should be $200.

Federal Information Processing Standards (FIPS): Publicly announced standards issued by NIST, after approval by the Secretary of Commerce pursuant to the Federal Information Security Management Act, for use in computer systems by non-military government agencies and government contractors.

Financial Accounting Standards Board (FASB): Private, not-for-profit independent board (consisting of a seven-member independent accounting professionals) who establish/improve, and communicate standards of financial accounting and reporting in the United States. FASB standards, known as generally accepted accounting principles (GAAP), govern the preparation of corporate financial reports and are recognized as authoritative by the Securities and Exchange Commission (SEC).

Financial Statements: A formal record of the financial activities and position of a business, person, or other entity. Financial statements for a business typically include: Balance Sheet, Income Statement, Statement of Owners' Equity, and Statement of Cash Flows. The balance sheet is reported at a particular point in time. The remaining three financial statements are prepared and reported for a specific period of time.

Finding (or Audit Finding): Result from a process that evaluates audit evidence, such as records, factual statements, and other verifiable information that is related to the audit criteria being used, and compares it against audit criteria. Audit criteria may include policies, procedures, and requirements. Audit findings can show that audit criteria are being met (conformity) or that they are not being met (nonconformity). They can also identify best practices or improvement opportunities.

Firewall: A part of a computer system or network that is designed to block unauthorized access while permitting outward communication.

Footing: Refers to the final balance when adding the debits and credits on an accounting balance sheet. Footings are commonly used in bookkeeping to determine final balances to be put on the financial statements.

Forensic Specialist: Also known as "Crime Scene Investigators" or "Crime Scene Technicians" are responsible of collecting evidence from crime scenes. They sometimes take direction from detectives at the scene, but officers also rely on their judgment and expertise.

Fourth-generation Language: Also known as "4GL" is a computer programming language that assimilates the human language better than other typical high-level programming languages, such as, 3GLs or 2GLs. Most 4GLs are used to interact with databases with query commands like "DELETE ALL RECORDS WHERE LAST NAME IS 'MURPHY'" or "FIND ALL RECORDS WHERE FIRST NAME IS 'CHRISTOPHER'."

Fraud: Wrongful or criminal deception intended to result in financial or personal gain.

Fraud Detection Software: Tool designed to test and compare all types of organizational data (e.g., financial, operational, etc.). Fraud detection software is highly effective in profiling and monitoring day-to-day activities to search for patterns in data, or instances of fraudulent activity, such as cash skimming, money laundering, and employees posing as vendors, among others. Tool typically used by auditors and forensic accountants to examine large volumes of transactional data so they can detect and prevent corporate fraud.

Fraud Hotline: Confidential and anonymous whistleblower reporting service for potential fraud, ethical issues, and other concerns.

Full System Backup (or Full Backup): Complete copy of all data files and folders to another set of media (e.g., tape, disk, CD, DVD, cloud, etc.). Full system backups are typically used as an initial or first backup followed by subsequent partial backups.

Functional Requirements: Functional requirements specify something the system should do (i.e., a function). Functions describe inputs, behavior, and outputs. A functional requirement may be for a system to compute the sales price for a customer invoice, or to add vendor contact information. An additional example of a functional requirement could be the generation of a reports, say a sales report for a particular period, a cost/benefit analysis, etc.

Fuzz Testing: Software testing technique used to discover programming errors and security gaps in a system (e.g., operating systems, networks, etc.). With this technique, massive amounts of random data are entered into the system in an attempt to stop the system from functioning properly.

Gateway: A gateway is hardware (i.e., routers or computers) that connects two networks that use different protocols.

General ledger: Provides a complete, sorted, and summarized accounting record of all balance sheet and income statement transactions. Examples of general ledger accounts include Cash, Accounts Receivable, Accounts Payable, Revenues, Expenses, etc.

Generally Accepted Accounting Principles (GAAP): A collection of commonly followed accounting rules and standards for financial reporting. GAAP specifications include definitions of concepts and principles, as well as industry-specific rules.

Generally Accepted Auditing Standards (GAAS): A set of systematic guidelines used by auditors when conducting audits on companies' finances, ensuring the accuracy, consistency and verifiability of auditors' actions and reports.

Generic Accounts: Computer account that is shared or is not uniquely tied to a specific user. For good control purposes, the use of generic accounts should be reduced. User accounts should be uniquely owned so that users are accountable for their own activities.

Go-live Date: The date a computer application system can be used or is ready to start operating as designed. The go-live date typically follows successful test results and formal approval from management.

Hash Total: Sums a nonfinancial numeric field, such as the total of the quantity-ordered field in a batch of sales transactions.

Hole: Weakness built into a program or system that allows programmers to enter through a "backdoor," bypassing any security controls.

Incremental Backup: Considered a partial backup. It copies only data items that have changed since the last backup operation of any type (i.e., partial or full). Since incremental backups copy smaller amounts of data than full backups, they are executed as often as needed, completed faster, and require less media to store the backup.

Independence: The independence of the (internal or external) auditor from parties that may have a financial interest in the business being audited. Independence requires integrity and an objective approach to the audit process.

Integrity: Being honest and forthright when dealing and/or communicating with others.

Internal Control or Internal Control Structure: Systems and procedures designed to ensure that all employees perform their duties ethically and honestly. Accounting controls deal specifically with the integrity of internal financial information and the accuracy of financial reports provided to outsiders.

International Accounting Standards Board (IASB): Independent, private-sector body that develops and approves International Financial Reporting Standards (IFRS). The IASB operates under the oversight of the IFRS Foundation. The purpose of the IASB is to develop a single set of high-quality, understandable, enforceable, and globally accepted financial reporting standards based upon clearly articulated principles.

International Financial Reporting Standards (IFRS): Set of accounting standards developed by the IASB that is becoming the global standard for the preparation of public company financial statements.

ISO/IEC 27001: British Standard International Organization for Standardization (ISO)/ International Electrotechnical Commission 27001. A well-known security standard that helps organizations manage the security of assets such as financial information, intellectual property, employee details or information entrusted by third parties. The ISO/IEC 27001 standard provides requirements to implement an information security management system for sensitive company information related to, for instance, personnel records, processes, and IT systems, among others.

IT Control: A procedure or policy that provides a reasonable assurance that the information technology (IT) used by an organization operates as intended, that data is reliable and that the organization is in compliance with applicable laws and regulations.

Iterative Approach: Method that divides the software development of applications into parts. When using the iterative approach of software development, code is designed, developed, and tested in repeated cycles. With each cycle (or iteration), additional features can be designed, developed, and tested until there is a fully functional software application ready to be implemented into the live environment.

Job Scheduling: Process of allocating system resources to many different tasks by an operating system. The operating system prioritizes job queues, and determines which job(s) is/are to be executed and the amount of time allocated for the job. Job scheduling enables an organization to schedule, manage, and monitor computer jobs (e.g., payroll program, etc.).

Just-In-Time Inventory: Refers to an inventory strategy or management system employed by companies to decrease inventory time and costs by having inventory readily available to meet demand, but not to a point of excess to create extra products. Goods are received only as they are needed.

Liquidity Risk: The risk that companies may be unable to meet their short-term financial demands. Risk may result from the inability to convert a security or hard asset to cash without a loss of capital and/or income in the process.

Malware: Malicious code that infiltrates a computer. It is intrusive software with the purpose of damaging or disabling computers and computer systems.

Management Letter: A management letter is an auditor-created document provided to management that identifies findings or areas/procedures to be improved or redesigned (also often called deviations, control deficiencies, exceptions, etc.), as a result of the audit. The letter may also include the risks resulting from those findings, as well as auditor recommendations describing a course of action by the company's Management to restore or provide accuracy, efficiency, or adequate control of audit subjects. Both risks and recommendations should be provided by the auditor for each audit finding for the report to be useful to Management. Management letters typically contain no financial information.

Margin: Also known as "gross profit." Difference between a product or service's selling price and its cost of production.

Market Risk: Also called "systematic risk." Risk to investors that the value of their investments will decrease (i.e., experience losses) due to factors that affect the overall performance of the financial markets.

Marketable Securities: Debts that are to be sold or redeemed within a year. These are financial instruments that can be easily converted to cash such as government bonds, common stock or certificates of deposit.

Materiality: The threshold above which missing or incorrect information in financial statements is considered to have an impact on the decision making of users. Materiality is sometimes construed in terms of net impact on reported profits, or the percentage or dollar change in a specific line item in the financial statements.

Material Misstatement: Material misstatements refer to those misstatements (or instances where a financial statement assertion is not in accordance with the criteria against which it is audited; for example, GAAP) that if present in the financial statements may affect the economic decisions of the users of such financial statements. Materials misstatements may be classified as fraud (intentional), other illegal acts such as noncompliance with laws and regulations (intentional or unintentional), or errors (unintentional).

Material Weaknesses: A deficiency, or a combination of deficiencies, in internal control over financial reporting, such that there is a reasonable possibility that a material misstatement of the company's annual or interim financial statements will not be prevented or detected on a timely basis.

Mole: A mole enters a system through a software application and enables the user to break the normal processing and exits the program to the operating system without logging off the user, which gives the creator access to the entire system.

Money Laundering: Concealment of the origins of illegally obtained money, typically by means of transfers involving foreign banks or legitimate businesses.

Online Processing: Automated way to enter and process transactions (data, reports, etc.) continuously into a computer system in real time. Bar code scanning is a good example of online processing.

Operational Risk: Risk means a company undertakes when it attempts to operate within a given field or industry. Operational risk is the risk not inherent in financial, systematic, or market-wide risk. It is the risk remaining after determining financing and systematic risk, and includes risks resulting from breakdowns in internal procedures, people, and systems.

Outsourcing: A practice used by different companies to reduce costs by transferring portions of work to outside suppliers rather than completing it internally. Effective cost-saving strategy when used properly. It is sometimes more affordable to purchase a good from outside companies than to produce them internally.

Pay-back Period: The length of time required for an investment to recover its initial outlay in terms of profits or savings.

Penetration Testing: Also referred to as pen testing, is the process of attempting to gain access to computer resources (e.g., application system, network, Web application, etc.) to find security weaknesses and vulnerabilities that can be exploited. The simulated and authorized attack replicates the types of actions that a malicious attacker would take in order to gain access to the system's features and data. Results provide organizations a more accurate representation of their security position at any given point in time.

Personally Identifiable Information (PII): Information that can be used on its own or with other information to identify, contact, or locate a single person, or to identify an individual in context.

Production Environment (or Live Environment): Describes the setting or environment where the actual software and hardware are installed in order to operate as intended by end users. A production environment, or a live environment as it is also referred, is where the real-time execution of software programs occur. A production environment includes the personnel, processes, data, hardware, and software necessary to achieve daily tasks and operations.

Pro Forma Financial Statements: Pro forma financial statements are prepared incorporating certain assumptions, projections, or hypothetical conditions that may have occurred or which may occur in the future. These statements are used to present a view of corporate results to outsiders, perhaps as part of an investment or lending proposal. For example, a corporation might want to see the effects of three different financing options. Therefore, it prepares projected balance sheets, income statements, and statements of cash flows.

Pseudocode: Informal, high-level notation (description) used by programmers to develop algorithms.

Public Company Accounting Oversight Board (PCAOB): A non-for-profit corporation instituted by Congress to oversee the audits of public companies in order to protect the interests of investors and further the public interest in the preparation of informative, accurate, and independent audit reports.

Ransomware: Malware (mainly in the form of a denial-of-access attack) that prevents computer users from having access to files and records. Ransomware installs secretly on a device (e.g., computer, smartphone, wearable device) and either holds the data hostage, or threatens the victim to publish such data until a ransom is paid. Simple ransomware may lock the system and display a message requesting payment to unlock it. More advanced ransomware encrypts files or the computer's hard drive making them inaccessible, and demands a ransom payment from the victim in order to decrypt them.

Reasonable Assurance: Level of confidence that the financial statements are not materially misstated that an auditor, exercising professional skill and care, is expected to attain from an audit.

Receivable (or Account Receivable): Outstanding invoices a company has or the money that the company is owed from its clients. Refers to accounts a business has a right to receive because it has delivered a product or service. Receivables essentially represent a line of credit extended by a company and due within a relatively short time period, ranging from a few days to a year.

Reliability: The extent to which an experiment, test, or measuring procedure yields the same results on repeated trials.

Return on Investment (ROI): ROI, usually expressed as a percentage, measures the gain or loss generated on an investment relative to the amount of money invested. It is typically used for personal financial decisions, to compare a company's profitability or to compare the efficiency of different investments.

Risk: A situation which exposes to danger, harm, or loss.

Risk Management: The forecasting and evaluation of financial risks together with the identification of procedures to avoid or minimize their impact.

Sarbanes-Oxley Act of 2002 (SOX): Act passed by U.S. Congress in 2002 to protect investors from the possibility of fraudulent accounting activities by corporations. The SOX Act mandated strict reforms to improve financial disclosures from corporations and prevent accounting fraud. The SOX Act was created in response to accounting malpractice in the early 2000s, when public scandals shook investor confidence in financial statements and demanded an overhaul of regulatory standards. Key provisions of SOX are Section 302 and Section 404.

Scope (or Audit Scope): Amount of time and documents which are involved in an audit, is an important factor in all auditing. The audit scope, ultimately, establishes how deeply an audit is performed. An audit scope can range from simple to complete, including all company documents.

Securities and Exchange Commission (or SEC): Agency of the United States federal government. It holds primary responsibility for enforcing the federal securities laws, proposing securities rules, and regulating the securities industry, the nation's stock and options exchanges, and other activities and organizations, including the electronic securities markets in the United States.

Skepticism: Doubting as to the truth of something.

Software Piracy: Illegal and unauthorized copying, distribution, or use of software.

Solid-state Drive: An electronic storage device considered an alternative to a hard disk. They are faster than hard disk drives, and are employed in many products, including mobile devices, cameras, laptops, and desktop computers, among others.

Source Code Audits: Examination of the source code in a programming (software) project to attempt to reduce errors before the software is ultimately released to the users. Source code audits concentrate on discovering errors, flaws, failures or faults in a computer software program, as well as security breaches or violations of programming conventions.

Spamming: Disruptive online messages, especially commercial messages posted on a computer network or sent as email.

Stakeholder: A person with an interest or concern in something, especially a business.

Strategy: A plan of action or policy designed to achieve a major or overall aim.

Structure Query Language: Standard computer language used in relational database management and data manipulation. Structure Query Language or SQL is used to query, insert, update, and modify data.

Subsidiary: A company controlled by a holding company.

Subsidiary Ledger: Designed for the storage of specific types of accounting transactions. A subsidiary ledger groups similar accounts whose combined balances equal the balance in a specific general ledger account. After transactions are grouped in the subsidiary ledger, they are summarized and posted to the general ledger. Information from the general ledger is then used to prepare the financial statements of a company.

Substantive Test (or Substantive Testing): Audit procedure that examines the financial statements and supporting documentation to see if they contain errors. These tests are needed as evidence to support the assertion that the financial records of an entity are complete, valid, and accurate.

Tape Cartridge (or Data Cartridge): Device used in tape library units to store digital data (e.g., corporate data, audio/video files, etc.) on magnetic tape.

Terrorism: The use of violence and intimidation in the pursuit of political aims.

Time Bomb: Code that is activated by a certain event such as a date or command.

Trend Analysis: Analysis conducted to identify patterns in problems or solutions. A common purpose of trend analysis is, for instance, to attempt to predict the future movement of a stock based on past data.

Trojan Horse: Piece of code inside a program that causes damage by destroying data or obtaining information.

Validity: The state or quality of being valid (effective or legally binding).

Version Control: Procedure to track and monitor files associated with a version. Version control procedures also assist in coordinating programmers working on various aspects or stages of the system. There are programs and databases available that can automatically store the various versions of files.

Virus (or Computer Virus): Piece of program code that contains self-reproducing logic, which piggybacks onto other programs and cannot survive by itself.

Vulnerabilities (or Vulnerability): Quality of being easily hurt or attacked. The state of being open to injury, or appearing as if you are.

Waterfall Approach: Systems development life cycle model that is linear and sequential. The waterfall approach has distinct goals for each phase of development. Once a phase of development is completed, the development proceeds to the next phase (without turning back).

White-collar Crime: Financially motivated nonviolent crime that typically involves stealing money from a company and that is typically done by business and government professionals (holding high positions within the organization).

Work Paper (or Working Paper): Formal collection of pertinent writings, documents, flowcharts, correspondence, results of observations, plans and results of tests, the audit plan, minutes of meetings, computerized records, data files or application results, and evaluations that document the auditor's activity for the entire audit period. A complete, well-organized, cross-referenced, and legible set of working papers is essential to support the findings, conclusions, and recommendations as stated in the Audit Report. Typically, copy of the final Audit Report is filed in the working papers.

Worm: Independent program code that replicates itself and eats away at data, uses up memory, and slows down processing.

Index

Page numbers followed by "e" indicate exhibits.